Handbook of
Project Finance
for
Water and Wastewater
Systems

Michael Curley

CRC Press
Taylor & Francis Group
Boca Raton London New York

CRC Press is an imprint of the
Taylor & Francis Group, an **informa** business

CRC Press
Taylor & Francis Group
6000 Broken Sound Parkway NW, Suite 300
Boca Raton, FL 33487-2742

© 1993 by Taylor & Francis Group, LLC
CRC Press is an imprint of Taylor & Francis Group, an Informa business

First issued in paperback 2019

No claim to original U.S. Government works

ISBN-13: 978-0-36-744999-5 (pbk)
ISBN-13: 978-0-87-371486-0 (hbk)

Visit the Taylor & Francis Web site at
http://www.taylorandfrancis.com

and the CRC Press Web site at
http://www.crcpress.com

Library of Congress Cataloging-in-Publication Data

Curley, Michael.
    Handbook of project finance for water and wastewater systems /
Michael Curley.
        p.   cm.
    Includes bibliographical references and index.
    ISBN 0-87371-486-5
    1. Waterworks--Finance.   2. Municipal water supply--Finance.
3. Water supply--Finance.   4. Sewage disposal--Finance.   5. Water
treatment plants--Finance.   6. Sewage disposal plants--Finance.

Dedication

To my mother, Lucy Alford Curley, who for twenty years was the Treasurer of the City of Buffalo, New York, and who taught me the rudiments of public finance as well as the importance of sound financial practices.

Acknowledgment

First and foremost, I must thank George Ames of the United States Environmental Protection Agency (EPA). While on a partial sabbatical from EPA, during which he founded the Council of Infrastructure Financing Authorities (CIFA), George asked me to make several presentations on financing for small water and wastewater systems. He then asked if I would commit the presentation to writing. This became <u>Financial Alternatives for Small Water and Wastewater Utility Systems</u>, which was published as a monograph by CIFA in January, 1990, and which is the source document for this book. If it were not for George, this book would never have been written.

Next, I would like to thank the publisher, Brian Lewis, and most especially my editors, Skip DeWall and Elise Hoffman, for their gracious encouragement and their patience.

I must thank my partner, Kenneth S. Hall, and my long-time colleague, Helen C. Robertson, for their encouragement, their advice and their commentary. I must also thank my sister, Lucy Curley Haley, and my brother, D. Patrick Curley, both of whom are financial consultants, who gave me many perspectives from the users' point of view. I thank Michael Deane of EPA for his advice and assistance on the State Revolving Fund program. I need also to thank my friend and fellow author, Joe Mysak, whose advice steeled me for the rigors of authorship.

I thank my wife, Barbara Ann Curley, who is the former general counsel of Industrial Development Bond Insurance Managers, Inc., for her many lawyerly admonitions and her editorial skills.

Last but not least, I must thank my children, Andrew and Elisabeth, who at this writing are three and five years old, respectively, for graciously allowing me the use of my computer, in peace, so to get the job done.

M.C.
New York, New York
October, 1992

The Author

Michael Curley is an attorney by profession and an investment banker. He is a partner in Hall, Curley & Co., Inc., a private merchant bank specializing in economic development and environmental infrastructure.

In 1981 Mr. Curley founded Industrial Development Bond Insurance Managers, Inc., one of the pioneer companies in the financial guaranty insurance industry (IDBI) which was sold to a major bank in 1988. He was also a partner in the New York law firm of Shea & Gould.

Prior thereto, Mr. Curley served in government in several capacities, including president of the New York Job Development Authority, deputy commissioner and counsel of the New York State Department of Economic Development, and counsel to the New York State Science & Technology Foundation.

Mr. Curley has served as an Adjunct Professor of Banking and Finance at New York University and has taught at the University of Oklahoma. He currently teaches in the environmental studies program at the Johns Hopkins University in Baltimore, Maryland.

Mr. Curley is a member of the Environmental Financial Advisory Board (EFAB) of the United States Environmental Protection Agency (USEPA), where he chairs the International Committee and is a member of the Executive Committee. He is also a vice chairman of the American Economic Development Council and has served on several boards and industry committees.

Introduction

This book is written for the non-financial executive. It is designed for the directors and managers of water and wastewater systems, who do not deal with project finance on a day-to-day basis.

Many people think that finance is a subject akin to black magic. Rubbish. There is no magic to it at all. It is a straightforward science, and a fairly simple one at that. Even for people who think they are "not good with numbers."

Being elected to the Board of Directors of a water or wastewater system is, indeed, an honor. Likewise, being chosen to manage such a system is also an honor. But this honor carries with it a mandate. Each member of the board of directors and each manager of a water or wastewater system has a firm obligation to the system's users to provide quality service at reasonable rates. This mandate implies financial determinations. It applies to the day-to-day operation of the system. It also applies to capital improvement projects undertaken by the system. You can't run a water or wastewater system without being able to make sound financial judgements.

Reckoning the quality of day-to-day service against its cost is a fairly straightforward proposition. One needn't be a chemist, an engineer, or a financial expert to be an effective board member or general manager. Reckoning the benefits and cost effectiveness of a major capital project, on the other hand, can sometimes seem absolutely daunting. But it is not. It lends itself readily to the type of analysis which even "non-financial types" can easily get comfortable with.

The subject of the book is the funding of capital projects, which are projects to improve, expand or renovate the physical facilities of a system. Collectively, these will be called: capital improvement projects.

Major system improvements do not happen every day. The programs and procedures for funding such projects are somewhat abstruse and technical. Directors who are elected to govern a water or wastewater system and even the professional managers responsible for the day-to-day operations of the system can't possibly have firsthand familiarity with the intricacies of capital funding programs. Nor are they expected to. After all, systems normally only embark on such projects every five to ten years. No one can be expected to be up to date on matters of such infrequent occurrence. This book is, therefore, designed to give those busy executives the necessary background to be able to analyze and understand the various capital funding programs available to water and wastewater systems, to be able to compare them, and to be able to make sound financial decisions concerning such projects.

Necessarily, this book will probably be of greater value to the management of small water and wastewater systems. They will generally not have the luxury of staff whose only responsibility is finance, much less project finance. Also, because of the small size and relative infrequency of projects, these managers will not be the subject of the intense affection and interest which bankers and Wall Street types traditionally lavish on the managers and directors of large systems when they sense the possibility of a big fee. Small systems don't produce big fees. This is, indeed, unfortunate because a great deal of free information and advice can be had from the financial community, which could be of great help to small system managers.

In addition, the book should also be of good service to the directors of larger systems. The problems are the same, and the analyses are certainly the same for both big and small systems. Directors of larger systems may find this book useful in analyzing financial options put forth by their staffs and in posing questions to their financial advisors, consultants, investment bankers, and the like. They have the same duty to system ratepayers as do directors of smaller systems. To the extent that this book can help directors of large systems come to an informed judgment, it will facilitate the discharge of that duty. For, indeed, the only thing small about small water and wastewater systems is their size. The responsibilities of the system's management are the same whether a system has 100 or 1,000,000 users. Directors and managers have the same legal and moral mandate to provide their users with quality service at reasonable rates.

When it comes to capital improvement projects, the management of water and wastewater systems have four discreet obligations to their ratepayers. First, the project must be effective; i.e., it must solve the problem at hand or meet the intended need. Second, the project must be done in a workmanlike manner so that it will last, in good working order, over its intended service life. Third, the project cost must include only necessary elements and those must be at reasonable prices. And, fourth, it must be financed in the most cost effective manner in order to minimize the actual financial impact on the system's ratepayers.

This book deals with this latter duty: to finance the system's needs in the most cost-effective manner, so that the impact on system ratepayers is absolutely minimized.

The book is divided into three major sections. The first section will present the two tools which are necessary for any type of analysis of capital project funding. These are: the methods of comparing costs, and the method of converting dollars into rates, and vice versa. The first section will conclude by presenting the two most commonly used (and most sensible) methods of comparing financing programs: annual cost and total cost.

The second major section deals with seven factors which can have a dramatic effect on annual cost and total cost. Annual cost can be defined as how much system ratepayers have to pay for their project each year. Total cost has to do with how long they have to pay for it. These seven factors are:

1.) Term of payment. The number of years over which the loan principal for a project is repaid.
2.) Rates. The interest rate at which the bond is issued or the loan is made.
3.) Financing costs. The up-front costs which must be added to the bond or loan principal, and the annual charges which are added to the interest rate.
4.) Delay. The effect of delays on both project cost and financing cost.
5.) Imposed costs. These costs are those additional charges which are extrinsic to the project but are required by the terms of a particular financing program.
6.) Ineligibility. The costs which cannot be financed by the loan or bond, and, hence, must be paid for otherwise.
7.) Coverage. The amount, if any, by which the ratepayers' annual payments must exceed the annual debt service due.

It is important to realize that there are three ways in which a water or wastewater system can be adversely affected by the particular financing approach chosen.

First, project costs may be inflated. Factors three, four and five, above, address this point. Sometimes additions to project cost are absolutely necessary and sometimes they may not necessarily result in either higher annual debt service payments nor in higher charges to system users. But additions to project cost must be accurately identified and factored into the decision making process.

Second, true annual debt service costs must be rigorously ascertained. As you will see below, sometimes a higher project cost will not result in higher annual debt service payments. This is due to differences in the Rates (Item #2, above) and the Terms (Item #1, above) of particular financing programs.

Third, both higher project costs and higher annual debt service payments may still result in lower annual charges to system users. This may be due to the fact that less expensive programs may exclude certain costs, which must then be financed by other more expensive means, making the less expensive program more costly overall. Factors six and seven, above, discuss this possibility.

All of these circumstances -- where higher really means lower -- may seem at first paradoxical. But they are not. They become obvious once you analyze all financial alternatives using the same approach and the same criteria in each instance.

The third section of the book presents the four most frequently used programs for financing the capital improvement projects of water and wastewater systems. These are:

1.) Municipal Bonds.
2.) Commercial Bank Loans. Both taxable and tax-exempt.
3.) The Water and Waste Disposal Program (WWDP) of the Farmers Home Administration (FmHA) of the U.S. Department of Agriculture (USDA).
4.) The U.S. Environmental Protection Agency's (EPA) State Revolving Fund (SRF) Program for wastewater projects only.

As each program is presented, it will be fully described. Then it will be evaluated in terms of the seven cost factors presented in section two.

Table of Contents

Section I

Section 1

Chapter 1

Cash Available for Debt Service

This section of the book deals with what many people think is the black magic aspect of finance.

Several years ago, the Council of Infrastructure Financing Authorities (CIFA) and the U.S. Environmental Protection Agency (EPA) sponsored a conference on the financing of small water and wastewater projects. Among the presentations at the conference was one on the FmHA "buy-back" program. This program derived from the Reagan Administration's effort to reduce the federal deficit. In this case, local water and wastewater districts were offered the opportunity to "buy back" loans which they had obtained from the Farmers Home Administration (FmHA) of the U.S. Department of Agriculture. The idea, of course, was that the government would accept a discount, i.e., less than full value on the loans. This would, in turn, produce a savings which was the presumed inducement for the small systems to buy back the loans.

A distinguished investment banker was making a presentation on how systems could use tax-exempt municipal bonds to effect their buy-back. During the question and answer period at the end of the presentation, a representative of the American Water Works Association (AWWA) asked an amazing question. She first called the audience's attention to the fact that, in the banker's presentation, the old payment on the FmHA loan was precisely the same amount as the new payment on the bonds issued to buy back the FmHA loan. If the two payments were equal, she queried, where was the savings? Apart from the obviously improved age and demeanor of the questioner, the circumstance was not at all unlike the famous Wendy's television commercial of the old lady staring at the hamburger asking: "Where's the beef?"

If the water system owed $8,000 a month before the buy-back and still owed $8,000 a month after the buy-back, where was the benefit to the system?

In point of fact, both the presenter and the questioner were right. The difference was in their analyses. The presenter's example had demonstrated a savings in total cost; while the questioner was looking at annual cost. Unfortunately, neither recognized the other's position in time to avoid a breakup of the session in a haze of confusion.

As we will see in the next few chapters, the wizards at the U.S. Treasury who designed the FmHA buy-back program realized that they needed a financial inducement for the small systems to buy back their loans. They could easily have structured the inducement to provide a savings in annual cost. But this would have meant less cash for the federal government. Since the goal was to raise as much money as possible, they structured the inducement, instead, to provide a savings in total cost. To the naked eye, and from the questioner's perspective, this presents no real savings at all, because the savings are not in today's payments. But, in fact, there is a savings and it is real. The investment banker was also right. On a total cost basis, the system would pay fewer dollars for the same project. So, the savings would be there; but just not in terms of what the system had to pay today.

It is precisely because of differences such as these that finance is thought of as black magic. How can less be more, or the same?

Less is not more. But it can appear to be when only looking at one aspect of a transaction. There is no black magic here, only a difference in perspective. Both annual cost and total cost can easily be measured. Once you are able to do so, the reality of the situation becomes plain. In the example above, on an annual cost basis, the FmHA loan and the bond were the same (as the questioner rightly observed); but on a total cost basis (which the investment banker presented) there really was a true dollar savings to the water district.

In the same vein, the Chairman of the Board of a major metropolitan water system recently confided his unease over a proposal brought to him by his agency's investment bankers. They proposed that his system refund some of its outstanding tax-exempt bonds. Refunding generally means replacing bonds. This can be done by either calling, i.e., redeeming, the outstanding bonds or by placing the proceeds of the new bond issue in a special account where the principal and interest, together, to be earned on the new bonds precisely equals the principal and interest, together, to be paid on the old bonds. This latter method results in an event known by the lawyer-thrilling term "defeasance."

What was bothering the Chairman of the water system was that his bankers' proposal called for the issuance of more new bonds than were then outstanding on the old bonds. (This is not unusual.) But, at the same time, his agency's annual payments on the larger amount of new bonds were actually going to be **smaller!**

This is almost the opposite of the FmHA question. There, the annual cost was the same both before and after the transaction but the total cost was lower. Here, the annual cost is lower and the total cost is higher. The Chairman's system will wind up paying less each year; but they will continue paying for a greater number of years.

There is certainly a relationship between annual cost and total cost, but it is not linear. When one goes one way, the other does not necessarily follow.

Anecdotes like these tend to make people think finance really is black magic. Increasing the loan amount and decreasing the loan payment at the same time just doesn't seem right. It doesn't seem to make sense. There appears to be something sinister about it. But there isn't. There is a simple and straightforward answer to both the Chairman's question and the FmHA problem.

To answer these questions requires being able to compare the costs of alternative methods of financing a given project. If two investment bankers approach you with two different proposals for funding the same project, the only way the proposals can be compared is by evaluating the annual cost of each <u>or</u> the total cost of each. You cannot compare an annual cost with a total cost. That is where the Chairman's question and the FmHA problem come from. Annual cost and total cost are a true example of apples and oranges. They simply cannot be compared.

To properly evaluate either the annual cost or total cost of a project we must first gather some analytical tools which we will need.

The first thing we will need to learn is what funds (of the water or wastewater system) are available to pay for the project. Current funds may be insufficient. But we will only know the precise amount by which they are deficient after a thorough analysis. In addition, by identifying those exact funds which can be used to pay for projects, we are, of course, at the same time specifying either the sources of income which must be increased or the expenses which must be curtailed in order to be able to pay for the project.

Next, and most importantly, we will need to know how much we have to pay each year for the project. Notice we used the phrase "how much we have to pay" instead of the simpler "how much it costs." This is because the term "cost" or, more commonly, "project cost" is often used in so narrow a sense that it is frankly misleading. When you hear used car dealers talk about their "low-cost" pre-owned models, you may get the uneasy feeling that the monthly payment they are talking about will be substantially different from the amount you actually wind up making the check out for each month. Rightfully so. And, unfortunately, this is often the case in the project finance business as well.

To determine how much we have to pay for a project we will need to be able to convert annual costs or charges into their one-time equivalents. We will also have to convert one-time charges or fees into annual costs. For example, if a lawyer charges a one-time fee of $10,000, and the fee is added to the amount of a loan to pay for the project, how much will the one-time legal fee add to the annual loan payment? Conversely, a lender may charge an annual inspection fee of $1,000 per year. How much does this add to the <u>total</u> project cost? To be able to convert from an annual cost to a one-time cost is not as simple as adding up all of the annual payments. Nor is converting from one-time cost to total cost as easy as dividing the one-time cost by the number of years of the loan. Unfortunately, these processes are complicated by the fact that you cannot add a dollar paid in one year to a dollar paid in another year. Here, again, is an apple and oranges problem. This leads us to the concept of discounting, by which dollars paid in different years can be reduced to the concept of "present value" dollars and can now be added readily. This subject will be covered in the next chapter. For now, let us begin by identifying the financial concepts necessary to analyze water and wastewater project costs.

*   *   *   *   *

The first concept involves determining what funds are available to pay for projects. Fortunately, the number of elements necessary to make sense out of this concept are relatively few. They are basically divided into three areas: Regular Income, Cash Expenses, and Cash Available for Debt Service.

<u>Regular Income.</u> The first major area to look at is income. From the point of view of project financing, the best way to look at income is as

regular income. By regular income we mean income that arises, or will arise, from the day-to-day delivery of water or wastewater services. The word "services" here is most important. Income from services means income derived from the delivery of water or the use of a sewer. Its principal characteristic, for financial purposes, is that it tends to be extremely regular and predictable.

Once a user is connected to a system, that user generally establishes a pattern of use which subsequently varies little from year to year. The more users a system has, the more reliable are the patterns of use. User payments, too, tend to be equally as predictable. Thus, a water or wastewater system with a large, stable user base develops a very predictable and highly reliable record of payments over time. This is extremely important because the more stable a data base is, the more it can be relied upon to predict how much money will be available for annual payments on a new project. History does not guaranty future events but it is certainly a good indicator of what will happen. Shakespeare is right: "What is past, is prologue."

We refer to this income as "regular" because payments are made for day-to-day service as opposed to any single or extraordinary event. Regular income is a critical element of project financing because it can be used statistically to predict future income. Irregular or, more correctly, non-recurring income cannot be used to predict future income. This is true whether the non-recurring income derives from the sale of a piece of property, or from a government grant, or from connection fees.

An important element of many systems' income derives from connection fees. Under such circumstances, the water system can be said to be underwriting part of the cost of the growth of its community. In areas where water and wastewater systems have historically issued bonds or taken out loans to finance the infrastructure of undeveloped areas, such fees may have always been considered "regular" because the system experienced regular or steady growth. Those days are probably over. Recent experiences in Texas, Louisiana and California have taught lenders that major growth cannot simply go on ad infinitum. Lending against future connection fees based on large-scale development has proven to be a risky business, but lending based on historical payment patterns for normal use is very sound. Please note that we are referring to only "major" growth here. This is because marginal growth can always be absorbed in a strong system. Systems experiencing growth of a few percent a year pose no difficulty; however, connection fees for such systems should still not be considered regular income. The acid test of whether any growth is marginal, in the sense used here, is whether the cost of such growth can be absorbed readily by the system's current users out of regular system income. In other words, can the system still support the debt even if no new connection fees are earned by the system?

The emphasis, again, is on completely regular income. Income which is either not going to recur or is otherwise contingent on an event, such as the sale of a home or lot, is not regular income. To the extent that a system has such income, it is certainly helpful to pay the bills, but is of little or no value in financing improvements to the system.

A word about interest income. Interest income is that derived from the earnings on any funds which the system has under its control. The question then arises: Should interest income be considered "regular" income for our purposes? The answer is relative. If the fund or account on which a system is earning interest is permanent and will remain stable for at least the first few years after the new project has been built, then yes. Otherwise, no. For example, a system may have a Repair & Replacement Reserve Fund. If this fund is constantly replenished out of normal user service charges, then the expectation is clearly that the fund will always be there. In such case, a lender should be able to rely on that fund earning at least a modest amount of interest each year. If, on the other hand, a system has a fund intended to be used, for example, to extend its mains to a new subdivision, for example, then the interest on this fund should not be included in regular income because it will presumably be used for its stated purpose. Therefore, it will not be available to make annual payments on the system's new project.

We can summarize the regular income concept with a simple formula:

$$\begin{array}{l} \text{Total income} \\ - \;\underline{\text{Non-recurring income}} \\ \;\;\text{Regular income} \end{array}$$

Therefore, the first element of project finance is to ascertain regular income by subtracting non-recurring income from total income.

Cash Expenses. Cash expenses are, as you might guess, the expenses you have to pay for in cash. The term "cash expense" is emphasized here, rather than just "expense," because the concept of "expense" has come to be a term of art in accountancy, generally meaning any sum which may rightfully be deducted from income. The term "cash expense" is used here to differentiate that which may be deducted from that which must actually be paid. In other words, a system may have expenses which do not result in cash expenditures. The most common expense which is not actually paid in cash is depreciation. Another is amortization.

The concept of cash expenses is important because cash expenses will almost always be less than total expenses. And, as anyone who sits down with a checkbook at the end of the month knows, the fewer the expenses, the more money that is left over. It is these funds -- the money that is left over after all cash expenses are paid -- that are used to pay debt service.

Cash expenses are total expenses minus non-cash expenses, or:

$$\begin{array}{l} \text{Total expenses} \\ - \ \underline{\text{Non-cash expenses}} \\ \text{Cash expenses} \end{array}$$

Once we have ascertained both regular income and cash expenses we can easily calculate Cash Available for Debt Service. As you might guess, Cash Available for Debt Service equals regular income minus cash expenses, or:

$$\begin{array}{l} \text{Regular income} \\ - \ \underline{\text{Cash expenses}} \\ \text{Cash Available for Debt Service} \end{array}$$

Cash Available for Debt Service is a very important concept. It is imperative that anyone involved in the financing of any new project for a water or wastewater system know exactly what income can be counted on in the future to pay for the project. This is Cash Available for Debt Service.

We capitalize the first letters of the words "Cash Available for Debt Service" because it has come to be a term of art, with a very specific meaning, among credit analysts. It is not simply a group of words. It is neither a general nor a casual concept. It is a highly specific term which is defined as the remainder of only certain defined expenses subtracted from only certain defined income. Knowing which types of income can be used in calculating a project financing is critical.

Knowing, as well, the precise expenses which can be deducted from this type of income is equally critical. Together, these concepts comprise Cash Available for Debt Service, which are the only funds which can be used to pay for a new project.

Before leaving this topic, it is important to note that the funds we have defined as Cash Available for Debt Service are the only ones which may be used for calculating or estimating project costs and payments. When it actually comes to making a payment, any cash which the system has on hand, and which is not legally barred from being used for such purpose, may be used for making debt service payments. In other words, if a system has connection fees on hand, and is not legally barred from disbursing those funds to make annual debt service payments, it may use such funds for that purpose. It is just that you don't use them in estimating income, because they are not a regular source of income from year to year.

Knowing what funds can be used to pay for a new water or wastewater project is the first major step involved in analyzing a total project financing. Now that we know what funds can be used to pay for a project, the next question, which will be addressed in the succeeding chapter, is: what is a project payment?

Chapter 2

Annual Debt
Service Payments

The preceding chapter dealt with the concept of "cash available for debt service." This is the specific term used to identify that portion of a water or wastewater system's net income which can be used to estimate, and pay for, project payments. Such project payments, on an annual basis, are usually called "annual debt service" or "annual debt service payments." This chapter will deal with the concept of "debt service" itself. As you will see, the notion of debt service can mean considerably more than the simple principal and interest normally involved in a loan transaction.

There are three components of "annual debt service payments": principal, interest and fees.

Fees.

Fees are not as easy to discuss as they might seem. This is because there are many types of fees. There are one-time only fees. There are annual fees which are expressed in flat dollar amounts, like $250 a year. And, there are fees which are expressed as a percentage of the outstanding principal balance of the loan.

You cannot simply add up these different kinds of fees. Here you get into the old apples and oranges problem again. The only way to deal with this problem is either to convert the one-time fees and the flat dollar fees into numbers which are based on outstanding balance, or to convert both types of annual fees into numbers which are the equivalent of one-time fees.

All of this seems complicated. And it is a bit complicated. But it is also important. Converting one-time fees to annual, and vice versa, will be dealt with in Chapter 5. Actually identifying and quantifying these fees will be covered in Chapter 11.

Interest.

The interest component of annual debt service is the simplest and most straightforward subject to explain and calculate. No matter what type of loan you are dealing with, and no matter how long the term of the loan is, the amount of interest due and payable in any year is calculated by simply multiplying the outstanding principal balance of the loan at that time by the rate of interest. Thus, if, in year x of a loan, the outstanding principal balance at that time is, say, $63,000, and the interest rate on the loan is 7%, then the next annual debt service payment will contain an interest component of $4,410, which is equal to $63,000 x .07.

That is all there is to say about interest.

The problem, as you might guess, is: how to figure out what the outstanding principal balance is in year x? This leads us to the discussion of principal, which is by far the most complicated aspect of dealing with annual debt service payments.

Principal.

We will begin with a definition of principal and a discussion of the several ways in which principal can be repaid.

The best definition of principal is that it is the amount of money which, with interest, your system must pay for its project.

You may think this is a querulous definition. After all, couldn't principal also be defined as the amount you borrow? Or, couldn't it also be defined as the total project cost? The answer to both of these questions is yes, most of the time, but not always. The definition which is true all of the time is that it is the amount of money which you must pay, with interest, for your project.

If you are now beginning to suspect that you may sometimes wind up paying back more than you borrowed, you are right. This actually happens, sometimes.

There are generally three ways in which principal is paid. They are: the level payment method, the level principal method, and a method in which nothing is level, which we will call the irregular method.

To emphasize the fact that the phrases Level Payment Method and Level Principal Payment Method are terms of art, with very specific meanings, we will henceforth capitalize the initial letters of these terms. We will not so dignify the irregular method, however, precisely because it is irregular, and is not a term of art.

The Level Payment Method is, by far, the most common and the most popular. Virtually all home mortgages are calculated on the Level Payment Method. We will look at this method first.

THE LEVEL PAYMENT METHOD

As you would clearly imagine from its title, the Level Payment Method means that the total amount of principal and interest, combined, which you pay every year is the same. In other words, it is where your principal and interest payments each year might be, for example, $75,000. The key point here is that the payment is exactly the same year in and year out.

Why is the Level Payment Method the most popular formula for making project payments? There are three reasons. First, it is almost always the cheapest. Second, it greatly facilitates planning. And, last but by no means least, it is the fairest. Let us examine these principles.

The first is the concept of how the Level Payment Method results in the cheapest means of financing a project. This principle involves an understanding of "discounting" or the "time/value theory of money." These topics are covered in Chapter 3 below, as is a thorough illustration of how the Level Payment Method can (and usually does) result in the lowest overall cost to ratepayers.

The second principle is that the Level Payment Method greatly facilitates fiscal planning for the system. You will recall a few paragraphs ago we said that debt service payments had three components: principal, interest and fees. As such, the first two, principal and interest, comprise the great majority of the payment. Fees are generally small in relation to principal and interest, combined. Therefore, if principal and interest, combined, comprise the great majority of the payment, and if principal and interest, combined, are always the same from year to year, then the fiscal planning of the system is greatly simplified.

Needless to say, implicit in this facilitation of the financial planning process is the notion that the Level Payment Method will also greatly facilitate the ratepayer relation process. As bad as it is to have to explain to ratepayers why their water or sewer payments must increase to pay for a new project, it would be far worse to have to do this every year, as would be the case if their rates were bobbing up and down as their annual debt service payments fluctuated from year to year.

This leads us to the third principle which is that the Level Payment Method is the fairest.

The payments for a project should, as far as is practicable, be spread over the same period of years in which the project will be used by the system. This principle is more fully discussed below in Chapter 9. As such, each ratepayer will pay his equitable share of the cost of the project.

Let us consider a project, the annual debt service payment on which is $75,000 per year for forty years. Let us also say that the project involves improvements to the system which will last forty years, at least. Finally, let us say that there are 1,000 users and that each pays pro rata. So each ratepayer will pay $75 per year for forty years.

In such circumstances, let us take the example of a ratepayer who sells his home after one year and moves out of the district. The ratepayer, as one of 1,000 users, gets 1/1,000th of the benefit of the project per year. Since the project (and the payments therefor) will last 40 years, each year he will actually get 1/40th of 1/1,000th of the benefit, or 1/40,000th of the benefit. By the time he moves out of the district at the end of the first year, he will have had the benefit of 1/40,000th of the project and he will have paid $75, which is exactly 1/40,000th of the total project cost. (Total project cost is 40 years x $75,000 or $3,000,000. 1/40,000 x $3,000,000 equals $75.)

This same process can be applied with the same results to someone who buys a home in the district and becomes a ratepayer in the 39th year of the project. He pays his fair share, just like the fellow who moved out after one year. The process also applies with the same inherently fair results to one who is a ratepayer for the full forty years. It is also true whether the district expands or contracts with fewer or more ratepayers from one year to the next.

In short, the Level Payment Method means that every ratepayer will pay his proportional share of the project cost regardless of how many years he enjoys the benefits of the project.

ILLUSTRATION OF LEVEL PAYMENT METHOD

As you know, the amount of interest you pay on any loan of any kind in a given year is directly proportional to the outstanding principal balance on the loan. Let us say you borrowed $200 from a friend and agreed to pay him back $100 plus 10% interest at the end of the first year and $100 plus 10% interest at the end of the second year. At the end of the first year, you would pay your friend interest of $20, which is 10% of $200. You would also pay $100 of principal. Thus, at the beginning of the second year, the outstanding balance on the loan is only $100. So, at the end of the second year you only pay $10 interest, which is 10% of $100. In other words, you only pay interest on the amount of money you owe; so as you pay off the loan, the amount of interest you must pay each year gets smaller.

Now, consider this: If the amount of interest declines each time a principal payment is made, how can <u>any</u> series of loan payments be constant? The answer is very simple. <u>As the amount of interest declines with every payment that is made, the amount of principal increases by the same amount</u>.

We all know intuitively that every loan payment is composed of principal and interest. Another way of saying this is that Principal (P) plus Interest (I) equals the Annual Payment (AP). The arithmetic expression for this sentence is:

$$P + I = AP$$

Now, as we said before, the Level Payment Method means, by definition, that all Annual Payments are equal. To thoroughly understand what is going on here, let us look at an Annual Payment schedule for a 5-year loan of $100,000 at an interest rate of 10%, as shown in Table 2-A, on top of next page.

Table 2-A

Year	Principal	+	Interest	=	Annual Payment
1	$16,380	+	$10,000	=	$26,380
2	18,018	+	8,362	=	"
3	19,820	+	6,560	=	"
4	21,802	+	4,578	=	"
5	23,980	+	2,400	=	"
	$100,000				

Please notice, before anything else, that all of the Annual Payments are the same. Please also note that as the amount of Interest decreases each year, the amount of Principal increases by exactly the same amount. Figure 2-A, below, shows this relationship in graphic form.

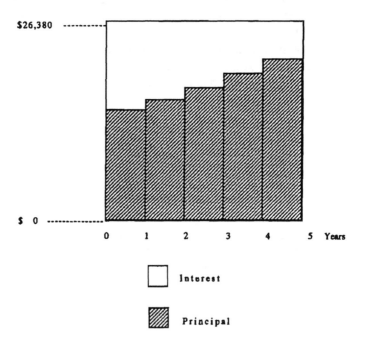

Figure 2-A (graph of Table 2-A)

The importance of the Level Payment Method is that the Annual Payment is always the same from year to year. This is achieved by increasing the Principal by the precise amount of the decrease in the Interest.

Now, the only other number which ought to ring a bell in the above example is the first year's Interest ($10,000). The reason this should look familiar is that it clearly represents 10% Interest on the first year's Principal of $100,000.

If you look carefully again at Year 1, there should be two things which bother you. How do we know the Annual Payment is $26,380? And, how do we know the first Principal payment is $16,380?

The answer to the first question is that we calculate the Annual Payment by using a formula. If you would like to see the formula and how it works, please turn to the section in Chapter 9, entitled "Calculating Annual Debt Service Payments for Loans Using the Level Payment Method."

(The equation in Chapter 9 is the same one which is used for calculating the monthly payments on a home mortgage. If you have a calculator, you can calculate your monthly mortgage payment by making only four keystrokes. These are listed in Appendix B. If you do not have a calculator and if you had a thirty-year mortgage, you would have to make over 360 [12 months x 30 years = 360 payments] individual calculations!)

For those unaccustomed to working with calculators, a table of Annual Payments for various bond terms and at various interest rates has been included in Appendix C.

CALCULATING ANNUAL PRINCIPAL AND INTEREST PAYMENTS: THE ANNUAL DEBT SERVICE PAYMENT SCHEDULE

Once you know the Annual Payment, calculating the amount of the first year's Principal payment is quite easy. You simply subtract the Interest ($10,000) from the Annual Payment ($26,380), as follows:

$$\$26,380 - \$10,000 = \$16,380$$

Now, once you have calculated the first year's Principal payment, you can easily calculate all of the other Interest and Principal payments. This is possible because the Annual Payment is always the same.

For example, what are the second year's Interest and Principal payments? We can figure this out by simple logic.

We know that the original Principal amount was $100,000, and we know that in the first year we paid off $16,380 of Principal. So, at the beginning of the second year, the following amount of Principal was outstanding:

$$\$100,000 - \$16,380 = \$83,620$$

Now that we know how much Principal is outstanding, we can easily find the second year's Interest by taking 10% of that number, as follows:

$$\$83,620 \times .1 = \$8,362$$

To calculate the amount of Principal paid in the second year, we use the same procedure used in the first year, viz., we subtract the Interest payment ($8,360) from the Annual Payment ($26,380), as follows:

$$\$26,380 - \$8,362 = \$18,018$$

To obtain the Principal balance at the beginning of the third year, we repeat the process by subtracting the second year's Principal payment ($18,018) from the outstanding Principal balance at the end of the first year ($83,620), as follows:

$$\$83,620 - \$18,018 = \$65,602$$

You can repeat these procedures each year to obtain the precise Interest and Principal payments as well as the outstanding principal balances for each year of the loan.

Here, in Table 2-B on the next page, is an annual debt service payment schedule for the five year loan in our example using the Level Payment Method:

(N.B. The information contained in the following Table usually appears on an "amortization schedule." Technically speaking, an amortization schedule is only a list of the principal amounts to be repaid in each year of a loan term. See Chapter 9. To preserve this distinction, we will refer to the more complete types of schedules, such as that in Table 2-B, below, as annual debt service payment schedules.)

Table 2-B   Annual Debt Service Payment Schedule, Level Payment Method
(Loan: $100,000, 5 years, 10% interest)

Year	Annual Payment	Interest	Principal	Outstanding Principal Balance
1	$26,380	$10,000	$16,380	$83,620
2	"	8,362	18,018	65,602
3	"	6,560	19,820	45,782
4	"	4,578	21,802	23,980
5	"	2,400	23,980	0
			$100,000	

Please note once again, with the Level Payment Method:

1.) The Annual Payments are always the same. This is the definition of the Level Payment Method.

2.) The amount of Principal paid each year increases by the exact amount in which the Interest payment decreases.

3.) The total of all of the annual Principal payments equals the original Principal amount of the loan.

RULES FOR CREATING ANNUAL DEBT SERVICE PAYMENT SCHEDULES FOR THE LEVEL PAYMENT METHOD

Even with a hand-held calculator, creating an annual debt service payment schedule requires listing all of the elements for each year. Here are the rules for creating such schedules without the use of a calculator: (Please note that we are assuming a loan in which Principal payments are made annually. If Principal payments are made more frequently, such as semi-annually, quarterly, or monthly, then the calculations described in steps 4, 5, and 6 will have to be performed for each Principal payment.)

1.) Once you know the Principal amount of the loan and the rate of Interest, determine the Annual Payment by either using the instructions in Appendix B, or by looking it up on the table and performing the simple calculation as described in Appendix C.

2.) Calculate the first year's Interest payment by multiplying the original Principal amount by the rate of Interest.

3.) Obtain the first year's annual Principal payment by subtracting the first year's annual Interest payment from the Annual Payment.

4.) Obtain the outstanding Principal balance for the next year by subtracting the first year's annual Principal payment from the original Principal amount of the loan.

5.) Obtain the next year's annual Interest payment, by multiplying the preceding year's outstanding Principal balance by the rate of Interest.

6.) Obtain the next year's annual Principal payment by subtracting that year's annual Interest payment from the Annual Payment.

7.) Obtain the next year's outstanding Principal balance by subtracting that year's annual Principal payment from the preceding year's outstanding Principal balance.

8.) Repeat steps 5, 6, and 7, for each year of the loan.

As you can see, if you have a forty-year loan, you have your work cut out for you. And, if you have a thirty-year mortgage and would like to create an annual debt service payment schedule for your home, you really have your work cut out for you, because you must do each of the calculations for each month. Since there are three steps required for each payment and since there are 360 monthly payments on a thirty-year loan, you will have 1,080 calculations! Now, you know why your banker didn't create your home mortgage payment schedule while you were sitting there in his office. You also know why computers are so popular among financial people.

THE LEVEL PRINCIPAL PAYMENT METHOD

Now that you know that the Level Payment Method means that the Annual Payments are always the same, you might correctly assume that with the Level Principal Payment Method, the annual Principal payments are always the same. There is an important corollary to this assumption.

With the Level Payment Method, the Annual Payment is a constant and both the Interest payment and the Principal vary each year.

In the Level Principal Payment Method, the annual Principal payment is a constant and both the Annual Payment and the annual Interest payment vary each year.

We can begin to examine the Level Principal Payment Method by first ascertaining what, in fact, the level Principal payments will be for each year of the loan. We do this simply by dividing the original Principal amount of the loan ($100,000), by the number of years in the term of the loan (5), or $100,000 / 5 = $20,000. We now know that we will pay off $20,000 of Principal in each year of the loan.

Here, in Table 2-C below, are the annual Principal payments for the five-year loan of $100,000 at 10% in our example:

Table 2-C

Year	Principal Payment
1	$20,000
2	20,000
3	20,000
4	20,000
5	20,000
	$100,000

Please note, above all, that the annual Principal payments equal the original Principal amount of the loan.

Next, referring back to our corollary, we must realize that if the annual Principal payments are constant, then both the Annual Payment and the annual Interest payment will vary from year to year. Let us see.

In the first year, all $100,000 of the original loan Principal is outstanding, so we can calculate the first year's annual interest payment by multiplying the original Principal balance ($100,000) by the rate of Interest (10%, or .1), as follows:

$$\$100,000 \times .1 = \$10,000$$

Now that we have both the first annual Interest payment ($10,000) and the level Principal payment ($20,000), we can readily see that the first Annual Payment is $30,000.

$$\$10,000 + \$20,000 = \$30,000$$

Thus, we now have both the first year's Interest payment ($10,000), the level Principal payment ($20,000), which is not only paid in the first year but in every year of the loan, as well as the first year's Annual Payment which is the sum of the two, or $30,000. Now, to calculate the succeeding years' Annual Payments and annual Interest payments.

In order to do so, we must first calculate the outstanding Principal balance at the end of the first year. This is done by subtracting the level Principal payment ($20,000) from the original Principal balance ($100,000), as follows:

$$\$100,000 - \$20,000 = \$80,000$$

With the outstanding Principal balance at the end of the first year, we can now calculate the annual Interest payment and the Annual Payment for the next year. The annual Interest payment for the second year is obtained by multiplying the outstanding Principal balance ($80,000) by the rate of interest (10%, or .1), as follows:

$$\$80,000 \times .1 = \$8,000$$

With the second year's annual Interest payment ($8,000) we can now calculate the second year's Annual Payment by adding it to the level Principal payment ($20,000), as follows:

$$\$20,000 + \$8,000 = \$28,000$$

Notice that the second year's Annual Payment is lower that the first year's Annual Payment.

By repeating the above process for every year of the loan term, we can construct an annual debt service payment schedule for our five-year loan based on the Level Principal Payment Method, as is shown in Table 2-D, below.

Table 2-D Annual Debt Service Payment Schedule, Level Principal Payment Method (Loan: $100,000, 5 years, 10% interest)

Year	Annual Payment	Interest	Principal	Outstanding Principal Balance
1	$30,000	$10,000	$20,000	$80,000
2	28,000	8,000	"	60,000
3	26,000	6,000	"	40,000
4	24,000	4,000	"	20,000
5	22,000	2,000	"	0
			$100,000	

Please note that with the Level Principal Payment Method:

1.) The annual Principal payments are the same in each year. This is the definition of the Level Principal Payment Method.

2.) Both the annual Interest payment and the Annual Payment decrease each year.

3.) The total of all of the annual Principal payments, of course, equals the original Principal amount of the loan.

RULES FOR CREATING ANNUAL DEBT SERVICE PAYMENT SCHEDULES FOR THE LEVEL PRINCIPAL PAYMENT METHOD

Creating an annual debt service payment schedule with the Level Principal Payment Method does not involve an elaborate equation or a table. It can be done with the aid of a simple adding machine or just a pencil and paper.

1.) Once you know the original Principal amount of the loan and the number of years in the term, you can calculate the level Principal payment, which will be the annual Principal payment for each year of the loan, by dividing the original Principal amount by the number of years in the term of the loan.

2.) Calculate the first year's annual Interest payment by multiplying the original Principal amount of the loan by the rate of interest.

3.) Obtain the first year's Annual Payment by adding the level Principal payment to the first year's annual Interest payment.

4.) Obtain the outstanding Principal balance at the end of the first year by subtracting the first year's level Principal payment from the original Principal amount of the loan.

5.) Calculate the next year's annual Interest payment by multiplying the outstanding Principal balance from the preceding year by the rate of Interest.

6.) Calculate the next year's Annual Payment by adding that year's annual Interest payment to the level Principal payment.

7.) Calculate the next year's outstanding Principal balance by subtracting the current year's level Principal payment from the preceding year's outstanding Principal balance.

8.) Repeat steps 5, 6, and 7, for each year of the loan term.

THE IRREGULAR METHOD

Unlike both the Level Payment Method and the Level Principal Payment Method, which have at least one constant, the irregular method of debt service payment has none. As its name suggests, it is totally irregular. Before looking at the irregular method, it will be worthwhile to explore the circumstances in which it is most often used.

The irregular method of payment is used, in short, when neither the Level Payment Method nor the Level Principal Payment Method will do for factual reasons.

There are actually very few sets of circumstances which come to mind where the irregular method might be preferable.

The first might be where a system issued a bond or obtained a loan prior to the construction of the project.

In such case the project might require two years or more to complete, and the system's board might not want (or might not be permitted) to begin charging the users for the project until it was on line and the benefits of the project were available to all of the ratepayers.

In such case, you would most likely see a payment schedule which provided for no principal payments at all for the first three years. During this time, the bondholders or the bank would only receive interest on the bond or loan.

As you know from the preceding paragraphs, however, in the first few years of the loan, interest payments can be substantial. Using the level payment method, interest makes up the great majority of the Annual Payment in the first several years of the loan. Thus, if the goal of the board of directors is to minimize the cost to the ratepayers prior to the actual operation of the project, then, in most circumstances, not only will the directors elect not to have any principal payments in the first three years, but they will also elect not to have any interest payments during that period, as well!

How so? Wouldn't we all like to elect not to make either principal or interest payments on our loans? Yes, of course, we would; it is just that average citizens are seldom given the opportunity to make such an election. Water districts, on the other hand, may have this luxury. It all depends on the financial history of the district and the current status of its funds. Financially strong systems will be offered the opportunity to defer such payments. Weak systems will not.

How is it done? Well, of course, if the system had issued a bond, the bondholders would still get their interest payments every six months.

Interest would actually be paid by the system. Yes, but from what funds? The answer is: from borrowed funds.

In short, a system which borrows in advance of charging its users will have to get the money to pay interest to the bondholders, during such time, from somewhere. That "somewhere" is the bond itself. In other words, the system must borrow more than it needs for the project. In fact, it overborrows the precise amount it needs to make up for the missing user fees.

Suffice it to say that when a system elects to defer principal payments, it will have an irregular payment schedule. If it elects to defer both principal and interest payments, it will not only have an irregular payment schedule but it will have an inflated total project cost, as well. The problem of inflated project costs will be dealt with in Chapter 14 below. The irregular payment schedule will be dealt with here.

Let us assume that a system issued a $1,000,000 bond for a forty-year term at an 8% rate, with no principal payments for the first three years. Assume, as well, that in year four, principal begins to be amortized according to the Level Payment Method. As such, the first ten years of the annual debt service payment schedule would be as presented, below, on Table 2-E.

Table 2-E

Year	Interest	Principal	Annual Payment
1	$80,000	$ 0	$ 80,000
2	"	"	"
3	"	"	"
4	"	4,900	84,900
5	79,608	5,292	"
6	79,185	5,715	"
7	78,727	6,173	"
8	78,234	6,666	"
9	77,700	7,200	"
10	77,124	7,776	"

As you can see, with the irregular method neither the Interest, nor the Principal, nor the Annual Payment remain constant over the entire term of the bond. As you can also tell by simple visual inspection, there is very little to be gained by this method. The difference between the first three years' payments and the remaining payments is only a little over 6%.

I want to point out that it would not be correct to compare the annual debt service payment schedule presented above with an annual debt service payment schedule for a bond of identical tenor but which uses either the Level Payment Method or the Level Principal Payment Method. This is because, for the reasons discussed above, in order to defer both Interest and Principal, a system would have to increase the amount it had to borrow. Thus in a fair comparison, both the Level Payment Bond and the Level Principal Payment Bond would be for $1,000,000, but the comparable irregular method bond would be for more than $1,000,000.

Another circumstance which might warrant the use of the irregular method is where a system might have a realistic expectation of future income. By "realistic expectation," I mean such circumstances as where a system is extending its lines to a group of existing homes which, for one reason or another, must connect to the system. I specifically do not mean the circumstance whereby a system builds lines out into unoccupied land and then prays that someone will buy a lot, build a home on it, and eventually connect to the system.

The additional income from the annual service charges -- not the non-recurring connection fees -- for the real, additional users can be calculated precisely using today's rates. A reliable schedule of when that additional income can be expected to reach the system's coffers can also be developed. It seems not only reasonable but laudable that the system would opt for the irregular method under these circumstances. After all, if it did not, the system would find itself in a few years with an excess of income (from the additional user fees) and its directors would be open to the charge of fiscal mismanagement.

This what might be termed "impending additional income" situation could just as easily arise from other types of circumstances. For example, a system may be planning to sell some of its property. It will certainly be able to reasonably estimate the amount of gain it will have from the sale and determine that at least a portion of that gain will be set aside to pay off loan principal. This would warrant using the irregular payment method with principal payments skewed to those years immediately after the planned sale date of the property.

The final circumstance which might also warrant the use of an irregular payment method, arises not from within the system, but from without the system. It actually arises from within the bond market.

In such circumstances, there are simply an inordinately large number of buyers willing to buy bonds of certain terms (e.g., five years, or ten years) but not other terms (e.g., six years, or nine years). Because there are a large number of buyers, the interest rates at these maturities should be very favorable to the water system which is borrowing. (Please see discussion of Rates vs. Maturities in Chapter 10, below.) In such case, a water system's financial advisor or investment banker might well recommend an annual debt service payment schedule with Principal payments skewed to those favorable maturities to take advantage of the more favorable rates.

In summary, there are certain factual situations which might warrant the use of the irregular payment method. If this method is recommended to you, you can construct an annual debt service payment schedule, albeit you must do so manually.

RULES FOR CREATING ANNUAL DEBT SERVICE PAYMENT SCHEDULES USING THE IRREGULAR METHOD

Creating an annual debt service payment schedule for the irregular method must be done manually. As you know, with the irregular method the annual Principal payments are given, literally. This means that either your financial advisor or investment banker will give you a schedule with the proposed Principal payments for each year. In some cases (where they are targeting certain maturities to get better rates), they will also give you the Interest rate for each particular maturity. This is where the calculation gets a little complicated. The reason for this is that you must actually make up a mini-schedule for each maturity, and then add all of them together. This is not as onerous as it sounds because each of the maturities has only one Principal payment; and, therefore, all of the interest payments in the intervening years are the same. In short, the mini-schedule for each of the maturities actually looks like a small bond with only one Principal payment. Let us look at an example.

Let us take the case of a loan or small bond issued for only seven years to purchase some machinery and equipment for your water or wastewater system. The bond is for $1,000,000 and your financial advisor recommends paying off 40%, or $400,000 at the end of the fifth year and the balance, 60% or $600,000, at the end of the seventh year. He makes this recommendation because he believes the five-year maturity will carry an Interest rate of 6% and the seven-year maturity will carry an Interest rate of 8%.

As indicated before, to create an annual debt service payment schedule for the whole bond we must first create mini-schedules for each of the maturities, and then add them together. Here, in Table 2-F below, is a schedule for the year five maturity at an Interest rate of 6%:

Table 2-F

Year	Interest	Principal	Annual Payment
1	$24,000	$ 0	$ 24,000
2	"	"	"
3	"	"	"
4	"	"	"
5	"	$400,000	$424,000

Now, here, in Table 2-G below, is the schedule for the final maturity in year seven at an Interest rate of 8%:

Table 2-G

Year	Interest	Principal	Annual Payment
1	$48,000	$ 0	$ 48,000
2	"	"	"
3	"	"	"
4	"	"	"
5	"	"	"
6	"	"	"
7	"	$600,000	$648,000

Now we can complete the annual debt service payment schedule for the whole bond by adding together the two schedules above, as shown on top of the next page in Table 2-H:

Table 2-H

Year	Interest	Principal	Annual Payment
1	$72,000	$ 0	$ 72,000
2	"	"	"
3	"	"	"
4	"	"	"
5	"	400,000	472,000
6	48,000	0	48,000
7	"	$600,000	$648,000

That is what an annual debt service payment schedule looks like when the irregular payment method is used. It is not pretty; but, at least, it is not difficult to create.

Now, before moving on, there are two minor points which need to be made.

Going back to our example, above, let us assume for the moment that your financial advisor is right, that you are able to get these rates, and that you are satisfied you are saving your system some money; there is still one thing which should be bothering you.

Look at the payment schedule. How do you think your ratepayers will react in year five when their rates are adjusted to accommodate a 656% increase (from $72,000 to $472,000) in annual debt service payments? Or, even better, how will they feel the following year, when they are adjusted again to accommodate a 90% decrease (from $472,000 to $48,000) in annual debt service payments?

The answer to this is obvious. The ratepayers would be furious. You and your fellow members of the system's management would undoubtedly find yourselves unflatteringly, but prominently, displayed on the front pages of the local newspaper.

You do not have to put your ratepayers through the agony of such a teeter-totter payment schedule. You do not have to collect the precise amount of annual debt service payments due each year from the ratepayers. You can collect any reasonable amount; and, under the circumstances, one of the more reasonable amounts to collect would certainly be the average of all the payments.

Place a piece of paper below the Annual Payment column in the final annual debt service payment schedule and add them up. They will total $1,456,000. Then divide by the number of years of the bond (7), to get the average of $208,000. If you collected $208,000 each year from your ratepayers, there would be no shock from radically changing rates, and you would certainly have sufficient funds to make all of the Annual Payments.

If you did, however, collect the $208,000 each year from your ratepayers, you would be collecting too much.

In year one you would collect $208,000 but would only pay out $72,000. At the end of the year you would have a balance of $136,000 which you would not need for several years. You would, of course, invest those funds. You would do the same thing every year, and at the end of the bond term in year seven, you would have a sizeable sum remaining after all of the bonds were paid off. This amount would constitute an overcollection from your ratepayers.

Now you may be able to brush such a matter under the rug, especially if it is small. But more than likely you will be open to the charge of fiscal mismanagement for not taking into account at least some investment earnings before you determined how much the ratepayers had to pay each year for the project.

A conservative and highly acceptable way of estimating these investment earnings is to assume that the funds will be invested from the date of receipt until they need to be disbursed. A conservative interest rate should also be used to estimate earnings. In this regard, assuming you are in a stable and otherwise normal interest rate environment, it is always prudent to use a rate slightly lower than the prevailing rate in making your estimates. In Chapter 4, we will discuss the pitfalls of choosing the wrong interest rate to estimate savings. Suffice it to say here that a rate slightly lower from the prevailing rate will save you from appearing in the headlines again -- this time for causing a default on your system's loan because you over-estimated your interest income.

In summary, with the irregular method, annual debt service payment schedules are created manually from information provided by your financial advisor or investment banker concerning the annual Principal payments and the Interest rates at various maturities over the term of the loan. Here are the simple rules:

1.) Obtain the annual Principal payments and the Interest rates applicable thereto for each maturity from the system's financial advisor or investment banker.

2.) For a particular maturity, calculate each of the annual Interest payments for each maturity by multiplying the amount of

Principal to be paid at that maturity by the rate of Interest applicable to that maturity.

3.) For the same maturity, calculate each of the Annual Payments by adding the annual Interest payment to the annual Principal payment, if any, for each year.

4.) Repeat steps 2 and 3, creating a schedule for each maturity.

5.) Add all of the schedules for all of the maturities.

There are times when using the irregular method can be very advisable for a water or wastewater district. As with any method, or with any financial program or technique, the real question is whether it is the most cost effective means of financing the project for the ratepayers. If it is, good. If it is not, there should be an extremely compelling argument as to why any method which is not the most cost effective should be used.

Chapter 3

Evaluating Annual Project Payments

This chapter deals with the problem of dollars which are paid or earned in different years.

The problem arises because the value of money changes over time.

More specifically, a dollar to be paid in twenty years is not the same as a dollar which must be paid today, or next year, or in any other year.

We can illustrate this point with a brief example.

Let us say your water system needs to finance a project and receives two offers:

Offer A requires the system to pay $190,000 for ten years.

Offer B requires the system to pay $100,000 for twenty years.

Which is less expensive for your system?

(There are important considerations as to whether you should make your decision on an annual or total cost basis. These matters will be discussed in Chapters 6 and 7, below. For now, however, we will consider only the total cost basis.)

The natural approach to the problem would seem to add up the total payments and see which is cheaper.

Offer A requires 10 payments of $190,000 each, or $1,900,000 (10 x 190,000 = 1,900,000).

Offer B requires 20 payments of $100,000 each, or $2,000,000 (20 x 100,000 = 2,000,000).

One would, therefore, conclude, by subtracting the sum of the payments in Offer A from the sum of the payments in Offer B (2,000,000 - 1,900,000 = 100,000), that Offer A is cheaper by $100,000.

The one who reached this conclusion, however, would be wrong.

Offer A is not $100,000 cheaper. Offer B is cheaper. And, it is $220,907 cheaper!

The reason why Offer B is cheaper than Offer A is because the value of money changes over time. The real value of each payment is different in each year. And, the sum of the real values is the real cost.

I honestly believe that one of the major reasons why many people think that finance is black magic is that they fundamentally do not believe that the value of a dollar changes over time.

In the last ten years I have taught what is called the "time/value theory of money" to somewhere between twenty-five hundred and three thousand students. Of these, I would estimate that between twenty and twenty-five percent did not believe that the dollar changed value over time and would not be convinced of it. And, another forty to fifty percent probably had at least some doubts.

I strongly suspect that one of the major reasons for this almost endemic phenomenon lies in the childishly simple notion that the actual dollars, i.e., the bills themselves, change very little over time.

As a matter of fact, as far as the notes themselves are concerned, apart from the changes of the names of the Treasurer of the United States and the Secretary of the Treasury, the only other change I can think of in the last thirty-five years is that the words "silver certificate" were replaced with "federal reserve note" on the front of the one dollar bill. Other than that, the dollar has not physically changed at all. If you put a fistful of dollars in a time capsule fifty years ago, they would still look the same. Furthermore, if, your parents put the exact amount of their first mortgage payment in a tin box and buried it in the back yard on the day they bought their home thirty years ago, they would find as they dug it up thirty years later that they had enough money to make the final payment. If that is true, how could the value of the dollar have changed?

The Time/Value Theory of Money.

We must begin by pointing out that the "Time/Value Theory of Money" is not a theory at all. It is fact. Absolute fact.

The word "theory" has nothing to do with whether the value of money changes over time; rather, it has only to do with how the value of money changes over time.

Having said this, we are now going to present a series of examples to illustrate just how the value of money does change over time.

The first, and perhaps, most familiar is the savings account.

The second deals with a U.S. Savings Bond and how the surrender value is calculated. The third and fourth examples are more abstruse.

The third deals with changes in the price of milk and how that relates to the change in the value of money over time.

The fourth example uses a thirty-year, fixed-rate mortgage to show the subtle changes which are actually occurring when neither the home nor the payment appear to change at all over the thirty-year period. Next, we will do some actual illustrations of how the change in the value of money over time relates to the payments a water or wastewater system makes on behalf of its projects. And, last, the chapter will end with the solution to the problem posed at the beginning of the chapter: how twenty years of $100,000 payments can actually be cheaper than ten years of $190,000 payments.

Now, for the first example, the savings account.

Let us say that you have $1,000 which you want to put into a safe investment account to have for your retirement in twenty years. Let us say your local bank will offer you a guarantied rate of 5%, compounded annually. How much money will you have in twenty years?

Before we calculate the answer, we have to get clear on some terminology to avoid confusion later on.

Present Value and Future Value.

In our example, the $1,000 you have today is called the Present Value.

The amount of money you will have in twenty years is called the Future Value.

Please note that the terms Present Value and Future Value are relative terms. They have nothing to do with today's calendar. The best rule to follow is that when you have two sums of money separated by time the Present Value is the earlier amount and the Future Value is the later amount.

Discounting and Compounding.

The process of going from Present Value to Future Value is called Compounding. This is the process by which the value of money increases over time.

The process of going from Future Value to Present Value is called Discounting. This is the process by which a known sum of money in the future can be reduced to today's terms.

As you can readily see, Discounting is the opposite of Compounding.

The Interest Rate is the factor by which you either Compound a Present Value to reach a Future Value, or the factor by which you Discount a Future Value to reach a Present Value. Thus, when going forward from Present to Future Values, it is called a Compounding Rate; and, when going backwards from Future to Present Values, it is called a Discount Rate.

The Period is the number of times you either Compound a Present Value to reach a Future Value or the number of times you Discount a Future Value to reach a Present Value. In most instances, you will be told what the Compounding or Discounting Period is. For example, in our problem above, we said that our $1,000 would Compound annually. For twenty years, this means that we would Compound twenty times.

In loan transactions, the Period is the number of times a Principal payment will be made. For example, most home mortgages have monthly Principal payments. This means that for a thirty-year home mortgage with monthly payments, there would be 360 Periods (30 years x 12 months per year = 360 Periods).

Most large loans, such as those made to water or wastewater systems, have annual Principal payments. So, this book will assume annual Compounding or Discounting, unless specifically noted otherwise.

Please be sure to note as well that the term Period refers to Principal payments only. It is common that Interest payments may be made more frequently during the year. For example, Interest is often paid quarterly or semi-annually, while Principal is only paid annually.

CALCULATING FUTURE VALUES

Now, here are the procedures for compounding or calculating Future Values:

To Calculate Future Values (Compounding)

1.) Convert the Interest Rate, or Compounding Rate (e.g., 5%), to a decimal (e.g., .05) and add the number 1 (1 + .05 = 1.05). We will call this number the Compounding Factor.

2.) Multiply the Present Value (e.g., $1,000) by the Compounding Factor as many times as there are Periods (e.g., 20). The result will be the Future Value.

Here is what the calculation of the Future Value of your $1,000 retirement account would look like, Figure 3-A.

The Future Value of your pension fund will be $2,653.30 in twenty years.

The Present Value is $1,000. The Compounding Rate is 5%, or .05. And, the number of Periods is 20. Thus, according to our formula, we first find the Compounding Factor by adding the number 1 to the decimal expression (.05) for the Compounding Rate (1.05). Then, we multiply the Present Value by the Compounding Factor 20 times, which is the number of Compounding Periods. The result is the Future Value.

This process is extremely important. The terminology may seem unwieldy at first, but it is worth getting used to, since it is widely used in comparing project costs not only for water and wastewater projects but for all types of financing.

CALCULATING PRESENT VALUES

Finding a Present Value, when you know the Future Value, involves precisely the opposite process.

Let us use a U.S. Savings Bond as an example. Let us say a rich aunt gives your newborn baby a $1,000 Savings Bond as a present. Such Bonds generally mature in twenty years. For the sake of continuity, let us also say that the Savings Bond carries an Interest Rate of 5%.

Figure 3-A

$1,000.00	
x1.05	(1)
1,050.00	
x1.05	(2)
1,102.50	
x1.05	(3)
1,157.63	
x1.05	(4)
1,215.51	
x1.05	(5)
1,276.28	
x1.05	(6)
1,340.10	
x1.05	(7)
1,407.10	
x1.05	(8)
1,477.46	
x1.05	(9)
1,551.33	
x1.05	(10)
1,628.89	
x1.05	(11)
1,710.34	
x1.05	(12)
1,795.86	
x1.05	(13)
1,885.65	
x1.05	(14)
1,979.93	
x1.05	(15)
2,078.93	
x1.05	(16)
2,182.87	
x1.05	(17)
2,292.02	
x1.05	(18)
2,406.62	
x1.05	(19)
2,526.95	
x1.05	(20)
$2,653.30	

Figure 3-B

```
$1,000.00
   x.9524   (1)
   952.38*
   x.9524   (2)
   907.03
   x.9524   (3)
   863.84
   x.9524   (4)
   822.70
   x.9524   (5)
   783.53
   x.9524   (6)
   746.22
   x.9524   (7)
   710.68
   x.9524   (8)
   676.84
   x.9524   (9)
   644.61
   x.9524   (10)
   613.91
   x.9524   (11)
   584.68
   x.9524   (12)
   556.84
   x.9524   (13)
   530.32
   x.9524   (14)
   505.07
   x.9524   (15)
   481.02
   x.9524   (16)
   458.11
   x.9524   (17)
   436.30
   x.9524   (18)
   415.52
   x.9524   (19)
   395.73
   x.9524   (20)
  $376.89
```

We are now prepared to find out just how much your rich aunt actually paid for the Savings Bond. We do so by finding how much it is worth today.

In this example, we know the Future Value ($1,000). We know the Interest Rate (5%), which, because we are going backwards from Future to Present Values, is a Discount Rate. We also know the Discounting Period, which is 20 years.

Here is the procedure for discounting or calculating the Present Value.

To Calculate Present Values (Discounting)

1.) Convert the Interest Rate, or Discount Rate (e.g., 5%) to a decimal (e.g., .05) and add the number 1 (1 + .05 = 1.05). Next, divide this number into the number 1, to obtain the Discount Factor: (1 / 1.05 = .9524). Please note there is an extra step here.

2.) Multiply the Future Value (e.g., $1,000) by the Discount Factor as many times as there are Periods (e.g., 20). The result will be the Present Value.

Here is what the calculation of the Present Value of the Savings Bond would look like, Figure 3-B.

The value of the 20-year, $1,000 Savings Bond today (i.e., its Present Value) is $376.89.

The Future Value is $1,000. The Discount Rate is 5%, or .05. And, the number of Periods is 20. Thus, according to our formula, we first converted the Discount Rate (5%) to decimal form (.05), and added it to the number 1, to get 1.05. Then, we divided this number into the number 1, to obtain the Discount Factor (.9524). Finally, we multiplied the

Future Value by the Discount Factor 20 times, which is the number of Discounting Periods. The result is the Present Value.

The third example for us to consider concerns the changes in the price of milk.

THE TIME/VALUE OF A QUART OF MILK

I remember growing up in Buffalo, New York, almost forty years ago, when my mother would give me a quarter and watch out the window as I crossed Linwood Avenue and walked the half block to Mr. Whitcomb's Delicatessen. For my quarter, I would get not only a quart of milk, but also two penny candies, which, of course, was the reason I volunteered to go to the store in the first place.

In the last few years, since the Governor broke the dairy cartel in New York City, the price of milk has been about ninety-five cents a quart.

The milk certainly has not changed; it has not improved. The dollar bill with which I pay for the milk has not changed. What has changed is the price.

This is my point. People don't believe that the dollar has changed because it hasn't. But they do believe that prices change and that they are changing every day.

What do we really mean when we say prices change? The simple answer, in terms of milk, is that it takes more dollars to buy the same amount of milk. The quantity of the goods (milk) remains the same but the quantity of dollars changes.

But what if we were to theoretically reverse the situation? What if we keep the number of dollars constant and, instead, change the quantity of milk? For example: How many ounces of milk could you buy for twenty-three cents in 1952? How many ounces of milk can you buy for twenty-three cents in 1992? The answer in the first instance is 32 ounces and in the latter case about 7 3/4 ounces. Here the number of dollars is constant and the quantity of goods which the dollar can buy changes.

This is a good example of Discounting. The change in the amount of milk is the discount. More accurately, the change in the amount of milk divided by the number of years is the average discount. Even more accurately, the discount rate equals one minus the nth root of the old milk divided by the new milk, where n is the number of years in between. Now I know you are really sorry you began reading this book.

What the above convoluted sentence is really saying is the following: By what number can I multiply the new amount of milk -- once for each year -- so that it will equal the old amount of milk?

In the context of our example, the precise question would be: By what number can I multiply 7.75oz. -- 40 times -- so that it will equal 32 ounces? The answer is 1.0361. Does this number look familiar? It should. It is the Compounding Factor. It is the number by which you multiply the original quantity (Present Value) the number of times in the Compounding Period (40 years) to reach the Future Value (32 ounces).

(N.B.: Please note that in this example the Present Value is the value in 1952 and the Future Value is in 1992. This illustrates the principle that the concepts of Present Value and Future Value are relative to any given example and have absolutely nothing to do with today's calendar.)

Now, to continue, we subtract the 1 (since that represents the original quantity [7.75 ounces] itself) and make a percentage out of the remainder: .0361 = 3.61%. That is the annual rate at which the amount of milk changed. That is the Compounding Rate.

By now, you undoubtedly feel some unease talking about an "annual rate at which the amount of milk changed." This is not a natural way to describe things. Amounts of milk don't change. The containers do not get smaller. What actually changes is the price. It is much easier to visualize the change in the amount of something like milk, but it is far more commonplace to talk about change in price than the change in the amount of a commodity.

In terms of price or dollars, the math is the same except we must reverse the time sequence. (When price is constant, the old quantity is large and the new quantity small. When quantity is constant, the old price is small and the new price is large.)

In other words, the proper question would now be: By what number would I multiply 23 cents -- 40 times -- so that it would equal 95 cents? Not surprisingly, the answer is again 1.0361, or 3.61%.

The number 1.0361 is the Compounding Factor. The number 3.61% is the Compounding Rate. They are called the Compounding Factor and the Compounding Rate because we are going from a Present Value to a Future Value. The earlier amount is the Present Value. The later amount is called the Future Value.

In the example about Mr. Whitcomb's milk, we have both changing quantities of goods (the milk) and changing quantities of money (the price).

Changing the quantity is certainly easy to visualize. Changing the price is also readily comprehensible. What is usually difficult to grasp, however, is the change which occurs when both the quantity and the price remain the same. This leads us to our fourth, and final, scenario. The best example of this scenario is the home mortgage payment.

THE TIME/VALUE OF A HOME MORTGAGE

Let us assume a thirty-year, fixed-rate home mortgage. Let us say you took out the mortgage in 1980 for $75,000 at a rate of 10%. As such, your first monthly payment in 1980 was $658.18, and your last monthly payment in 2009 will also be $658.18. As you can see, there is no change in the dollars. Also, there is obviously no change in the "goods." Your home is the "goods," and it doesn't change at all. The only thing that changes in this example is the time. If the value of money is supposed to change over time, where is this change manifest?

The answer in this case is quite subtle. There really is change. And, the change really does occur in your home!

How can this be so? Obviously, the structure does not change. No, of course not. What changes is the value! In the example about Mr. Whitcomb's milk, the amount of milk was constant and the amount of dollars changed. Then we reversed the example and kept the amount of dollars constant which caused the amount of milk to change.

Here, too, the amount of dollars, $658.18, is constant, but the house -- or, more correctly, the value of the house -- is changing.

In support of this contention, let me pose two questions. How much do you think the home you bought in 1980 will be worth in 2009? Or, conversely, how much of a home do you think you will be able to buy in 2009 with a $658.18 monthly mortgage payment? Each person reading this book may give a different answer to these questions, but I am willing to bet that 99.9% of the respondents will say that their 1980 home will be worth substantially more in 2009, and that they will not be able to get a very nice new home in 2009 with a monthly payment of only $658.18.

Now, we have established more or less intuitively that the value of the home is changing, just like the amount of milk did in the previous example. But unlike the milk, we cannot measure the amount of change. You can estimate how much you think your house is worth today, but you will never know for sure unless you sell it. Even more ephemeral is the value of your home in 2009. The best you can do is make an educated guess as to what the value will be in the next nineteen years based on what changes have occurred in comparable values over the last eleven years.

If you paid $100,000 for the home in 1980 and you think it is worth $175,000, then your home has increased in value by about 5.2% a year. You can check this with the formula for calculating the Future Value when the Present Value is known.

First, we convert the assumed Compounding Rate (5.2%) to a decimal (.052) and add the number 1 (.052 + 1 = 1.052), which is the Compounding Factor. Next, we multiply the Present Value ($100,000) by the Compounding Factor (1.052), the number of times in the Compounding Period (11), as shown in Figure 3-E, below:

Figure 3-E

```
$100,000
  x1.052  (1980)
$105,200
  x1.052  (1981)
$110,670
  x1.052  (1982)
$116,425
  x1.052  (1983)
$122,479
  x1.052  (1984)
$128,848
  x1.052  (1985)
$135,548
  x1.052  (1986)
$142,597
  x1.052  (1987)
$150,012
  x1.052  (1988)
$157,813
  x1.052  (1989)
$166,019
  x1.052  (1990)
$174,652
```

There you have it: $174,652, or call it $175,000. This is the Future Value.

Now, if you want to find out how much your home will be worth in 2009, you simply continue the process another 19 times, as shown in Figure 3-F.

There you have it. This is the Future Value of the house for which you paid, or the Present Value of which was, $100,000 in 1980. So, in the year 2009, the home you paid $100,000 for in 1980 will be worth the grand sum of $457,554.

This is quite impressive. But before you start counting your money, please remember back a few pages to our example about the quart of milk. You will recall that today the milk costs 90 cents. Well, if the cost of milk increases as fast as the value of your home, you will be paying $2.36 for that same quart of milk in 2009! Furthermore, if in 2009 your children buy the exact same quality of home as you bought in 1980, they will have to pay $457,554 for it!

In our example, we assumed you got a $75,000, or 75%, mortgage in 1980. If your children get a 75% mortgage in 2009, it will be for $343,166!

Now, before we leave this example, there are a couple of additional questions which can shed some light on this difficult concept.

How much were you earning the day you bought the home in 1980? How much are you earning now? How much do you think you will be earning in 2009?

If you have had a reasonably normal life, you are earning more now than you were in 1980 and you can expect to be earning considerably more in 2009.

Figure 3-F

```
$174,652
  x1.052  (1991)
$183,734
  x1.052  (1992)
$193,288
  x1.052  (1993)
$203,339
  x1.052  (1994)
$213,912
  x1.052  (1995)
$225,036
  x1.052  (1996)
$236,738
  x1.052  (1997)
$249,048
  x1.052  (1998)
$261,999
  x1.052  (1999)
$275,623
  x1.052  (2000)
$289,955
  x1.052  (2001)
$305,033
  x1.052  (2002)
$320,894
  x1.052  (2003)
$337,581
  x1.052  (2004)
$355,135
  x1.052  (2005)
$373,602
  x1.052  (2006)
$393,029
  x1.052  (2007)
$413,467
  x1.052  (2008)
$434,937
  x1.052  (2009)
$457,554
```

Let us say you were earning about $32,000 a year in 1980. Your monthly payment of $658.18 means you were paying $7,898.16 a year.

Divided by your salary, this means you were paying 7,898.16 divided by 32,000, or about 25% of your income for your home. Let us say you are now earning $60,000 per year. You are still paying $658.18 a month, or $7,898.16 a year; but now your mortgage payment is only 7,898.16 divided by 60,000, or about 13% of your salary.

Now, for the good part. If you keep getting raises at the same rate, in 2009 you will be earning about $160,000! At that point, when you make your last mortgage payment, it will constitute only 7,898.16 divided by 160,000, or just about 5% of your income.

The entire point in this exercise involving mortgage payments, the value of your home, and the amount of your income, is to illustrate that the value of money changes over time. Here is the "time/value theory of money" again. But as you can readily see, it is not the least bit theoretical. It is fact. And very tangible fact, at that.

Now, let us go from examples such as the price of milk or the value of a home -- where the change in the value of money over time is quite apparent -- to the world of financing water and wastewater projects, where it can be less obvious.

THE TIME/VALUE OF PROJECT PAYMENTS

Perhaps the best way to envision the change in the value of money over time when it comes to such matters as water and wastewater improvement projects is to think of it in terms of the purchasing power of the ratepayers. This is very similar to the example about the difference between the mortgage payment and the amount of the homeowner's income. The amount of the mortgage payment remains the same but the homeowner's salary keeps rising. Thus, the value of the mortgage payment keeps declining in relation to the homeowner's income.

By the same token, when the board of directors of a water system elects to finance a project over a number of years, it can be sure, whether the loan payments are the same amount every year or different, that whatever the loan payment amount is in a future year, the actual impact on the ratepayer will be less because his income, or buying power, has increased.

Whenever you begin to have doubts, just remember that you will be earning $160,000 in the year 2009, and your $658.18 monthly mortgage payment will have declined from 25% to only 5% of your total income.

Let us say that a water system is adding a substantial amount of new mains, replacing some old mains and installing a new standpipe to increase storage capacity. Let us say the project will cost $1,000,000.

Now, you could -- theoretically -- bill your users for the full $1,000,000 today as a surcharge. If you had a large system with 100,000 or so users, you would probably get away with it. But if your system had 1,000 users, you might, indeed, have a rebellion on your hands. In any event, if you did surcharge the users, there would be no doubt that the real cost of the project would be $1,000,000, and no more.

Let us say, as an alternative, that you consulted three local bankers, each of whom was willing to make your water system a loan of $1,000,000 to pay for the project, as follows:

Banker A will lend for five years at 8% interest with level principal payments.

Banker B will also lend for five years at 8% interest but with level payments.

Banker C will lend for ten years but at 9% interest and also with level payments.

Now, let's look at the five Annual Payments on the loan from Banker A:

Year 1 - $280,000
Year 2 - 264,000
Year 3 - 248,000
Year 4 - 232,000
Year 5 - 216,000

We know from our examples about the milk and the mortgage payment that the value of money changes over time. In the milk example, we could measure the change in the value of money by measuring the change in the amount of milk. In the home mortgage example, we could measure the change in the value of money either by measuring the change in the value of the home or by measuring the change in the homeowner's income relative to his mortgage payment. Here, the value of the new mains and the standpipe doesn't change at all (leaving aside the concept of depreciation). So, what we are effectively doing is measuring the amount of the payment in terms of the ratepayer's rising income or buying power.

The actual measurement of the change in the value of money in this example involves the use of a parameter. A parameter is a constant number which is, itself, made up of variables. The parameter we can use here is the <u>rate of inflation.</u>

ANNUAL INFLATION RATE AS DISCOUNT FACTOR

The next chapter, Chapter 4, deals specifically with the determination of which (of several) discount factors to use in a given instance. It also illustrates the pitfalls which arise when an inappropriate discount rate is used. For our purposes here, in this chapter, we will assume that the annual rate of inflation as expressed by the Consumer Price Index (CPI) is the appropriate discount rate to use.

The annual inflation rate is published by the federal government in the form of the rate of change of the "Consumer Price Index." The CPI is the cost of a large group of goods and services weighted for the proportion which those particular goods and services make up out of the average household's annual budget. In short, the CPI measures buying power.

Using the year 1967 as a base year, it measures the cost of the same group of goods and services every year and expresses them as a percentage of the cost of the same goods and services in 1967. The rate of change of the annual CPI is widely regarded as the annual rate of inflation. This is a very reasonable number to use in our example for two reasons. First, a homeowner's buying power will remain steady as long as his or her income increases at the same rate as the inflation rate. Second, it is precisely this number, the CPI, which is used in many, many businesses as a basis to determine wage and salary increases. For this reason, the CPI has an almost built-in accuracy and stability factor in that last year's CPI is used to calculate this year's wage increase, which is, of course, reflected in this year's CPI, which is used to calculate next year's wage increase. And so on.

In summary, wages are likely to increase at the same rate as the CPI, or better; so the CPI is a good number against which the cost of a water project reflected in annual water rates can be measured.

As you can imagine, the rate of inflation is a very specific number (and a very important one). You can actually look up the rate for every year going back to 1967, and even back hundreds of years, using other measurements similar to the CPI. But, unfortunately, you cannot look up the CPI for the <u>next</u> five years. Since most financial matters relate to future payments, what can you do? The answer is: guesstimate.

This is not as frivolous as it sounds for two reasons. First, pinpoint accuracy is not necessary. Remember, what we are doing is comparing the cost of alternative loans. We are not trying to put an astronaut on the rim of a specific crater on the moon. We need to be accurate enough to measure which loan costs less, but it would be both unnecessary and frivolous to try to figure it down to the penny.

The second reason is that there is at hand a formal and acceptable method of projecting CPI increases by what statisticians call a "linear regression analysis." Such an analysis involves ascertaining the trend of the CPI over the last certain number of years and then projecting this trend forward. A less formal way of accomplishing the same end would be to take the average of the last five years' CPI's. This, however, would be reasonable only if -- and this is most important -- there were no major and unique economic event which would significantly skew the average of the last five years.

So, as I am writing this chapter, my best recollection is that the CPI's for the last five years or so have been in the 4-6% range. Furthermore, I can't think of any economic event which has happened or is likely to happen which would have the effect of either increasing or decreasing the CPI over the next five years. Under those circumstances, I would feel very comfortable estimating an annual inflation rate of 5% for the next five years.

Now I have made two fundamental decisions. I have decided that I will determine the change in the value of money by measuring it against the rate of inflation. I have also decided that the annual rate of change which I will use is 5%. This number, 5%, is the discount rate.

DISCOUNTING LOAN A

Let us now proceed to discount the loan from Banker A. Since we know precisely how much we will have to pay each year, we must work backward to measure the change in value. In other words, each Annual Payment in a future year will be treated as a Future Value.

Since we know the Future Values of each of the Annual Payments, having calculated them above, we can now proceed to calculate the Present Values for each of them, by Discounting. To do this we first convert our Interest, or Discount Rate, 5%, to a decimal, .05. Then we add the number 1, to get 1.05. Next, we divide 1.05 into the number 1, to obtain .9524, which is the Discount Factor. Finally, we multiply each Future Value by the Discount Factor, the number of times in the

Discount Period. Note that for each Annual Payment, the Discount Period is the number of years from the Present Value until that particular Annual Payment. In other words, it is the number of years from the time the loan is made until that particular loan payment is made. Let us illustrate this point.

If someone agrees to pay you $100 in one year (Future Value), then it is worth $95.24 today. We know this by multiplying $100 by .9524 ($100 x .9524 = $95.24). If someone agrees to pay you $100 in two years, then it is worth $90.70 today. We know this by multiplying $100 by .9524 twice, once for each year between the time the loan is made and the payment is made ($100 x .9524 x .9524 = $90.70). For three years, you multiply three times ($100 x .9524 x .9524 x .9524 = $86.38), and so on. In like manner, we can now determine the real, or discounted, value of the five loan payments in our example, as is shown in Figure 3-G, below:

Figure 3-G

Year	Annual Payment		Discounted Value
1	$280,000 x.9524	= $	266,672
2	264,000 x.9524 x.9524	=	239,465
3	248,000 x.9524 x.9524 x.9524	=	214,245
4	232,000 x.9524 x.9524 x.9524 x.9524	=	190,882
5	216,000 x.9524 x.9524 x.9524 x.9524 x.9524	= $	169,259
			$1,080,523

We can now compare the discounted values of the loan payments with the stated values, as shown in Figure 3-H, below:

Figure 3-H

Year	Stated Value	Discounted Value
1	$ 280,000	$ 266,672
2	264,000	239,465
3	248,000	214,245
4	232,000	190,882
5	216,000	169,259
	$1,240,000	$1,080,523

What useful conclusions can we draw from this exercise?

The most valuable conclusion is that it will only cost $80,523 to spread the payments out over five years. If the decision is whether to levy a $1,000,000 surcharge on the hapless ratepayers all at once or to spread the cost over a number of years, then it is useful to know that the real cost of doing so is really only $80,523. How so?

First, we know that we are borrowing -- and hence must repay -- $1,000,000. By subtracting the loan <u>Principal</u> of $1,000,000 from both the stated value and the discounted value, we can calculate the total amount of interest paid. In stated value, it is $240,000; in discounted value, it is $80,523. This means that although the amount paid will be $240,000, the actual economic value of these payments (or the real cost to the ratepayers in terms of their buying power) will only be $80,523. In other words, it will cost the ratepayers $80,523, which is only about 8% of $1,000,000, for the privilege of paying for the project over five years instead of one.

The most important lesson from the above exercise, however, is the simple fact that the discounted value, or Present Value, of Loan A is $1,080,523.

DISCOUNTING LOAN B

To obtain the Present Value of Loan B, the first thing we must do is calculate the Annual Payments. We can do this by performing the operations on a calculator according to the instructions in Appendix B, or by looking up the factor on the tables in Appendix C and performing the appropriate calculation.

No matter which method you use, you will find that the Annual Payment for Loan B, with $1,000,000 of Principal at an Interest rate of 8% for a term of five years, is: $250,456. Note that Loan B is a Level Payment Loan. This means that all of the Annual Payments in each year will be equal. Finding the discounted value, or Present Value, of these Annual Payments is easy using the same method as we used above for Loan A. This is demonstrated in Figure 3-I, on the next page.

Figure 3-I

Year	Annual Payment		Discounted Value
1	$250,456 x.9524		= $ 238,534
2	"	x.9524 x.9524	= 227,180
3	"	x.9524 x.9524 x.9524	= 216,366
4	"	x.9524 x.9524 x.9524 x.9524	= 206,067
5	"	x.9524 x.9524 x.9524 x.9524 x.9524	= 196,258
			$1,084,405

In short, the Present Value of Loan B is $1,084,405.

DISCOUNTING LOAN C

Loan C is also for $1,000,000, but for a 10-year term and at an Interest rate of 9%. Like Loan B, Loan C is also a level payment loan; so, again, all of the Annual Payments will be equal.

By again consulting either Appendices A, B, or C, we can calculate an Annual Payment of $155,820. To find the discounted value, or Present Value, of these payments we use the same procedure as we used above for loans A and B. The discounted values are (without showing all of the steps), as indicated in Figure 3-J, below:

Figure 3-J

Year	Annual Payment	Discounted Value
1	$155,820	$ 148,403
2	"	141,339
3	"	134,611
4	"	128,204
5	"	122,101
6	"	116,289
7	"	110,754
8	"	105,482
9	"	100,461
10	"	95,679
		$1,203,323

Now, let us compare the discounted values of all the loans, as shown in Figure 3-K, below.

Figure 3-K   Discounted Values

Year	A	B	C
1	$266,672	$238,534	$148,403
2	239,465	227,180	141,339
3	214,245	216,366	134,611
4	190,882	206,067	128,204
5	169,259	196,258	122,101
6	-	-	116,289
7	-	-	110,754
8	-	-	105,482
9	-	-	100,461
10	-	-	95,679
	$1,080,523	$1,084,405	$1,203,323

These numbers all look very nice. But what is their importance?

The single most important aspect of these Discounted Values is that they are now all the same. They are all Present Values. They can now be compared with each other.

As such, they unequivocally prove that the loan with the lowest cost to the system's ratepayers is Loan A. Some numbers may lie; but these don't. The loan with the lowest Present Value costs the least.

We have already mentioned the old saying, that you cannot add apples and oranges. I don't particularly like this analogy because it might leave people with the lingering notion that since apples and oranges are both fruit, there is, in fact, a common basis for adding the two together. After all, two apples and three oranges do equal five pieces of fruit. As far as adding together the Annual Payments in, say, years four and five of any loan transaction are concerned, this absolutely cannot be done. There is no common basis whatsoever. For this reason, I prefer using the phrase, "cabbages and kings," from Lewis Carroll's Through the Looking Glass. Nobody is going to mistakenly try to add cabbages and kings.

Loans A, B, and C, as they were originally stated, cannot be compared. They are cabbages and kings. They cannot be compared with each other. Even when you calculate the Annual Payments for each, they are still cabbages and kings. They still cannot be compared.

Only when all of the Annual Payments are reduced to Present Values -- when all of the loans are cabbages -- can they be compared.

The purpose of this chapter was to demonstrate how different financing options can be compared by reducing all future loan payments to Present Values. Once they are all stated in Present Value terms, they can be readily compared. Now that we know what the actual cost of each loan is, we have accomplished the first step in thoroughly evaluating the financing options available to water and wastewater systems. Chapters 4 and 5 will add two more necessary steps. Chapters 6 and 7 will return to examples such as those above, examples to illustrate how a full and totally comprehensive evaluation of different financing options can be made on either an annual cost basis or a total cost basis.

*   *   *   *   *

Before leaving this chapter, however, we have a piece of unfinished business.

We began this chapter with a question of which was cheaper: a financing (Offer A) with $190,000 Annual Payments for 10 years; or a financing (Offer B) with $100,000 Annual Payments for twenty years?

By adding the stated values of the Annual Payments, Offer A requires $1,900,000 in total payments, and Offer B requires $2,000,000 in total payments. It seems, therefore, that Offer A might be $100,000 cheaper. But then we said no, Offer B was cheaper by some $220,907. We then left the matter hanging.

By now you realize that the answer lies in the Present Value of each of the loan payments.

You now know that you cannot add up each of the Annual Payments in any loan. Annual Payments are cabbages and kings. You cannot add up the 10 Annual Payments of $100,000 in Offer A and get $1,900,000; because the number $1,900,000 has no meaning at all. It is as if you tried to add six cabbages and four kings. The answer is nonsense.

In order to compare Offer A and Offer B, we must reduce all of the annual payments in each loan to their Present Values.

To do this, we begin by realizing that each Annual Payment is a separate Future Value. We now know that we must discount these Future Values to reach their Present Values. To do so, we take a Discount Rate of 5% (for the reasons mentioned previously), convert it to a decimal (.05), add the number 1 (1.05) and divide it into the number 1 to get the Discount Factor (1 / 1.05 = .9524). Next, we multiply each Future Value by the Discount Factor the number of times in the Discount Period for each Annual Payment. Without showing all of the multiplication, here are the Present Values in Figure 3-L, below:

Figure 3-L   Present Values of Offer A and Offer B

Offer A		Offer B	
Stated Value	Present Value	Stated Value	Present Value
$190,000	$180,956	$100,000	$95,240
"	172,342	"	90,707
"	164,138	"	86,389
"	156,326	"	82,277
"	148,885	"	78,360
"	141,798	"	74,630
"	135,048	"	71,078
"	128,620	"	67,695
"	122,498	"	64,472
"	116,667	"	61,404
		"	58,481
		"	55,697
		"	53,046
		"	50,521
		"	48,116
		"	45,826
		"	43,644
		"	41,567
		"	39,588
		"	37,704
	$1,467,278		$1,246,442

Now, by subtracting the sum of the Present Values of all of the Annual Payments in Offer B from the sum of the Present Values of all of the Annual Payments in Offer A, as follows,

$$
\begin{array}{r}
\$1,467,278 \\
- \ 1,246,442 \\
\hline
\$ \ 220,836
\end{array}
$$

you can see that Offer B is the better offer by a margin of $220,836. As the mathematicians say: Q.E.D.

Chapter 4

Determination of a Discount Rate

Under most circumstances, the determination of the appropriate discount rate to use in any given instance should not occupy an entire chapter in a book on water project financing. But the purpose of this book is to give the non-financial director or manager of a water or wastewater system the background he needs to understand what the system's bankers, investment bankers and financial advisors are telling him. In this regard, one of the most important lessons to be learned is that discounted numbers may not be what they seem. They can be manipulated to arrive at different conclusions. If this happens, the manipulation will occur in the selection of a discount rate. For that reason, the selection of the appropriate discount rate takes on a somewhat added significance.

The Use of Present Values and Future Values in Determining the Discount Rate.

To illustrate how Present Values and Future Values affect determinations of the discount rate, let us look at two simple examples of Present Value and the same examples as Future Values.

What is the Future Value of $100 in one year at a compound rate of 3%?

The answer is: $100 x 1.03 = $103.

What is the Future Value of $100 in one year at a compound rate of 10%?

The answer is: $100 x 1.1 = $110.

Notice that the Future Value is higher in the second example where the compound rate is higher. What you are seeing here is a general rule:

The higher the compound rate, the higher the Future Value.

Now, let us look at the same set of numbers from the Present Value perspective.

What is the Present Value of a payment of $100 in one year at a discount rate of 3%?

As you recall from Chapter 3, to obtain the Present Value of a future payment without a calculator, you first take the discount rate (3%) converted to decimal form (.03) and add the number one (.03 + 1 = 1.03). Next find the reciprocal by dividing this number into the number one (1 / 1.03 = .9709). Then multiply the future payment ($100) by this number (.9709) as many times as there are years to the future payment (1), as follows:

$$\$100 \times .9709 = \$97.09$$

Now let us perform the same operation for the $100 future payment at a discount rate of 10%.

First, we convert 10% to decimal form: .1, then we add 1 to get 1.1. Next we find the reciprocal by dividing 1 by 1.1, where we get .9091. Finally, we multiply the future payment ($100) by this number (.9091) as many times as there are years (1) to the future payment, as follows:

$$\$100 \times .9091 = \$90.91$$

Now we can compare the Present Values of a future payment of $100 at two discount rates.

We can see that the Present Value with the 3% discount rate is $97.09.

And we can see that the Present Value with the 10% discount rate is $90.91.

Clearly, the Present Value is much larger with the 3% discount rate than it is with the 10% rate. Again, what we are witnessing is the operation of a general rule:

The lower the discount rate, the higher the Present Value.

Why is it important to know that the higher the compound rate the higher the Future Value, and the lower the discount rate the higher the Present Value?

It is important to know these rules when someone is trying to convince you of something. The "something" they will usually be trying to convince you of, is that you can either make a lot of money or save a lot of money for your system by doing whatever that person suggests.

Manipulating and Distorting Real Values.

Two extremely clever ways of doing this are by showing you how small your project payments or how large your savings on the project payments will be.

Project payments can be calculated by the Level Payment Method, the Level Principal Payment Method or the irregular method, as we saw in Chapter 2. If there are two sets of project payments to be compared, the "savings" can be calculated by subtracting the larger payment from the smaller one in each year. Once these amounts have been ascertained, their Present Value can be obtained by discounting.

Let us say that you receive a proposal from Banker A which will require your system to pay out $10,000 a year for five years. As part of his sales pitch, the banker tells you that although the actual cash outlay will be $10,000 for five years, or $50,000, on a fully discounted basis it will only cost your ratepayers $30,000.

The next day you are sent another proposal, this time from Banker B, for an investment plan for your reserve account where your system will receive $10,000 a year for five years, for a grand total of $50,000. Now, you keep asking Banker B what the discounted value of the investment earnings is. He keeps dodging the issue because the discounted number is always a smaller number, and he wants to keep you focused on the full amount, despite the fact that you will not receive most of it for a few years. Finally, he gives in and tells you that the fully discounted value is $49,000.

Now, let us say you make the inexcusable mistake of calling a public hearing to get public reaction to these proposals. And, unfortunately, not only are over 90% of your ratepayers political activists who like to show up at every meeting, but 90% of them are also CPA's. You have just created what might well be your worst nightmare.

The problem you will encounter centers around the subject matter of this chapter: the Determination of the Discount Rate. Or, more appropriately, the Determination of the Correct Discount Rate.

Let us assume, after thoroughly reading this chapter, you independently concluded that the correct discount rate to use for the next five years is 5%. Table 4-A, below, shows the discounted values for $10,000 a year for the next five years.

Table 4-A

$10,000	-	$9,524
$10,000	-	$9,070
$10,000	-	$8,638
$10,000	-	$8,227
$10,000	-	$7,835
		$43,294

As you can see, the discounted value of $10,000 for five years is not what either banker said it was. Here is what happened.

Banker A wanted to emphasize how little your ratepayers would pay. How small their annual payments would be. Banker A, having read this book himself, and having remembered the axiom that the higher the discount rate, the lower the Present Value, chose a discount rate of 10%. Table 4-B, below, shows the discounted values of five $10,000 payments at a discount rate of 10%.

Table 4-B

$10,000	-	$9,091
$10,000	-	$8,264
$10,000	-	$7,513
$10,000	-	$6,830
$10,000	-	$6,209
		$37,907

If you tell your ratepayers that the real value of project payments your system must make is only y; and the discount rate you used to calculate that value is unrealistically high, then you will be open to criticism for underestimating the cost of the project.

Now let us look at Banker B's proposal.

Banker B has an investment proposal. He is trying to persuade you to invest a fund which your system controls. In short, Banker B wants to show you the largest return possible on your investment, the highest numbers possible. When you asked him what the discounted value of the earnings on the investment was, he was reluctant to tell you because discounted numbers are, by definition, lower. When you finally got him into a corner, however, he did remember reading this book, and this chapter in particular. He also remembered the axiom: the lower the discount rate, the higher the Present Value. Keeping that in mind, when you pressed him on it, he finally did give you a discounted value; but the discount rate he used was 1%. (What is the matter with 1%? After all, 1% is a nice, low, conservative number, isn't it?) Table 4-C, below, shows the discounted values of the five $10,000 payments at a discount rate of 1%.

Table 4-C

$10,000	-	$9,901
$10,000	-	$9,803
$10,000	-	$9,706
$10,000	-	$9,610
$10,000	-	$9,515
		$48,535

If you tell your ratepayers that a course of action you propose will result in a savings to their system of x; and the discount rate you used to achieve the projected savings is unrealistically low, then you will be open to criticism for overestimating the savings.

Now that you realize what Bankers A and B have done, and you realize what the angry hall full of CPA-ratepayers is going to do, here is how to handle the situation.

Stride imperiously to the front of the room. Raise your arms to silence the din. Tell the crowd that you called the meeting because two Bankers had attempted to hoodwink your system -- by underestimating costs and by overestimating income -- into taking some ill-advised financial actions; and that you wanted to have a meeting to publicly denounce the Bankers and ban them from any future dealings with your system. As they carry you from the hall on their shoulders amidst a thunderous hail of grateful applause, don't forget you owe it all to the two axioms: the higher the discount rate, the lower the Present Value; and the lower the discount rate, the higher the Present Value.

Whenever confronted with a proposal involving a series of future payments which allegedly will mean either a greater savings or a lower cost to your system, I would strongly recommend testing the value of those savings by doing the Present Value calculation for at least several different interest rates. Which exact discount rates you might want to use will be discussed below.

A much simpler variation of this same problem arises on the question of investing funds of the system. This is simpler because everyone knows that the higher the rate of interest on your account, the more money you will earn. Another way of saying this is that the higher the rate of interest on your account, the more money you will have when you go to withdraw it. Please note that both of these expressions are simply a restatement of the general rule: the higher the compound rate, the higher the Future Value.

By reference to the example above, if you invest $100 of system funds for one year at a rate of 3%, you will have $103 at the end of the year.

If you invest the $100 at 10%, you will have $110 dollars at the end of the year.

In other words, the Future Value will be higher because the compound rate is higher.

The point here is to beware of the use of discount rates and compound rates in the hands of anyone trying to sell you anything or persuade you to do anything. The best way to guard against being fooled by this type of manipulation is to do the numbers yourself at several different rates. Whether you use a calculator or not, it is clearly worth the small effort of the hour or so it would take to do even thirty or forty years of calculations by hand, just for the peace of mind of knowing what the real values and the real costs of a project actually are.

SELECTING A DISCOUNT RATE

There are really only three sensible discount rates for a water system to use in doing the financial calculations for its projects. In technical terminology these are: the CPI, the reinvestment rate and the cost of funds rate. The CPI is, of course, the Consumer Price Index. The reinvestment rate is the rate, either short-term or long-term, at which the system could invest its funds. The cost of funds is the rate of interest at which the system could borrow, again, either short-term or long-term.

Time Frame of Reference.

All of these rates imply a time period of reference. Therefore, not only do you have to choose which index you will use, but you have to select the time frame of reference as well.

For example, the CPI is published monthly, quarterly, and annually. Which one should you use? Or, should you use some combination or average of the CPI at several intervals?

The reinvestment rate is even more varied than the CPI. You can invest funds on a daily basis, for a week, a month, any number of months, a year and any number of years up to forty years. For each time period, there is a different rate. For an analysis of how individual rates relate to such time periods, please turn to the last section of Chapter 10, where the "yield curve" is discussed.

The cost of funds rate generally has two, sometimes three, but always a very limited number of expressions. There are two basic reference points. The first is a floating rate, which is used for short-term borrowings, and is expressed as a function of the prime rate. If your system is a unit of government and is tax-exempt, you will be able to borrow at rates below your local bank's prime rate. Rates of 50%, 60%, 75%, up to 90% of prime are common.

If your district is a cooperative or is a privately owned utility, then your short-term rate will be something over the prime rate, such as prime plus one (P + 1), or prime plus two (P + 2), i.e., one or two percentage points over the local bank prime rate. If you are a very old, large district with a long and excellent credit history you will be able to borrow at the prime rate itself.

Whether you can borrow at 50% or 90% of prime, if you are a unit of government, or at P or P + 2, if you are private, is a function of two factors: first, the credit history and strength of your system; and, second, the amount of competition there is among local banks for your business. Please see Chapter 18 for a discussion of the prime rate as well as the effect of competition on bank rates.

The second rate is generally a fixed rate which is used for longer-term borrowings. It is expressed simply as a number. Again, it will be different depending on whether your system is a tax-exempt borrower. For example, if a bank were willing to lend for, say, ten years to a tax-exempt borrower at 8%, it might charge a taxable borrower, of comparable credit quality, 11%.

For the reinvestment rate and the cost of funds rate, the averages for the last year (or longer, if there has been an unusual economic event) will generally be fine.

The Consumer Price Index (CPI).

The Consumer Price Index is a measure of change in prices for a market basket full of a wide variety of consumer goods and services such as clothing, food, utilities, housing, transportation and entertainment. It is compiled and published on a monthly basis by the U.S. Bureau of Labor Statistics.

Assuming the CPI turns out to be the appropriate rate to use in a given instance, the question is: which expression of the CPI should be used? Last month's CPI? The average of the last six months? The CPI for the last year?

The average of the last five years' of the Consumer Price Index is to me the most sensible rate to use to discount future <u>costs</u>. If there has been a virtually unique economic event (e.g., the 21.5% prime interest rate spike in 1981 or the Arab oil embargo in 1974) in the last five years, then the reference period should be extended to ten years.

The reason for favoring the five-year average is that, although the last five years' CPI cannot flawlessly predict the next five years' growth in costs, there is no other set of numbers which can do so any better.

If you took the average change in the cost of Operations & Maintenance for your water or wastewater system for the last five years, the chances are good to excellent that it would fall within 10%, plus or minus, of the average of the CPI over the same period.

If you took the average change in those costs over the last forty years, it would as a virtual certainty fall within 10%, plus or minus, of the average of the CPI over the same forty year period.

The point here, of course, is that changes in costs do average out; and that they average out very well, indeed, over time. Why not, therefore, use a measurement which reflects this process. The CPI does just this.

Now, having established that the CPI does an excellent job of tracking <u>historical</u> costs, why, if at all, is this significant in predicting <u>future</u> rates?

We must begin any answer with the usual admonition that the future cannot be predicted. This is especially true when using past events to predict future events. Now, having begun with the proper note of caution, we can certainly say that the major reason the CPI is usually so reliable is that its use has become somewhat circular. By this we mean that the CPI has become so well known that it has come to be used to determine next year's rates for a wide variety of economic activities.

For example, many wage rate increases are pegged to the CPI. Many commercial rent rate increases are determined by the CPI. In short, many of the actual expenses which a system must pay, especially to third parties, increase, themselves, at a rate equal to the CPI. They do so not by accident, but rather because of agreement.

In other words, lease agreements between landlords and tenants provide for annual rent increases based on the prior year's CPI. If you are leasing space, therefore, you already know your rent will go up according to the CPI. If your system's workers receive annual wage increases based on the CPI, then you know that your labor costs, too, will increase according to the CPI. Thus, if you are making projections for your whole budget, you already know that the rent component and the labor component will increase at the rate of the CPI.

If these two components -- and any others equally affected by the CPI -- constitute a large enough part of a systems budget, then the entire budget itself will increase at the rate of the CPI. You can now see what we mean by the circularity of the CPI.

The CPI is a strong and broad measurement, on its own. The fact that it is so widely used as a basis for future cost increases, in and of itself, makes it in many ways a self-fulfilling prophecy. This circularity makes it all the more valuable in projecting cost increases into the future.

Please notice that throughout this subsection we have talked of the CPI in terms of its use as a predictor of cost. It is not only an excellent predictor of cost, but also of those other financial factors, such as rates, which are inextricably bound to underlying costs.

If you needed, for example, to predict the rates your system would putatively charge in the year 2025, probably the best place to begin would be to separate out the Operations & Maintenance cost and determine its Future Value by using the formula contained in the last chapter and a discount rate equal to the CPI for the last five years.

The CPI can also be used to predict certain replacement costs. For example, common machinery and vehicles need to be replaced. If a certain machine will have to be replaced in five years, then you might take today's cost as the Present Value and compute, using the formula in the last chapter, the Future Value by using a discount rate equal to the CPI for the last five years.

In addition, and as a final note about the CPI, you should be aware that because prognostication is not an exact science, or even a science of any kind, you will not be expected to be exact (nor is there any good reason why you should be) in your estimates. Therefore, if the Bureau of Labor Statistics reports that the average CPI for the last five years was 4.83862%, you can easily use 5% for five-year projections or less and 4.8% for longer projections.

Just remember that the CPI is a measurement of historical cost. It is extremely useful in estimating cost. But as you know, cost is just one component of most financial concepts. As far as your system is concerned, the cost of Operations & Maintenance is just one of the elements of the system's total income and expenses.

A final consideration about the use of the CPI is that it only works for cost estimates or for other estimates, such as rates, which are closely tied to cost. You should not, for example, try to project future earnings on a fund by using the CPI. For measuring numbers other than cost, one of the other indices, such as the reinvestment rate or the cost of funds rate, will be more accurate.

The Reinvestment Rate.

The reinvestment rate is the amount of money you could earn, if you invested money prudently. The "money" we are talking about is any fund or funds which a system has legally under its control and which does not need to be spent until some time in the future. By the word "prudently," we are referring to sane and normal investments -- the type which a "prudent man" would make when entrusted with funds belonging to someone else. Generally speaking, when it comes to public funds, investments are limited to obligations of the U.S. government, agencies of the United States, or other obligations which are guarantied by the U.S. government. (These matters are governed by state law. Please check with your system's counsel to ascertain the legal investment criteria for public funds in your state.)

Let us say that you currently have $50,000 in a particular fund which is earmarked for a new pumphouse. You plan on replacing the old one in five years. If you don't add any more funds to the account, how much will you have in five years?

This is, of course, a simple Future Value problem, the formula for which is in Chapter 3. The formula, however, requires the use of a discount rate. Which discount rate will you use?

As you might imagine, the reinvestment rate is designed specifically for situations such as this. But what actually is the reinvestment rate?

If you are talking about five years or less, you can ascertain the reinvestment rate by simply checking with two or three local banks. They will be able to tell you directly what the rates are for Treasury Bills and Notes in the maturities you are interested in. The will also be able to give you the rates for any government agency paper as well as the rates on their Certificates of Deposit (CD) for the appropriate time period. Certificates of Deposit of banks which are members of the Federal Deposit Insurance Corporation (FDIC) are insured by the United States government for amounts up to $100,000 per customer. (Note that the $100,000 limit is by customer, not by account.) They will also be able to give you rates for any other legal investment in your state.

Whatever category of legal investment in your state bears the highest rate of interest is the category you should use. And, the average of three banks' rates of interest for such instruments will very adequately serve as the reinvestment rate. Generally speaking, the CD rate will be the highest of the short-term rates for legal investments. So, in most cases, the average of the rates which three banks pay for their CD's for a given maturity will serve as the reinvestment rate.

For longer terms, banks do not have CD's. Nor are they usually willing to guaranty a rate of interest for a long period of time. Thus, in the case of longer maturities, the local banks may not be able to offer the best investment rate. But they will be able to tell you what the prevailing long-term rates are for the various kinds of legal investments. So, in such case, where different banks are not offering different rates on their own certificates, the highest interest rate on any legal investment for the desired maturity then becomes the reinvestment rate which you should use.

The Cost of Funds Rate.

The cost of funds rate is the rate at which a system borrows. That is all there is to it.

The cost of funds rate is, literally, used to calculate the cost of funds which a system borrows. It is also used in calculations involving funds which a system either has not yet borrowed, or is not going to borrow.

In other words, it is used to make projections and to make up hypothetical examples against which costs can be measured.

A good example might be an exercise to determine whether it is financially wiser to levy a surcharge for two or three years and invest the funds before and during construction, or to borrow the funds from the bank during the construction period, then issue long-term bonds to pay off the bank loan.

In this case you would first use the reinvestment rate to determine how much interest would be earned on the funds provided by the surcharges. Then, one borrowing rate would be used to determine at what rate the system could borrow from the bank during the construction period. Finally, another borrowing rate would have to be estimated for the long-term bond which would be issued to pay off, or take out, the bank loan.

The borrowing rate is the simplest rate to determine. You simply need to determine the maturity and then consult with one or more bankers and investment bankers. They will be able to give you the rate at which an institution (the bank) will lend to your system, but also the rate at which your system can borrow in the public market (the investment bank). The lowest rate will be your "borrowing" rate.

The borrowing rate for your system will be a reflection of two factors. The first is the rate at which banks will lend, for a given term, to their customers with the highest credit ratings. The second factor is the relative quality of your system's credit. In short, there is an absolute minimum rate at which a bank will lend to its best customers. All other borrowers must pay a premium.

*   *   *   *   *

As you know, the ability to calculate Present Values and Future Values is the key to being able to make meaningful comparisons among alternative financial proposals. To do so, we must use compound rates and discount rates. And, as you have seen in the chapter above, the determination of the correct discount rate is one of the most important factors involved in making the correct financial choices.

Chapter 5

Comparing Upfront and Annual Costs

At least once a month, the real estate section of most newspapers carries information on the availability of home mortgages as well as their rates. You will see such items as this:

Mortgage Bank A: fixed-rate loan, 30 year term, 10% interest rate, two points upfront.

Mortgage Bank B: fixed-rate loan, 30 year term, 10.75% interest rate, no points upfront.

First, in case you are not familiar with the terminology, a "point" is one percent. Two points are two percent. And so on. So, in the example above, the Bank A mortgage requires "two points upfront." This means that 2% of the mortgage amount, or $2,000 on a $100,000 loan, must be paid in cash at the closing of the loan.

Which of these two home mortgages would you prefer?

When we say, "Which do you prefer?," we are obviously asking which is the cheaper mortgage. As you can see, both mortgages have fixed rates. Both have thirty year terms. The only two aspects in which they differ are in their interest rates and in the fact that one requires two points to be paid while the other requires no points.

Mortgage A has a 10% rate but 2 points. Mortgage B has a 10.75% rate but no points.

In short, when you reduce the question to its most basic terms, what we are really asking is which is cheaper, the two points, or the extra .75% interest.

This question illustrates a problem which is common to many different kinds of financings. The problem arises from mixing annual costs with one-time costs which are commonly paid at the loan closing and which are generally referred to as "upfront costs." How can these two different types of costs be compared? In this chapter, we will analyze these types of costs and see if we cannot find a common basis for comparison.

Please be aware of the fact that when we say we are comparing loans with different upfront fees, what we are actually doing is comparing loans of different principal amounts.

As you will see below, a $10,000 loan where you have to pay the banker a 2%, or $200, fee at closing, can just as easily be considered a $9,800 loan. Another way of expressing this is by saying that the procceds of the loan are less than the loan itself. This is the scenario referred to in the "Principal" section of Chapter 2, where you wind up paying back more than you actually received.

In Chapter 3, we worked with home mortgages as an example. They are probably the most familiar type of financing, but they are surely the most tedious to work with. As you might have guessed from our work in Chapter 3, at least one of the methods of comparing costs in different years involves calculating Present Values. This, in turn, requires separate calculations for each Annual Payment. And, in the case of home mortgages, we would normally be working with thirty-year loans, which would require thirty long and tedious calculations. To avoid this needlessly repetitive work, we will use five-year Auto Loans instead of thirty-year mortgages as our example.

So, instead, let us say you turn to the automobile section of your newspaper, where two of your local banks have decided to publish the following information:

Bank A Auto Loan: Five year term, fixed-rate, 10% interest rate, two points upfront.

Bank B Auto Loan: Five year term, fixed-rate, 10.75% interest rate, no points upfront.

Now, which of these two auto loans would you prefer? Let us assume you want to borrow $10,000. The two points in Auto Loan A are, thus, 2% of $10,000, or $200.

Just as in the home mortgage advertisement, the two loans contain a mix of annual costs and upfront costs.

There are two basic methods of comparing these types of costs. The first, as you might have guessed from the references to Chapter 3, is the Present Value Method. The second is the True Interest Cost (TIC) Method. We will now analyze these two loans using each of these methods and see if we can develop a basis of comparison.

The Present Value Method.

The first step in comparing costs by the Present Value Method is to ascertain the method of payment for both loans. In our example, the payment method for neither loan was specified. We know, therefore, from Chapter 2, that when no payment method is specified, the level payment method is assumed.

We know that both loans have the same term (5 years). Only the interest rates are different. When only the interest rates differ and both loans use the Level Payment Method, the loan with the higher interest rate will have the higher annual debt service payment. Thus, we know that Auto Loan B will have the higher annual debt service payment. The question now becomes whether the difference between the annual debt service payments on Auto Loan B and Auto Loan A exceeds the $200 upfront fee on Auto Loan A.

We know that both Auto Loans have five annual debt service payments and, since both use the Level Payment Method, each annual debt service payment will be the same for each loan. Thus, the difference between the annual debt service payments for Auto Loan A and Auto Loan B will also be the same each year. So, we will have the same amount every year for five years.

Now, we know from Chapter 3 that we absolutely cannot add these amounts together because they occur in different years. We cannot simply find the difference between the annual debt service payments for Auto Loans A and B, multiply them by five, and then subtract them from the extra $200 upfront cost on Auto Loan A. Annual debt service payments in different years are cabbages and kings. They cannot be added together.

Chapter 3 provides a method for adding different costs or different payments which occur in different years. The method, of course, is to reduce future costs, or Future Values, to Present Values.

The difference between each annual debt service payment in Auto Loans A and B is a Future Value. To reduce these Future Values to Present Values we would multiply each Future Value by a Discount

Factor as many times as there are Discount Periods for that particular annual debt service payment. Thus, for example, we would take the difference in the two annual debt service payments in Year One and multiply it by the Discount Factor one time. We would then take the difference in the two annual debt service payments in Year Two and multiply it by the Discount Factor two times. And so on. In Chapter 3, we did this exercise for a thirty-year mortgage. It is a tedious process, indeed. At the end, we would add all of the Present Values for all of the annual debt service payments to arrive at the Present Value for all of the annual debt service payments. Then we would subtract this number from the $200 upfront payment in Auto Loan A. We will now do this exercise for the two Auto Loans.

The first step is to find the annual debt service payments for both loans. From Chapter 2, we know that we can find the annual debt service payment for a Level Payment Loan by either using the calculator steps in Appendix B, or the tables in Appendix C.

By whichever method you use, the annual debt service payment for Auto Loan A is $2,638 and the annual debt service payment for Auto Loan B is $2,689.

The second step is to find the difference between the annual debt service payments for the two loans. The difference between the two annual debt service payments is $51 per year (2,689 - 2,638 = 51).

The third step is to find the Present Value of the $51 difference in the two annual debt service payments in Years One through Five of the loan term. To accomplish this, we must determine a Discount Factor and to do this, we must settle on a Discount Rate.

Selecting a Discount Rate.

Since we are dealing with loans which have known rates of interest, should we use one or the other of the interest rates on the two loans? We also know from Chapter 4 that we are generally safe in using an approximation of the rate of inflation as a Discount Rate. We could, therefore, use 5%.

Or, is there any other rate which might be better to use?

As was discussed in Chapter 4, another rate which could be used is what we called the Reinvestment Rate. This is the rate at which you could invest the $10,000 for the same term, i.e., five years. The assumption here is that you have, in hand, the $10,000, and are going to make a decision whether you should spend the $10,000 on the car and not borrow at all, or whether you should invest the $10,000 and borrow another $10,000 to pay for the car. (As you can undoubtedly see, there may be tax ramifications here in that the interest you earn may be taxable and the interest you pay may be deductible. But that is not the subject under consideration here. So, for the purposes of our example, let us say that the question is tax neutral.)

At any given time, it is a good bet that if a bank is willing to lend you money in the 10% range for five years, then that same bank should not be willing to pay more than 7% for a five-year Certificate of Deposit.

In this case, since we are dealing with a personal loan, as we discussed in Chapter 4, the best rate to use is probably the Reinvestment Rate, or 7%.

Now that we have selected a Discount Rate, we can calculate the Discount Factor by first converting 7% to a decimal, or .07, then adding the number 1 to get 1.07, and then dividing this number into the number 1 to get .9346. This is the Discount Factor.

To get the Present Values for the differences in each of the five annual debt service payments on the two Auto Loans, we now multiply the difference in the annual debt service payments by the Discount Factor as many times as there are Discount Periods for each annual debt service payment, as follows:

Figure 5-A

Discount Period	Difference in Annual Payments	Present Values
1	$51	$47.66
2	"	44.55
3	"	41.63
4	"	38.91
5	"	36.36
		$209.11

We have now established that the annual debt service payments for Auto Loan B are $209.11 higher than the annual debt service payments for Auto Loan A. But since Auto Loan A requires an upfront payment of two points, or $200, we must now subtract the Present Value of the additional annual cost in Auto Loan B from the additional upfront cost in Auto Loan A. We can now do this <u>because both numbers are Present Values</u>.

We reduced the Future Values of the differences in the two annual debt service payments to Present Values, and then we added them up to get $209.11. The $200 upfront payment in Auto Loan A is, itself, a Present Value. Since they are now both Present Values we can now compare them. They are no longer cabbages and kings. They are either both cabbages or both kings, as you like.

So, we now subtract $200 from $209.11 to learn that the entire difference between Auto Loan A and Auto Loan B is $9.11.

If the difference between two $10,000 loans for five years is only $9.11, then you might as well make up your mind on the basis of which bank is closer, since you are likely to burn up more than $9.11 in gas in getting to the farther bank.

Before we leave the Present Value Method, however, there is one thing you should be aware of. You recall that a few paragraphs above we were deciding whether to use 5%, 7%, or 10% as the Discount Rate. Here is what would have happened had we chosen any of these Discount Rates:

Figure 5-B

Discount Rate	Present Value of Annual Payments
5.0 %	$220.80
7.0 %	209.11
10.0 %	193.33

Please note that if we use 5% or 7% as the Discount Rate, Auto Loan A winds up cheaper because the Present Value of the difference in the Annual Payments on the two loans exceeds the $200 upfront fee required on Auto Loan A. But if we had used 10% as the Discount Rate, then Auto Loan B would have been cheaper because the Present Value of the difference in the Annual Payments on the two loans is less than the $200 upfront fee on Auto Loan A.

This problem with the Discount Rate illustrates two things. First, as you have seen in Chapter 4, beware of people trying to sell you things based on Present Values. It is one thing when you, yourself, are doing calculations for your own information; but it is another thing entirely when someone is trying to sell you something. In other words, by choosing different Discount Rates, unscrupulous people can manipulate true value. Always heed the advice in Chapter 4. Check out the alleged savings (or benefit) yourself, by doing the calculation with at least a couple of different Discount Rates.

The True Interest Cost (TIC) Method.

The True Interest Cost (TIC) Method involves looking at the whole loan transaction is a slightly different way.

If, at the closing, you were sitting across the table from the banker on Auto Loan A, he would wind up giving you (or the dealer from whom you are buying the car) a check for $10,000, and you would give the banker a check for $200.

Now, isn't this exactly the same thing as borrowing $9,800 in the first place?

Of course it is. You are actually borrowing only $9,800, but you must repay $10,000, plus interest on the full $10,000. Your annual debt service payment on Auto Loan A would still be $2,638.

Now, hold onto your hat. Here comes the black magic.

What interest rate would you be paying if your annual debt service payment was $2,638, but the amount of the loan was only $9,800?

Auto Loan A requires you to make five annual debt service payments of $2,638 each, for a total of $13,190. Every loan can be broken down into principal and interest. If you are borrowing $10,000, you can determine the amount of interest you are paying by subtracting the principal from the total of the annual debt service payments. In this case, you would subtract $10,000 from $13,190, leaving $3,190. This sum, $3,190, is the total amount of interest you will pay on the loan.

But what if you think of the transaction as a $9,800 loan instead of a $10,000 loan? In that case, by subtracting $9,800 from the total of the annual debt service payments, which is still $13,190, you would find you are paying $3,390 , instead of $3,190.

If you spend a minute looking at the admittedly complicated procedure for calculating annual debt service payments in Appendix A, you can see that there are only four elements involved in the calculation. They are: the principal amount, the interest rate, the term, or period of years, and the answer, which is the annual debt service payment.

In our example, the annual debt service payment stays the same, $2,638. So does the term, five years. The principal <u>decreases</u> from $10,000 to $9,800. And, the interest <u>apparently increases</u> from $3,190 to $3,390.

In other words, all other elements of the transaction remain the same except that you are apparently paying <u>more</u> <u>interest</u> on <u>less</u> <u>principal</u>. Therefore, it is reasonable to assume that the interest <u>rate</u> must change. This makes perfectly good sense.

But... Does the interest rate really change? No. That is why we have been saying that the amount of interest <u>apparently</u> increases. In fact, we must <u>impute</u> a change in the interest rate, so that we can measure the change in the value of the transaction.

The <u>stated</u> interest rate on the loan doesn't change. As far as the banker is concerned, it is a 10% loan. The loan documents will undoubtedly say so. But your common sense will tell you that you can't decrease the principal by $200 and increase the interest by $200, without something happening to the rate.

The best way to express a solution to this problem is to create a concept, which, in this case, is called the True Interest Cost (TIC).

We said to watch out for the black magic. This is why. Calculating a TIC may seem, initially, like a lot of hocus pocus, but it is, in fact, a very powerful method for comparing loans when one has upfront costs and the other does not.

Now that we know what the TIC is, the next question is how to calculate it.

The smart-aleck answer is to say: Look at Appendix A and work backwards. But even the briefest look at Appendix A will indicate just how utterly daunting a task that would be.

To alleviate this problem, we have provided a series of tables in Appendix D. In addition to the tables in Appendix D, the steps for calculating TIC on a calculator are described in Appendix E. But for the purposes of thoroughly illustrating how the TIC method works, we will use the tables in Appendix E in our example.

The tables indicate the TIC for adding different numbers of points to a transaction for various maturities. In other words, the tables show the increase in True Interest Cost for adding from 1% to 10% in upfront costs to loans with interest rates from 5% to 10%, and with terms of 5, 10, 15, 20, 25, 30, 35, and 40 years. We can use Auto Loan A to illustrate how the tables work.

As you recall, Auto Loan A had a five year term, a 10% interest rate and two points, or 2%, in upfront fees.

Table D-6 is for loans with interest rates of 10%. The first column is for loans with five year terms. Line two is for the number of points, which is two. As you can see, according to the table, the answer is 10.78%. This means that the TIC on Auto Loan A is 10.78%.

To be absolutely precise, the TIC on a five-year loan, at a 10% interest rate, with 2% in upfront fees is 10.78%. This means that <u>paying two points on a five year loan at a 10% interest rate is the equivalent of paying an interest rate of 10.78% with no points</u>.

In other words, <u>the two points upfront will cost you the equivalent of a 0.78% increase in your Interest Rate</u>.

Now we can compare the two loans using the TIC method:

Auto Loan A has a TIC of 10.78%.
Auto Loan B has a TIC of 10.75%.

The difference between Auto Loan A and Auto Loan B is 0.03%.

As you can see, the two loans are virtually identical. This is the same conclusion we reached above, using the Present Value method.

Auto Loan A is marginally more expensive, but only by a factor of 0.03%. This number, 0.03%, or three one-hundredths of one percent, is the equivalent of about $3 a year on the $10,000 loan. So, as we said above, this is one loan decision that you don't have to make strictly on the economics, because economically speaking, the two loans are a wash.

To be able to financially differentiate two $10,000 loan offers to a tolerance of less than $3 a year, is a powerful analytical capability. The TIC method is not only a powerful, but also a highly exact means, of comparing loans with and without upfront fees, or loans where the upfront fees are different.

Comparing the Present Value Method and the True Interest Cost (TIC) Method.

We cannot press on without comparing the Present Value method and the TIC method. This is not just a mental exercise. It does absolutely no good to learn either method without also learning exactly when to use them. They are not interchangeable. Neither should be used 100% of the time. Nor, as you will see, is the decision as to which method to use based on any abstract rule. On the contrary, which comparative method to use is based entirely on the circumstances of the financial programs to be compared.

In our example above, depending on the method we used, we generated two entirely different kinds of answers. The Present Value method gave us a number which we could compare with the upfront fee.

As you will recall, the loan without the upfront fee has a higher annual debt service payment. So, the Present Value method takes the increase in the annual debt service payment on the loan without the upfront fee and finds the Present Value for each of such increase for each year of the loan, and then totals them. This number can then be compared directly to the dollar amount of the upfront fees to see which is less.

As you will also recall, the Present Value method left us with several options as to the Present Value of the increase in the Annual Payments, depending on which Discount Rate we used. The Present Values had a range of $27.47. In this particular example, the outcomes were so close that the selection of the Discount Rate actually became critical.

You can see from reviewing Figure 5-B, that by selecting a higher Discount Rate, Auto Loan B becomes the cheaper loan; whereas if a lower Discount Rate is chosen, Auto Loan A, with its $200 upfront fee, is the cheaper loan.

This is the problem with the Present Value method. The outcome may depend on the selection of the Discount Rate. This is precisely the situation to avoid, if possible.

The TIC method, on the other hand, produces a very exact answer. By using the TIC method, we calculate the TIC for Auto Loan A to be 10.78%, not 10.77% or 10.79%. There is no ambiguity here at all. It is very easy to compare the two auto loans using the TIC method. The TIC for Auto Loan A is 10.78% and the TIC for Auto Loan B is 10.75%. The difference between the two is unmistakably 0.03%; i.e., the TIC on Auto Loan A is 0.03% higher. Auto Loan A is more expensive than Auto Loan B.

In this case, it is very clear that the TIC method is the better method to use. It is easy to see that, in this particular example, at least, the TIC method is more accurate.

But are there instances where the Present Value method is preferable? The answer is yes. And, surprisingly so, in most instances.

As it turns out, the TIC method only works when the loans being compared have the same term. In all other instances, i.e., when you are comparing loans or financial programs with different terms, the Present Value method is the only method which can be used.

Although the TIC method has great accuracy, its use is, unfortunately, limited.

By using one or the other method, however, all financial programs can be compared, especially those which may, or may not, have upfront fees or one-time costs which are added to the cost of the project.

This leads us to a final note.

Please be aware that, as we said at the beginning of this chapter, loans with different upfront fees can, for analytical purposes, simply be considered loans with different principal amounts. By the same token, it must be said that a financing program that produces a total project cost of $9,800, or $98,000,000, is exactly the same thing as a $10,000, or $100,000,000, financing with a two-point fee upfront.

In other words, as you continue through this book looking at different factors which inflate project costs, you will see that they do so in either of two ways. They will either add to principal or they will affect the annual debt service payment. Points, as you will see in Chapter 11, are simply additions to principal.

So, what this chapter is really all about is converting changes in principal to changes in annual debt service payment, and vice versa, so that they can be compared from one financial program to another.

Chapter 6

The Annual Payment Method
of Evaluating Project Financing

The preceding five chapters have established the groundwork for evaluating alternative options to finance projects for water and wastewater systems. We know what annual payments look like, and what different types of annual payments there are (Chapter 2). We know how to reduce annual payments in different years to a common value by calculating their Present Values (Chapter 3). And, we know how to convert one-time costs to their annual cost equivalents. Now that we have covered the technical matters necessary to financially evaluate water and wastewater projects, we will turn to the methods of analysis, themselves.

There are two principal ways to financially evaluate projects. These are called the Annual Payment Method and the Total Payment Method.

There is absolutely no mystery here, at all. The Annual Payment Method tells you which of the alternative financial programs you are evaluating has the lowest Annual Payment, and the Total Payment Method tells you which financial alternative has the lowest total payments.

Which method you use under any given circumstance does not depend on any formula or set of events or circumstances. Which formula you use is subjective. As you will see, it is entirely up to you and the personal values you bring to the job.

A Matter of Principle.

Financing anything implies taking on debt. When it comes to debt, there seem to be two distinct kinds of people in the world: those who believe in debt, and those who shun debt.

Both views are supported by firm moral arguments.

Those who favor debt believe that one should pay for things over the same period of time that he uses them. Pay as you go, in other words.

Those who shun debt believe that one should simply not be beholden, period.

Both of these positions make eminently good sense. Both clearly lead, however, to two different financial conclusions. And, not unsurprisingly, they both lead to entirely different analyses in evaluating the cost of any project financing.

One thing to keep in mind as you read the next few pages is that this is a book about financing water and wastewater projects. The point is that these types of projects don't usually involve just a few hundred or even a few thousand dollars. They generally involve hundreds of thousands of dollars, at least; and sometimes tens of millions of dollars. In short, they involve real money.

Thus, there is seldom a question as to whether a project will, or will not, be financed. Other than for admittedly costly hook-up charges, homeowners might sit still for a one-time surcharge of a few hundred dollars to pay for an improvement project. But once that number starts to creep over the one thousand dollar mark, ratepayer resistance begins to increase exponentially. Likewise, if the surcharges start coming every other year or so.

So, almost all water and wastewater projects will be financed. The questions which boards and managers must face are -- and here is where the differences in personal philosophy come in: How much should be financed? And, for how long a period of years should it be financed?

Here is where the Pay As You Go people say as much and as long as possible; and the Not Beholden people say as little as possible and for as short a time as possible.

So, as you can see, the argument generally isn't whether to finance. It is how much and how long? These are the issues along which the battle lines form.

Whether you are a Pay As You Go-er, or a Not Beholden-er, you will want to be able to make your point forcefully and to be able to understand all the facts and figures behind the other argument.

This chapter will present the arguments in favor of the Pay As You Go position and the financial analysis which supports that position, which is the Annual Payment Method of evaluating project financing. The next chapter, Chapter 7, will present the argumentation and supporting analysis for the Not Beholden position and the financial analysis which supports that position, which is the Total Payment Method of evaluating project financing.

The Pay As You Go Theory of Project Finance.

The Pay As You Go side of the argument takes the position that the system itself should pay for things it needs over that period of time that it uses such things. The period of time over which something is used is called its "service life."

If you need pencils and paper, you should pay for them out of today's revenues. The reason for this is that the pencils and paper have a very short service life Basically, they should be used up within days of delivery. It is not wise to pay for something tomorrow which, by then, will be used up and gone. Everyone hates to pay for things they don't have the use of. This is doubly true when it comes to paying for goods and services which have already been used up.

In 1975, the City of New York went bankrupt. In the financial autopsy that was subsequently performed, it was disclosed that the City was financing the purchase of pencils and paper with the proceeds of twenty-year bonds. In other words, they were paying for the pencils and paper, which were going to be completely used up in 1975, over a twenty-year period. This means that as I am writing this page early in 1992, the people of the City of New York still have three more years to go to pay off the pencils and paper that were used seventeen years ago!

These are the types of nightmares which make Not Beholden types out of ordinary citizens.

But, as usual, it is the abuses which make the headlines. Once in a great while, especially before the municipal accounting rule reforms, some issuer would finance something, not over its real service life, but for as long a term as they could get away with.

The overwhelming majority of project financings do not suffer from such abuses. Projects are built, and goods and services are bought and paid for over their actual service lives. That is the philosophy of the true Pay As You Go proponent.

I say the true Pay As You Go person because there are always those who fundamentally don't ever want to pay for anything. These people aren't really Pay As You Go people; they are Never Pay people. They are just like the city officials in New York before 1975.

The true Pay As You Go person is governed by a sense of fairness and looks to the laws of economics to justify his position. The Never Pay people have no position at all, other than they don't want to pay. They have the laws of economics in their favor; but that is all they have.

The Never Pay person will advocate a five-, seven-, or ten-year auto loan, for example, utterly indifferent to the fact that the vehicle may only be used for three years; and that ratepayers or taxpayers four years hence will have to pay for two cars, the one they just bought and the one they no longer have.

As we said, however, the true Pay As You Go person's position is rooted in real principle. And that principle is that payment should follow use or enjoyment.

In addition to this principle, which is a logical principle, there is also the principle of fairness. Suffice it to say that the principle is rooted in the same facts as is the service life argument. The service life argument holds that a project should be paid for over the period of time in which it is used. The fairness argument holds that the ratepayers who get the benefit of the project (in every year over the project's entire service life) should pay pro rata for such benefits, over each year of the project's service life.

In other words, if a standpipe will last twenty years, the ratepayers in year one will get 1/20th of the benefit, and the ratepayers in year twenty will also get 1/20th of the benefit. Fairness dictates that each group of ratepayers pay their 1/20th of the total project cost. To do otherwise would simply not be fair.

(As you can see, the fairness argument not only supports the Pay As You Go theory, but it also supports the Level Payment Method of annual debt service payment, as well.)

In summary, the Pay As You Go theory of project funding says that a water project should be financed over the period of time during which its use will benefit the system's ratepayers. The theory is rooted both in logic and in fairness. As such, it is, by far, one of the most widely used principles in project finance.

When it comes to financing water or wastewater projects, the principle of Pay As You Go, or payment following use, begets two derivative principles:

1.) The term of the loan will coincide with the service life of the project; and,

2.) The proper way to evaluate the project is with the Annual Payment Method.

The principle that the term of the loan should coincide with the service life of the project simply means that the system should borrow for no longer a period of time than it will use whatever it is financing.

If a system is financing a car or truck, which management expects will have to be sold or traded in three years' time, then it should pay off any loan used to buy the truck over a three-year period. If a system is buying pumps or other heavy equipment, which it does not expect to have to replace for ten years, then it should pay off any loan taken out to buy such equipment over a ten-year period, as well.

This principle is simple enough, but it does have two drawbacks. The first drawback becomes apparent when you consider land, which has an infinite service life, or such things as mains or standpipes, which have extraordinarily long service lives. In such cases, loans which coincide with service life are impossible to obtain, or simply don't exist.

A loan with an infinite term is, by definition, not paid off. There is no such thing. As a matter of fact, when it comes to commercial banking, loans of more than twenty years are virtually impossible to obtain, and even 15 to 20 year loan terms are extremely rare. Large institutions, such as life insurance companies and pension funds, will make some loans of up to thirty years. And, about the only lender in the marketplace which will go out beyond thirty years is the Farmers Home Administration, which will lend for up to forty years. So, there is simply no one to go to, at all, for a loan of longer than forty years.

This is nothing peculiar to the water and wastewater industry. Private homes certainly have useful lives of longer that thirty years, yet a thirty-year mortgage is the longest term available on a home.

The conclusion, here, is that service life is not the only consideration. The financing term available under any particular set of circumstances is also critical. Thus, if a system is installing a quarter of a mile of new mains, which have a service life of one hundred years or more, and the maximum financing term available is forty years, then, obviously, the forty-year term controls.

This is often the case with shorter term assets as well. For example, a system might well take out a simple bank loan to pay for a new service vehicle with a realistic, and intended, service life of five years.

Nonetheless, the bank may be unwilling to lend for more than a three-year term. This is because the bank always wants the value of the collateral to, theoretically at least, substantially exceed the outstanding principal value of the loan. Thus, if a new vehicle costs $21,000, and the bank makes the system a Level Principal Payment Loan for a three-year term, just before the last payment is due the outstanding principal balance on the loan will be $7,000. If the system defaults at this time and the bank actually has to repossess the vehicle, the bank's assumption is that, since the vehicle should have two more years of service life to go, it should be able to get its $7,000 back when the vehicle is resold.

Those readers who are familiar with such matters as the cost of used equipment, may find the above scenario dubious. It is dubious. A three-year-old vehicle with an original purchase price of $21,000 may not really be able to bring $7,000, especially on a distressed sale where all the bidders know the bank wants desperately to get rid of the vehicle. That is why the word "theoretically" is underlined in the above paragraph.

On a straight line depreciation basis (which, of course, has little or nothing to do with reality), after three years a $21,000 vehicle with a five-year service life should still have a value of 2/5ths of its original value, or $8,400. So, on paper at least, the bank is in good shape. In the unlikely event the bank actually has to repossess the vehicle and auction it off, of course, it is likely to have at least some loss. But, then, the possibility of a water system defaulting on a loan for a truly needed service vehicle is also an "unlikely event." As a matter of fact, it is a highly unlikely event.

In any case, not only must you consider the actual service life of the asset to be financed, but also the term which any financial institution will allow on such asset.

Now, having discussed why assets should be financed over their service life, let us turn to the method of analyzing such financings.

The Annual Payment Method.

There are two variations for analyzing project financing costs by the Annual Payment Method. Which variation is used in any particular instance depends on whether the loan in question is a Level Principal Payment Loan or a Level Payment Loan. We will begin by comparing two Level Principal Payment Loans using the Annual Payment Method.

Comparing Level Principal Payment Loans by the Annual Payment Method.

Banker A offers your system a $20,000 Level Principal Payment Loan with a five-year term at a rate of 8.25% with no points.

Banker B offers your system a $20,000 Level Principal Payment Loan with a five-year term at a rate of 8%, with one point up front.

The problem here is cabbages and kings. As you can see, it is easy to figure out the first year's annual debt service payment on each loan. For Loan B, the first year's annual debt service payment is $5,600, composed of $4,000 of principal and $1,600 of interest. For Loan A, the first year's annual debt service payment is $5,650, composed of $4,000 of principal and $1,650 of interest.

Thus, on an <u>annual</u> basis, Loan B seems $50 cheaper.

The problem, of course, arises from the "one point" which is charged on Loan B.

As you know, one point equals 1%. This 1%, which is paid to the bank at the closing, has the effect of reducing the actual amount of the loan. In other words, 1% of $20,000 equals $200. This amount must be deducted from the principal amount of the loan. In other words, you are borrowing (and therefore must repay) $20,000, but you are only receiving $19,800. (This is the scenario alluded to in the section on "Principal" in Chapter 2, where you actually wind up paying back more than you receive.)

In order to accurately compare the two loans using the Annual Payment Method, we must convert our cabbages and kings to either all cabbages or all kings.

It is always easier to convert from the smaller to the larger. Hence, we will begin by determining, by percentage, how much larger Loan A is than Loan B. We can do this simply by dividing Loan A by Loan B, as follows:

$$\frac{20,000}{19,800}$$

The answer is 1.0101 etc. In other words, we can now make Loan B <u>equivalent</u> to Loan A by multiplying Loan B by 1.0101 etc., as follows: 19,800 x 1.0101 = 20,000.

Now, here comes the important part. We can also make the annual payments equivalent by multiplying the annual debt service payment on Loan B by 1.0101, as follows:

$$5,600 \times 1.0101 = 5,657$$

Now, the equivalent annual debt service payment for Loan B is $5,657 and the equivalent annual debt service payment for Loan A is $5,650.

Thus, in comparing Loan A and Loan B by the Annual Payment Method, Loan B is more expensive.

Now, let us compare the same loans, but this time using the Level Payment Method.

Comparing Level Payment Loans by the Annual Payment Method.

Comparing Level Payment Loans by the Annual Payment Method involves a determination of True Interest Cost, as was described in Chapter 5. Here, we will be able to compare our two loans by calculating the TIC for both.

Therefore, let us say Loan A-1 is a five year Level Payment Loan of $20,000 at an interest rate of 8.25%, with no points.

Loan B-1 is a five-year Level Payment Loan of $20,000 at an interest rate of 8%, with one point.

Now, it is possible to calculate the TIC by using the calculator steps as described in Appendix A or the tables in Appendix B, but his would be a torturous process, indeed. The easiest method would be simply to consult the TIC tables in Appendix E.

As you know, the TIC for Loan A-1 is 8.25%. This is because the proceeds of Loan A-1 are the full $20,000, or 100% of the loan amount.

For Loan B-1, however, the proceeds are only $19,800, which is 99% of the loan amount. Thus, the proceeds are 1% less than the loan amount.

As you recall from Chapter 5, to determine TIC we must first determine how much we must pay out of, or deduct from, loan proceeds, as a percentage of the loan amount. We have already determined that this is 1%. Therefore, we turn to Appendix D and look for the table which corresponds to the interest rate on the loan, or 8%. As you can see, this is Appendix D-4. Now, we look for the line which corresponds to the percentage by which the loan amount has been reduced (1%). Finally, we look for the column which corresponds for the loan term (5 years).

The point at which the line and the column intersect is the <u>addition</u> to the loan interest rate. As you can see from Table D-4, the number is: .38. This means that to ascertain the TIC you must add .38 to the loan interest rate of 8.00%. The TIC for Loan B-1 is, therefore, 8.38%.

We can now compare. The TIC for Loan A-1 is 8.25%. The TIC for Loan B-1 is 8.38%. Therefore, Loan B-1 is more expensive than Loan A-1.

This is how Level Payment Loans are compared by the Annual Payment Method.

Now, three final points; but they are brief.

For those readers who are comfortable with using calculators, the steps for ascertaining TIC are set forth in Appendix E.

In addition, there may be a question in some readers' minds as to why the quite simple method of comparison which is used with Level Principal Payment Loans cannot simply be used with Level Payment Loans, as well.

The answer is that to do so produces a slight inaccuracy; and, I emphasize slight. This is caused by the fact that different amounts of principal are repaid at different rates of interest, and the two do not correlate.

Now, just to demonstrate that the two methodologies are, indeed, similar, let us run through the exercise very briefly.

We can determine the annual debt service payment on Loan A-1 by using the calculator steps described in Appendix B, or by consulting the tables in Appendix C. The annual debt service payment on Loan A-1 is $5,042. For Loan B-1, the annual debt service payment is $5,009.

Since we have ascertained that Loan A-1 is 1.0101 times larger than Loan B-1, we can multiply the annual debt service payment on Loan B-1 by 1.0101, which equals $5,060 and compare them.

As you can see, at $5,060, Loan B-1 is more expensive than Loan A-1 at 45,042.

Now, for the third and final point, let us look at all four loans (A, A-1, B, and B-1) together.

Comparing Level Payment Loans and Level Payment Loans, Themselves, Using the Annual Payment Method.

As you have seen, the equivalent annual debt service payments for Loans A and A-1, respectively, are $5,650 and $5,042. The equivalent annual debt service payment on Loans B and B-1, respectively are $5,657 and $5,060.

The point here is that Loans A and A-1 are identical except for one factor. And, Loans B and B-1 are identical except for one factor. That one factor is that Loans A and B are level Principal Payment Loans. Loans A-1 and B-1 are Level Payment Loans.

Please note that the annual debt service payments are higher for the Level Principal Payment Loans than for the Level Payment Loans. In fact, what we have here is a general rule: the annual debt service payments for identical loans is always higher for Level Principal Payment Loans.

This means, of course, that from the point of view of this chapter, which is the Annual Payment Method, it is always preferable to have Level Payment Loans since their annual payments are always lower.

Chapter 7

The Total Payment Method
of Evaluating Project Financing

At the beginning of the last chapter, we pointed out that there are two distinct personal philosophies about how to pay for projects. These can be characterized at the Pay As You Go philosophy which holds that financing is proper, as long as the period of financing is confined to the service life of the project. In other words, if a new standpipe will last for twenty years without a major overhaul, then it is perfectly proper to issue a twenty-year bond or take out a twenty-year loan to pay for it.

The contrasting philosophy is known as the Not Beholden theory. Believers in the Not Beholden theory do not believe in debt, at all. And, if it is absolutely necessary to incur debt, then it should be for the smallest amount of money and for the shortest period of time.

We also noted that if one were an adherent of the Pay As You Go philosophy, the appropriate form of analysis of project financings would be with the Annual Payment Method, which was also described in the last chapter.

The opposite side of the coin, of course, is that if one is an adherent of the Not Beholden philosophy, then the appropriate form of analysis of project financings would be with the Total Payment Method, which we will now describe.

* * * * *

In the previous chapter, we also reviewed all of the philosophical reasons in favor of the Pay As You Go Method. We will now review the principles which underpin the Not Beholden theory, as well.

The Not-Beholden Theory of Project Funding.

The philosophical root of the Not Beholden theory of project funding is quite simple: aversion to debt.

Those who abhor indebtedness usually cannot be persuaded otherwise. The payment-according-to-service-life argument is not persuasive. Nor is the fairness argument: that today's ratepayers would be subsidizing future ratepayers.

Paying for a project according to its service life is eminently logical. To do otherwise is illogical.

Having ratepayers pay equally, each year, for their pro rata share of of a project, for as many years as the project lasts, is eminently fair. To do otherwise is patently unfair.

Proponents of the Not Beholden theory have no response to these charges of illogic and unfairness. Nonetheless, most people in the world are Not Beholden types.

This is undoubtedly because there is a primal instinct to own. Ownership means control. As long as you have a mortgage, there is a nagging feeling at the bottom of your soul that your home is really not your own: it is the bank's. When people pay off their mortgages they feel a great relief. This feeling is not just confined to mortals. I am told that even professors of economics, who surely must realize that paying off a mortgage is clearly an unwise move from the economic point of view, have this same feeling of freedom and relief on the day they burn their mortgage.

Even the expression "burning the mortgage," has an atavistic and primal ring to it; as if it were some ancient, holy ritual, like the sacrifices in the Old Testament.

In any event, notwithstanding either the laws of economics or the laws of metaphysics, most people loathe debt.

The problem with water and wastewater systems is the same one confronting householders: major projects, such as installing a mile of water mains or buying a new home, simply cost too much money. Nobody can pay in cash. Virtually everyone must finance.

Bearing this in mind, the Not Beholden types will opt for the financing program which lets them pay off the loan as soon as possible.

In some case, people will try to pay off mortgages so fast, which, in turn, eats up so much of their income, that it actually causes great hardship to their families.

Suffice it to say that the Not Beholden theory of debt requires that loans be paid off as quickly as possible.

This leads us to the actual topic of this chapter. Not Beholden types should favor Level Principal Payment loans, because, as you know, they repay principal faster than Level Payment Loans. The Not Beholden types will also certainly want to consider shorter terms - terms which are shorter than service life - even with Level Payment Loans.

These two types of financial structures - Level Principal Payment Loans and the comparison of Level Payment Loans with different terms - require the Total Payment Method of analysis.

The Total Payment Method.

The Total Payment Method involves ascertaining the cheapest possible way to pay for something. Period.

It contrasts with the Annual Payment Method which seeks the most painless way to pay for something. In short, with the Annual Payment Method you trade off lower yearly costs for higher total costs. With the Total Payment Method we are seeking the lowest total cost regardless of annual cost.

Unlike the Annual Payment Method which involved two different techniques to compare loans, the Total Payment Method involves only one. The key to the Total Payment Method is to determine the Present Values of each annual debt service payment of each loan.

In the last chapter, where we looked at the Annual Payment Method, we used two loans, in both the Level Payment format and the Level Principal Payment format, as illustrative examples. In this chapter, we will use the same two loans. And, we will begin, as we did in the last chapter, with the Level Principal Payment Method.

Comparing Level Principal Payment Loans Using the Total Payment Method.

Please recall that Loan A was a $20,000 Level Principal Payment Loan for a five year term with an interest rate of 8.25% and with no points.

Loan B was a $20,000 Level Principal Payment Loan for a five year term with an interest rate of 8% but with one point.

Now, as we said above, the key to the Total Payment Method is to determine the Present Value of the annual debt service payment for each year of the loan. To calculate a Present Value, we must discount. And, to discount, we need a Discount Rate.

In Chapter 5, we used a Reinvestment Rate of 7% as the Discount Rate. We will use the same 7% rate in this example for the same reasons.

Below, in Figure 7-A, are the Present Values for the respective annual debt service payments for Loan A.

Figure 7-A

Year	Annual Payment	Present Value
1	$5,650	$5,280
2	5,320	4,647
3	4,990	4,073
4	4,660	3,555
5	4,330	3,087
		$20,642

Now, on the next page, in Figure 7-B are the Present Values for the respective annual debt service payments for Loan B.

Figure 7-B

Year	Annual Payment	Present Value
1	$5,600	$5,234
2	5,280	4,612
3	4,960	4,048
4	4,640	3,540
5	4,320	3,080

STOP! What is wrong with the annual debt service payment structure for Loan B? We did not bother to add up the Present Values for the annual debt service payments because, as they are, they would be wrong. What is missing?

The point. The one point fee that was paid to the bank at the closing. That fee was surely a real payment. It was a cash outflow. And, to get an accurate valuation on the loan we must take into account all of the payments, not just the five annual payments. Here, in Figure 7-C, below, is precisely how we do so.

Figure 7-C

Year	Annual Payment	Present Value
0	$200	$200
1	5,600	5,234
2	5,280	4,612
3	4,960	4,048
4	4,640	3,540
5	4,320	3,080
		$20,714

Now, we have the correct Present Value of all of the payments made in respect of Loan B. (N.B. The Present Value of the $200 payment is $200 because this sum is paid immediately on closing. Its Present Value is its stated value.)

Now, we are in a position to compare it to Loan A. And, as you can see, by using the Total Payment Method of comparison, we can now conclude that Loan B is more expensive than Loan A. To be precise, it is exactly $72 more expensive.

Let us move on to Loans A-1 and B-1, which are Level Payment Loans and analyze them using the Total Payment Method.

Comparing Level Payment Loans Using the Total Payment Method.

As you recall, Loan A-1 is the same as Loan A, except that Loan A-1 is a Level Payment Loan. In specific, it is a $20,000 loan for a five year term at an interest rate of 8.25%, with no points.

Loan B-1 is the same as Loan B, except that it is a Level Payment Loan. In specific, it is a $20,000 loan for a five year term at an interest rate of 8%, with one point.

Here again, we need to determine the Present Values for each of the annual debt service payments. And, again, we will use a Discount Rate of 7%.

Figure 7-D, below, presents these data for Loan A-1.

Figure 7-D

Year	Annual Payment	Present Value
1	$5,042	$4,712
2	"	4,404
3	"	4,116
4	"	3,847
5	"	3,595
		$20,674

Now, before we present the same data for Loan B-1, we must remember that Loan B-1, like Loan B, above, has the problem of the one point paid to the bank on closing. We will solve this problem in the same way which we did on Loan B, by adding the 1%, or $200, payment at the very beginning of the loan. Figure 7-E, top of next page, shows these data.

Figure 7-E

Year	Annual Payment	Present Value
0	$200	$200
1	5,009	4,681
2	"	4,375
3	"	4,089
4	"	3,821
5	"	3,571
		$20,737

Having established the Present Value of all of the payments made in respect of Loan B-1, we can now compare it with Loan A-1.

As you can see, Loan B-1 is more expensive than Loan A-1. To be precise, once again, it is $63 more expensive.

Now, before leaving the Total Payment Method of evaluating project financings, we have one more topic to cover. Earlier in this chapter, we observed that the Total Payment Method was especially valuable in evaluating Level Payment Loans with different terms. We will now illustrate how this works; but, to do so we will need to hypothesize a third loan with a different term than the two loans we have used above.

Comparing Level Payment Loans With Different Terms Using the Total Payment Method.

Let us now create Loan A-2, which will be similar to Loan A-1, except that the term will be six years instead of five and the interest rate will be 8%, instead of 8.25%.

In other words, Loan A-1 was a $20,000 loan at an interest rate of 8.25% with a term of five years.

Loan A-2 is a $20,000 loan at an interest rate of 8% with a term of six years.

The Present Value data for Loan A-1 is shown above in Figure 7-D.

The Present Value data for Loan A-2 is shown below, in Figure 7-F, using 7%, once again, as a Discount Rate.

Figure 7-F

Year	Annual Payment	Present Value
1	$4,326	$4,043
2	"	3,778
3	"	3,531
4	"	3,300
5	"	3,084
		$17,736

As you can clearly see, Loan A-2 is considerably less expensive than Loan A-1. To be precise, it is $2,938 less expensive.

Section II

Section II

Chapter 8

Factors Which
Inflate Project Cost

The last seven chapters of this book, which constitute Section One, defined many of the terms which need to be understood in order to be able to fathom various financing programs available to water and wastewater systems. In addition, the two most important tools used to analyze capital project funding were also presented. These are: the methods of comparing costs, and the methods for converting onetime dollar costs or annual dollar costs to rates, and vice versa. Finally, the two most common methods of comparing different financing programs - the annual cost method and the total cost method - were also presented. In short, Section One is designed to give the reader the tools he needs to understand how various categories of project cost can impact the rates of his water or wastewater system.

Section Two will now present specific categories of cost which serve to inflate project costs and, hence, ratepayers' rates.

This chapter is going to introduce the seven factors which serve to inflate the cost of water and wastewater projects. The individual factors themselves will be examined in detail, one each in the next seven chapters of this book. These seven factors are:

1.) <u>Term</u>. The number of years, or Period of years, over which the Principal amount of a loan is repaid.

2.) <u>Rate</u>. The Interest Rate at which either the bond is issued or the loan is made.

3.) <u>Financing Costs</u>. Upfront costs which must be added to the bond or loan Principal as well as annual charges which serve to increase the True Interest Cost.

4.) <u>Delay</u>. The effect of delay on both project cost and on financing cost.

5.) <u>Imposed Costs</u>. These costs are those additional charges which are extrinsic to the project but are required by a particular financial alternative.

6.) <u>Ineligibility</u>. That portion of total project cost which cannot be financed through a particular financing program.

7.) <u>Coverage</u>. The amount, if any, by which the ratepayers' annual payments must exceed the annual debt service due.

These cost factors fall into four broad categories. The first five factors relate directly to the discussions of "Upfront costs vs. Annual Cost" in Chapter 5. The first two factors, Term and Rate, have a direct impact on the <u>annual</u> cost of a project; whereas the next three factors, Financing Costs, Delay and Imposed Costs relate directly to upfront costs. In other words, Financing Costs are, generally, upfront costs. They result in an increase in the amount to be borrowed or an increase in loan Principal. Delay results in a higher project cost, which also results in an increase in the loan Principal which system's must borrow. Imposed Costs are also additions to loan Principal.

These five factors are quite straightforward in their impact on the cost of a project to a water or wastewater system. The last two factors, Ineligibility and Coverage, have, however, a far more insidious effect. They are also very difficult to categorize.

Ineligibility, as you will see, means that the particular financial program you have chosen to fund your system's project will not permit certain categories of costs to be included. The implications of this, at first, can be quite subtle; but they can sometimes be quite dire.

But, in addition, they are difficult to define. Ineligible costs do not add to project costs. (They most often subtract costs.) They are not upfront costs and they are certainly not annual costs. But rather, in effect, they create a second project, at least from a financial perspective. Let us look at a brief, theoretical example.

Let us say, for example, that two years ago your system was offered an extensive piece of property, at an extremely attractive price, on which to drill a series of new wells. In addition to the land that was absolutely necessary, the system's board decided to buy a considerable amount of surrounding acreage. They felt this was the most prudent course of action. On the one hand, the additional acreage might be needed for future wells, if the system's needs expanded. But, on the other hand, even if the system didn't need any further wells, the additional acreage would assure good wellhead protection for the original well field. So the board went ahead and bought the property and financed it with a short-term bank loan, on which they only had to make quarterly interest payments.

The board's intention was to roll the acquisition of the land into a much larger project which would involve drilling the new wells, adding new storage capacity and pumping capacity, as well as connecting the new well field to the system itself. They have now completed all the planning and the engineering studies as well as the design of the new improvements. They are now ready to finance the project.

Unfortunately, the board now learns that most of the land cannot be financed through the program they have chosen. That program allows the system to pay for the project over a twenty-year period, but it will only permit land costs to be included in the project to the extent of the minimum acreage required for the wells themselves. In other words, no credit will be given for wellhead protection, and no credit will be given for future expansion.

At first this does not sound so drastic. After all, they already have interim bank financing. So what if it cannot be financed through the program the board had chosen. How hard can it be to get additional financing for the extra acreage?

Well, consider this. The bank might have been very happy indeed, to give you an interest-only bridge loan for a couple of years on the land. They did so in anticipation that the loan would be repaid in its entirety in a year or so when the full project was financed. Now the bank finds out that most of the land will not be financed with the project. Now they are being asked to finance it. And, what is worse, they won't even have a first mortgage on the full property, since part of the mortgage must be split off as security for the part of the land that is being financed through the other program.

The bank decides it won't give you a loan on the excess acreage for more than five years and at a very substantial rate. As you will see in Chapter 9, reducing an amortization schedule from 20 years to five years can have a drastic effect on the Annual Payments which the ratepayers must make. This is to say nothing of the effect of the increased rate, which will be covered in Chapter 10.

It is important to note that whenever you see the concept "Eligible Project Costs" in literature describing a financing program, you must be aware that it has a hidden counterpart called "Ineligible Project Costs." These ineligible costs still have to be paid for, even if they cannot be financed through a particular program.

The point here is that the ineligibility of a project cost means that the particular cost must be paid for by some other means. Perhaps, such costs can be paid for by another financial program. Perhaps, a bank will finance them. Perhaps, no one will finance them and they must eventually be absorbed directly by the ratepayers in the form of a one-time assessment.

The effects of an ineligible project cost on your system may be quite mild, especially if the cost is small and there is a reasonable alternative readily available. But, on the other hand, if the ineligible cost is large and there is no alternative available, you might wind up with some angry ratepayers who are forced to come up with a large sum of money on a one-time basis.

You can easily see what the real-world effects on ineligible costs may be. The question remains, however, as to how these costs can be measured and how they can be integrated into a total comparison for several financial programs. In other words, let us say in Program A, the land costs described above might be eligible; in Program B, however, they might be ineligible. In order to be able to compare Programs A and B, we would have to deem that Program B is really two programs: Program B1 and Program B2. Program B1 would contain all of the eligible costs, and Program B2 would contain all of the ineligible costs.

If the costs in Program B2 could be financed, then the annual payments in respect thereof could simply be reduced to Present Values and added to the Present Value of the annual payments in Program B1. In such case Programs B1 and B2 would contain the identical costs as Program A. Once all annual payments were reduced to their Present Values, they could be readily compared.

So, it is important to realize that ineligibility does not directly relate to either upfront costs nor annual costs. Rather, ineligible costs create a separate problem which must be dealt with by creating a subprogram. Ineligible costs must be treated separately. But, they can then be converted to Present Values, which can be added back to the original project costs. In such case, you will then be able to compare programs where the eligible (and ineligible) costs differ.

The seventh and last of the factors which inflate project cost is "coverage," which will be considered in Chapter 15.

The concept of coverage creates a unique type of cost for water and wastewater systems. At first look, the concept of coverage seems to involve a new type of annual cost. But, in fact, it is not a cost at all; it is simply additional funds which are collected from the ratepayers but not paid out in respect of any project.

Thus as far as the ratepayers are concerned, the concept of coverage means more money they will have to pay to their system in respect of a project; but the system will not pass along any of such excess payments to any lender.

As you will see, the concept of coverage, as an additional form of cost, simply refers to the fact that a system must collect a certain percentage more from its ratepayers than it needs for its annual debt service payments. The purpose of this overcollection is that such additional funds constitute additional security for the lender. In other words, if a bank were the lender, the bank might well set, as a condition of the loan, that cash available for debt service be at least 125% of annual debt service. This means that the system must collect 25% more than it needs from its ratepayers to pay for the project. Lenders make such demands in order to make sure that systems will have more than enough funds to make their debt service payments.

Lenders are not at all happy when the cash available for debt service of a water or wastewater system to which they were lending a substantial sum of money is only 100% or 101% of annual debt service. The obvious reason is that any slight slip up in the collection of system revenues will result in a shortfall on the loan repayment. For that reason, lenders often require that systems charge their ratepayers enough to create a cushion, which might be as high as 25% of cash available for debt service.

Now, it is important to note that nothing happens to the additional 25%. The system does not pay it to the lender, or to anyone else. The funds stay right there in the system and can be used for discretionary payments or for other small capital projects. In other words, although the system is required by the lender to collect 125% of annual debt service, the lender does not get the money.

For this reason -- that the extra money is not used to make annual debt service payments for the project -- it cannot be considered an annual project cost. But it certainly has the same effect on the ratepayers as an additional annual project cost. They must pay more, because of the project, regardless of where the money is actually going.

So, the concept of coverage is, indeed, a subtle form of cost. Though neither an upfront project cost nor an annual project cost nor even a real project cost, at all; it, nonetheless, requires higher annual payments from ratepayers. Thus, it is not a cost to the system, yet it is a cost to the ratepayers in the truest sense of the word.

Chapter 9

Term

The Impact of Term on Annual Debt Service Payments.

Of all the factors which inflate project costs, term is the one which is least understood. Term has an enormous impact on the annual payment.

What is the actual impact of term on annual debt service payments? As you will see in the next few pages, just this: the longer the term, the lower the annual debt service payment. And, the lower the rate of interest, the more so this is true.

Before we look into the impact of term on annual payment, however, let us make sure we are all speaking the same language.

Definitions: Term, Maturity, Amortization, and Amortization Schedule.

Term is the number of years in a loan during which principal is outstanding. Term is the most important concept relating to principal.

Another word which is also used to refer to principal is: maturity. Maturity is the time at which an installment of principal, including the final installment, is repaid. As you will note, a loan can have many maturities, but only one term. In common parlance, the word "maturity" is often used instead of the word "term." In such cases, the speaker is actually referring to the final maturity, which is, of course, synonymous with the term.

There is another word which relates both to time and to principal which is also used in conjunction with loans. This word is "amortization." <u>Amortization is the process of repaying principal</u>. The word "amortization" contains the element "mort," as in mortal, mortality, or mortuary. The root "mort" refers to death; and in the word "amortization," it is actually referring to the "death" of the debt. As principal is being repaid, the obligation is dying. Hence, the process of reducing the obligation is called: amortization.

Finally, there is also the phrase "amortization schedule." <u>The amortization schedule is a list of maturity dates and amounts of principal actually due and payable on those respective dates.</u>

Graphic presentations of loan terms.

Now, when you think of loans, try to picture them on a graph such as the one in Figure 9-A, as follows:

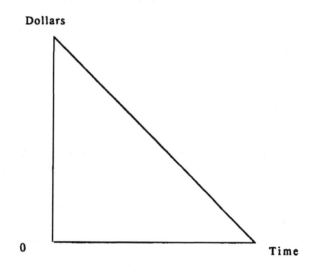

Figure 9-A

Think of the amount of outstanding principal as the number on the vertical, or y-axis. Then think of the amount of time during which principal is outstanding as the horizontal, or x-axis. Figure 9-B, below, shows a $100,000 loan for five years which is to be repaid at a rate of $20,000 per year plus interest.

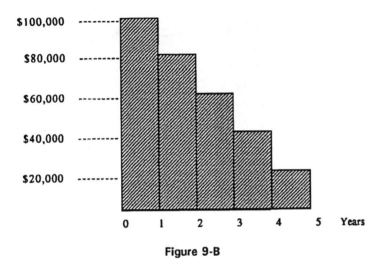

Figure 9-B

Along the vertical, or y-axis, you have the outstanding principal balance on the loan. Along the horizontal, or x-axis, you have the time period. So, for example, at the end of year three, the outstanding principal balance on the loan is $40,000.

There is another way of looking at the loan illustrated in Figure 9-B, above. Look at Figure 9-C, below.

Here, as in Figure 9-B above, the horizontal, or x-axis, represents the number of years during which loan principal is outstanding. But, in figure 9-C, the vertical, or y-axis, now represents the amount of loan principal repaid in each year. Please note that the loan described in both Figure 9-B and Figure 9-C is a Level Principal Payment Loan. (See Chapter 2.)

The number of years over which the principal of a loan is outstanding is, as you know, called the term. The number of years, however, during which the principal is scheduled to be repaid is called the "amortization period." Please note the careful choice of words. At first glance, these two concepts -- term and amortization period -- might seem identical. They are not. They are similar, but not identical; and the difference between them is important.

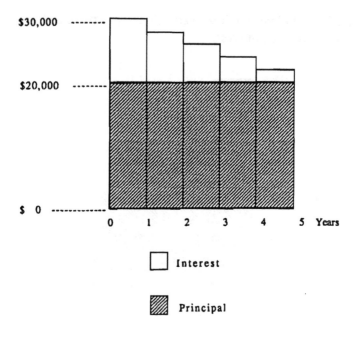

Figure 9-C

First, let us look at a loan where the term and the amortization period are identical. The loan described in Figures 9-B and 9-C, above, is just such a loan. The term is five years and the amortization period is also five years.

"Balloon" Payment Loans.

But there are also many types of loans where the terms and the amortization periods are different. The key to the difference between the two is the phrase "scheduled to be." There are many loans where, for example, the principal is "scheduled to be" repaid over a thirty-year period, but where the final principal payment is made at the end of the tenth year. Loans where the term and the amortization period are different are called "balloon" loans. As you might imagine, the final payment of such loans is called the "balloon payment."

Some people have "balloon" mortgages. You might hear someone say: "I have a thirty-year mortgage with a five-year balloon." This means that the loan has a thirty-year amortization schedule and a five-year term. What this means to the mortgagor is that the first five annual payments are identical to the first five payments on a thirty-year mortgage. (Remember from Chapter 2: Mortgage loans are level payment loans.) At the end of the fifth year, however, in addition to making annual payment #5, the hapless mortgagor must also pay the entire outstanding principal balance, which is then due and payable.

You might wonder what this poor soul's banker expects him to do at the end of the fifth year. How could anyone be expected to pay off twenty-five years of principal payments at one time? The answer, of course, is that no one expects the mortgage to be paid off at that time. It will simply have to be renegotiated. People often refer to this event as a "rollover." Technically, it is not; nothing is actually rolled over. The old mortgage is actually paid off, and a new mortgage is created. The borrower will have to pay a different rate of interest. He may well have a different term. And the amount of the new loan will be the outstanding principal balance of the old mortgage at the end of year five. The "balloon" becomes the initial principal amount of the new mortgage.

So, there are two types of loans. The first, and by far the most prevalent, are loans where the term and the amortization period are the same. The second, which is used mostly in the commercial real estate market, is the balloon loan where the amortization period exceeds the term. (Please note that there is no such thing as a loan where the term exceeds the amortization period. By definition, once the principal is repaid, there is nothing left upon which to charge interest; the loan is over.)

Figure 9-D presents a graph showing the outstanding balance of a thirty-year mortgage of $100,000 at a 10% interest rate.

Figure 9-E, below, shows the same loan, but this time the vertical, or y-axis, indicates the amount of principal repaid in each year.

Now, compare the above two figures with the two figures below, which show the same loan but with the added feature of a five-year balloon. In other words, the term is five years but the amortization period is thirty years.

Figure 9-F, below, shows the balloon loan with the outstanding principal balance indicated on the vertical, or y-axis.

As you can see, over 96% of the original principal remains unpaid. Imagine the surprise of the poor, unfortunate mortgagor, who didn't know what a "balloon" was, when he got his last annual payment notice.

Figure 9-G, below, shows the same loan with the amount of the annual principal payment indicated on the vertical, or y-axis.

Dollars

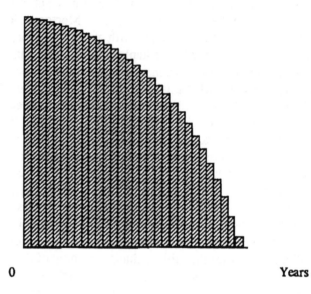

0   Years

Figure 9-D

Dollars

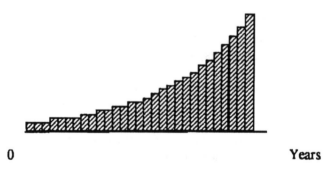

0   Years

Figure 9-E

Dollars

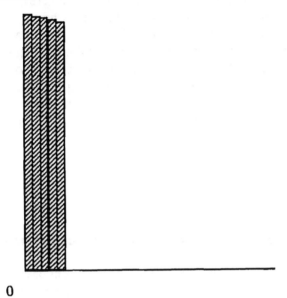

0   Years

Figure 9-F

Dollars

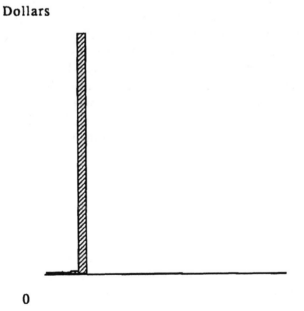

0   Years

Figure 9-G

The difference between term and amortization period is important. One should not be confused with the other. But from here on in, we are going to assume that we are dealing only with loans where the terms and the amortization periods are the same. Thus, for the balance of this book, when we speak of "term," implicit will be the fact that the loan to which we are referring has an amortization period coterminous with the term of the loan. So, when we speak of extending the term of a loan, say, from ten to twenty years, it will also be implicit that the amortization period is extended for the same period, as well.

*   *   *   *   *

The Impact of Term on Annual Debt Service Payment.

We must begin any analysis by saying that the term of a loan has an enormous impact on its annual debt service payment. This is true whether you choose the Level Payment Method or the Level Principal Payment Method.

You may recall that, in Chapter 2, we briefly discussed these two methods of repaying the principal of a loan. These concepts take on great importance when discussing the impact of term on annual debt service payments. As you will see, the impact of term on annual debt service makes the decision of whether to choose a loan with the Level Payment Method or the Level Principal Payment Method a critical judgment. For this reason, we will discuss the impact of term on annual debt service separately for each type of payment method. Then we will compare the two types of loan payment methods in terms of their respective impacts on annual debt service payment.

Characteristics of Level Principal Payment Loans; Problems with Fairness.

If the term of a loan is only one year, then 100% of the principal is due and payable at the end of that year. If the term of a loan is two years, then 100% of the principal must be repaid by the end of the second year.

Let us assume that we are dealing with a Level Principal Payment Loan, so that 50% of the principal balance is due and payable at the end of each year. Then, for a $100,000 loan at an interest rate of 7%, the annual debt service payment at the end of the first year would be $57,000, consisting of $50,000, or 50%, of the principal and $7,000 of interest, representing 7% on the $100,000 of principal which was outstanding during the first year of the loan.

The second annual debt service payment, however, which would be due at the end of year two, would only be $53,500. This payment is composed of $50,000 of principal, which is the remaining principal outstanding, and $3,500 of interest, which is 7% of the $50,000 of principal which was outstanding during the second year of the loan.

So, the annual debt service payment at the end of year one would be $57,000, and the annual debt service payment at the end of year two would be $53,500.

The difference between these two payments is only $3,500, which is about 7%. Even if the amount of the loan were increased from $100,000 to $1,000,000, the difference in the two annual debt service payments would be only $35,000, which is, of course, still only 7%.

A 7% difference in annual debt service payments cannot be considered significant. After all, annual debt service payments only make up a fraction of the total annual service charges to system users. For example, if the operation and maintenance expense for a water or wastewater system were only equal to annual debt service payments, then operation and maintenance expenses would be 50% of total system user charges and annual debt service payments would be 50% of total system user charges. In such case, the 7% difference in the two annual debt service payments in our example, above, would only necessitate a 3.5% difference in total system user charges, since 7% of 50% is equal to 3.5% of 100%.

In our example, the first year's annual debt service payment is $57,000. If the operation and maintenance expense that year were also $57,000, then, theoretically, the total system user charge for the year would be $114,000. Assuming the operation and maintenance expense stayed the same in year two, the second annual debt service payment would be only $53,500. Total system user charges for the year would be reduced to $113,500, which is about a 3.5% decrease.

A 3.5% difference in total system user charges from one year to the next is not really significant. Especially when it is a decrease.

When, however, we start talking about loans with terms of more than two years, the story begins to change radically.

Let us use the same $100,000 loan and the same 7% interest rate, but this time use a 20-year term. Let us look at the difference between the first annual debt service payment and the last annual debt service payment.

Assuming, once again, that we are dealing with a level <u>principal</u> payment loan, the amount of principal which would be repaid in every year would be 1/20th of $100,000, or $5,000. For the first year, the outstanding principal balance would be the original principal balance, or $100,000. The interest on $100,000 for one year at a rate of 7% would be $7,000. Since the principal payment at the end of the first year would be $5,000, and the interest payment would be $7,000, the total annual debt service payment for year one would be $12,000.

For year twenty, the outstanding principal balance would only be $5,000, since the other $95,000 had been repaid in the preceding 19 years. The interest, at a rate of 7%, on the outstanding principal balance of $5,000 would be $350. Therefore, the total annual debt service payment for year twenty would be $350 of interest and $5,000 of principal, or $5,350.

Now, compare the first year's annual debt service payment of $12,000 with the annual debt service payment in year twenty of $5,350. Here, the difference in the two annual debt service payments is $6,650, or a 55% decrease! I am certain that the ratepayers in year 20 would be pleased with this circumstance, but what about the ratepayers in year one who are getting the benefit of the same project?

In Chapter 2, we discussed Level Payment Loans and Level Principal Payment Loans. We observed that one of the key points in favor of Level Payment Loans involved the concept of fairness. But we did not go into the matter in detail. Let us illustrate this point now with an example.

As you will recall, a Level Payment Loan means just that: each annual debt service payment is equal to any other annual debt service payment. If the service life of, say, a new treatment facility is twenty years, then, using our example above of a $100,000 loan at an interest rate of 7%, the annual debt service payment under the Level Payment Method would be $9,439. This means that the annual debt service would be $9,439 in each of the twenty years the loan was outstanding. In such case the total debt service would be twenty times $9,439 (20 x $9,439), or $188,786.

A ratepayer in year one would pay his share of the $9,439, which is 1/20th of the total project cost. Assuming that the project -- say a new treatment facility -- had a service life of twenty years, that ratepayer would also enjoy 1/20th, or 5%, of the benefit of the project. In other words, in year one, he would pay 5% of the cost of the project and he would receive 5% of the benefit of the project. Fair enough.

Now, what if the project loan were a Level Principal Payment Loan instead of a Level Payment Loan. Table 9-A, below, sets forth the annual principal payments, the annual interest payments, and the annual debt service payments on a twenty-year, Level Principal Payment Loan of $100,000 at an interest rate of 7%.

Table 9-A

Year	Annual Principal Payment	Annual Interest Payment	Annual Debt Service Payment	% of Total Debt Service
1	$5,000	$7,000	$12,000	6.9%
2	"	6,650	11,650	6.7%
3	"	6,300	11,300	6.5%
4	"	5,950	10,950	6.3%
5	"	5,600	10,600	6.1%
6	"	5,250	10,250	5.9%
7	"	4,900	9,900	5.7%
8	"	4,550	9,550	5.5%
9	"	4,200	9,200	5.3%
10	"	3,850	8,850	5.1%
11	"	3,500	8,500	4.9%
12	"	3,150	8,150	4.7%
13	"	2,800	7,800	4.5%
14	"	2,450	7,450	4.3%
15	"	2,100	7,100	4.1%
16	"	1,750	6,750	3.9%
17	"	1,400	6,400	3.7%
18	"	1,050	6,050	3.5%
19	"	700	5,700	3.3%
20	"	350	5,350	3.1%
	$100,000	$73,500	$173,500	100%

Please note that the last column in Table 9-A, above, is captioned "Percentage of Total Debt Service." The percentages listed on each line represent the annual debt service payment for that year divided by the total debt service.

As you can see, with the Level Principal Payment Loan, the ratepayer in year one pays 6.9% of the project cost. The problem is that, since the service life of the project is only twenty years, he only gets the benefit of 1/20th, or 5%, of the project. So, he winds up paying for 6.9% of the project but getting the benefit of only 5% of the project.

By the same token, look now at the ratepayer in year twenty. He pays his proportionate share of the annual debt service payment of just $5,350, which is only 3.1% of total project cost. So, the year twenty ratepayer only pays for 3.1% of the project, but winds up getting the benefit of a full 5% of the project.

Now, 6.9%, 5%, and 3.1% look like pretty small numbers. But, in the next chapter we will discuss how absolute havoc can ensue from the way numbers can be <u>characterized</u>. If you want to characterize the annual debt service payments in years one and twenty, you can see that the difference between 6.9 and 3.1 is 123%. This means that the annual debt service payment in year one is 123% <u>higher</u> than the year twenty annual debt service payment.

Now, the differences between 3.1% and 6.9% are not much; but system ratepayers will certainly want to know why they are being charged 123% more, when they are only getting the same benefit out of the project!

Now, the theme of this chapter concerns the effect of rate on term. Lets look, without going into all of the detail, at the same two loans, if they had 40-year terms instead of twenty-year terms.

On the Level Principal Loan, the annual principal payment on a 40-year, $100,000 loan would be 1/40th, or 2.5%, or $2,500. The interest payment would be the same in year one, as for the 20-year loan (since the full $100,000 is outstanding in the first year), or $7,000. So, the annual debt service payment in year one for the Level Principal Payment Loan would be $9,500. For the fortieth year, only $2,500 would remain outstanding. The interest on this amount, at the 7% rate, would be $175. So, the fortieth year annual debt service payment would be $2,675.

The difference between the annual debt service payment in year one can be characterized as being <u>258% higher</u> than the annual debt service payment in year forty. How do you think your ratepayers would react to those numbers?

As we said before, the <u>characterization</u> of numbers will be discussed more fully in the next chapter. This chapter focuses on the effect of term on annual payment.

Now, that we have reviewed the impact of term on the annual debt service payment of a Level Principal Payment Loan, we will move on to the consideration of the impact of term on the annual debt service payments of a Level Payment Loan.

Calculating Annual Debt Service Payments for Loans Using the Level Payment Method.

The impact of term on annual debt service payments is as significant when the Level Payment Method is used as when the Level Principal Payment Method is used. With the Level Payment Method, however, there is no problem with fairness. There is no problem with an adverse characterization of the numbers.

For nine chapters so far, we have put off the task of calculating annual debt service payments on loans when using the Level Payment Method. We cannot go any further. To determine the impact of term on the annual debt service payments of loans using the Level Payment Method, we must first be able to calculate annual debt service payments themselves on loans using the Level Payment Method. Figure 9-H, below, shows the actual formula for deriving the annual debt service payment on a loan using the Level Payment Method. (Note that when discussing the Level Payment Method, we refer to the annual debt service payment, not payments; since all of the annual debt service payments are the same. The annual debt service payment on a loan using the Level Payment Method is frequently referred to simply as the "level payment.")

Figure 9-H

$$ADSP = P \times \cfrac{i}{1 - \left[\cfrac{1}{1 + i}\right]^n}$$

In this terribly imposing formula, ADSP is annual debt service payment, P is the original principal amount of the loan, i is the interest rate expressed as a decimal, and n is the term of the loan.

Gratefully, most of us are relieved of the burden of having to deal with this dreadful equation on a daily basis. There are a proliferation of books containing tables of annual debt service payments for Level

Payment Loans. There are also hand-sized, and smaller, calculators which can reduce this daunting host of symbols to <u>four simple keystrokes</u>. If your calculator has the following keys: <u>n</u>, <u>i</u>, <u>PV</u>, and <u>PMT</u>, you will be able to compute any level debt service payment in seconds, just by following the simple instructions in the manual.

Now, however, at the risk of alienating those who relied on the representation at the beginning of this book that it was written for non-financial executives, we will actually use the equation to calculate the level debt service payment. We will then go back and demonstrate, by using simple arithmetic, that the annual debt service payments determined by the formula are correct and that they produce annual debt service payments which are level, or equal. Mercifully, however, we will select a loan with only a two-year term as a means of abating the agony of going through the formula.

Figure 9-I shows how the formula works for a two-year loan of $100,000 at an interest rate of 7%. Translating into the equation: the original principal, or P, is $100,000, the interest rate, i, is 7%, or .07, and the term, or n, is 2 years.

Figure 9-I

$$ADSP = 100,000 \times \frac{.07}{1 - \left[\dfrac{1}{1 + .07}\right]^2}$$

As you will see, the calculation can now be done on an old-fashioned adding machine. The only real difficulty lies in remembering the sequence of operations from high school algebra. The first step is to begin with the bottom, or denominator, of the fraction in the parentheses. This simply involves adding 1.00 and .07, to get 1.07, as is shown in Figure 9-J, below.

Figure 9-J

$$ADSP = 100,000 \times \frac{.07}{1 - \left[\dfrac{1}{1.07}\right]^2}$$

Step 2 involves elimination of the fraction in the parentheses by dividing the number 1 by 1.07, to obtain .93458, as is shown in Figure 9-K. (Note the reason we are using so many decimal places is that we have a large [$100,000] loan. The rule of thumb to use is that you should have one less decimal places than you have digits in the original principal amount.)

Figure 9-K

$$ADSP = 100,000 \times \frac{.07}{1 - (.93458)^2}$$

In Step 3, we eliminate the parentheses by multiplying the number in the parentheses by itself. (The superscript 2 indicates the number of times the number should appear in the multiplication, i.e., .93458 x .93458.) The result is .87344, as shown in Figure 9-L.

Figure 9-L

$$ADSP = 100,000 \times \frac{.07}{1 - .87344}$$

Step 4 involves the simple subtraction now left in the denominator of the main fraction: 1 - .87344 = .12656. This is shown in Figure 9-M, below.

Figure 9-M

$$ADSP = 100,000 \times \frac{.07}{.12656}$$

In Step 5, we eliminate the main fraction by dividing the numerator, .07, by the denominator, .12656, to yield: .55309, as shown in Figure 9-N, below.

Figure 9-N

$$ADSP = 100,000 \times .55309$$

All that remains in Step 6 is to multiply the original principal amount, or 100,000, by .55309. The result is 55,309, or $55,309. **This is the annual debt service payment.**

Now, we can put an end to this brief exercise by doing a little arithmetic to prove that $55,309 is the exact amount of principal and interest due in both years of the loan. We do this by working backwards. We already have $55,309 which we believe to be the annual debt service payment. We also know that this amount is composed of principal and interest. We first need to determine how much of each: how much of the $55,309 is interest in the first year, and how much is principal.

We can begin by calculating the amount of interest. We can do this because we know the original principal amount and the rate of interest. Because the entire original principal balance had to be outstanding during the first year, we can calculate the amount of interest due at the end of the first year by multiplying the original principal amount, $100,000, by the interest rate, 7%, or .07. This is shown in Figure 9-O, below.

Figure 9-O

$100,000 x .07 = $7,000

Now that we know the amount of interest paid in year one, we can calculate the amount of principal paid in year one by subtracting the amount of interest paid, $7,000, from the annual debt service payment, $55,309, to yield $48,309, as shown in Figure 9-P, below.

Figure 9-P

$55,309 - 7,000 = $48,309

Now that we know how much principal was paid in year one, we can calculate the outstanding principal balance for year two, which can be used to calculate the interest due at the end of year two. We do this by subtracting the principal payment in year one, $48,309, from the original principal amount, $100,000, to obtain $51,691, as shown in Figure 9-Q, below.

Figure 9-Q

$100,000 - $48,309 = $51,691

Since we now know the outstanding principal balance for year two, we can calculate the amount of interest due in year two by multiplying the outstanding principal balance, $51,691, by the rate of interest, .07. The result is $3,618, as shown in Figure 9-R, below.

Figure 9-R

$$\$51,691 \times .07 = \$3,618$$

Now, the crucial part. We know the amount of interest due in year two, $3,618. We also know the outstanding principal balance for year two, $51,691. Since year two is the last year of the loan, the outstanding principal balance for year two must also be the amount of principal which must be paid in year two. Therefore, we can calculate the annual debt service payment due in year two simply by adding the amount of interest due in year two to the amount of principal due in year two, as is shown in Figure 9-S, below.

Figure 9-S

$$\$3,618 + 51,691 = \$55,309$$

As you can now see, the formula does work. The number, $55,309, is the annual debt service payment which we used in year one (Fig. 9-P) and which we calculated for year two (Fig. 9-S); and it is the same number predicted by the infamous formula as shown in Figure 9-N.

So, you can rest assured the formula works. And since it is the formula which is both built into hand-held calculators and the one from which all of the tables are derived, you, who tend to be financial Doubting Thomases, can now also be assured that your calculator and the tables are accurate.

Comparing Changes in Annual Debt Service Payment Caused by Variations in Term; and the Effect of Interest Rates on the Impact of Loan Term on Annual Debt Service Payments.

Let us begin with the point of departure between Level Payment Loans and Level Principal Payment Loans. That is the zero interest rate loan.

The amortization schedule for a zero interest rate loan, regardless of the term, is the same whether using the Level Payment Method or the Level Principal Payment Method. So is the Annual Debt Service Payment.

A Level Principal Payment Loan of $100,000 with a two-year term at an interest rate of 0% has an annual debt service payment of $50,000. So does a Level Payment Loan of the same amount for the same term at the same 0% rate of interest.

A Level Payment Loan of $100,000 with a twenty-year term at an interest rate of 0% has an annual debt service payment of $5,000. So, does a Level Principal Payment Loan of the same amount for the same term at the same 0% rate of interest.

With the Level Principal Payment Loan, all of the principal payments are equal. With the Level Payment Loan, all of the payments, themselves, are equal. With a zero interest rate loan, all of the annual debt service payments are composed of principal only. There is no interest component. So, here all of the annual debt service payments are equal; and, since they are composed of principal only, all of the principal payments are equal, as well.

The two-year loan has an annual debt service payment of $50,000, and the twenty-year loan has an annual debt service payment of $5,000. Here, a difference of 18 years in the principal repayment schedule results in one annual debt service payment's being ten times (10x), or 1,000%, higher than the other.

This is a good example of the impact of term on annual debt service payments.

This example is unique, however. It deals with 0% interest rate loans, which are so rare as to be almost hypothetical. It is also the only instance where the Level Payment Method is identical in effect to the Level Principal Payment Method in terms of annual debt service payment. That, however, is where the similarity between the Level Principal Payment Method and the Level Payment Method ends.

We will now look at the impact of term on the annual debt service payments of both Level Principal Payment Loans and Level Payment Loans where realistic rates of interest are involved. As you will see, the impact is similar on both types of loans, but not identical. This is because, as you know, annual debt service payments on Level Principal Payment Loans decline over the term of the loan. Thus, there are as many different annual debt service payments in a Level Principal Payment Loan as there are years in the loan term. On the other hand, with Level Payment Loans, the annual debt service payment is a constant over the entire term of the loan.

Level Payment Loans.

With a Level Payment Loan you can measure the impact of term on annual debt service by simply comparing the annual debt service payment of two loans with different terms, where each loan is for the same principal amount and carries the same rate of interest.

With a Level Principal Payment Loan, it is a little more difficult. Since the annual debt service payments on a Level Principal Payment Loan decline every year, the only way to obtain an accurate description of how term really affects annual payment is to compare the annual debt service payments at several intervals over the loan term. Again, we must make sure, when comparing these loans, that the principal amounts and the interest rates are the same for both.

Since they are easier to work with, we will begin by analyzing the impact of term on the annual debt service payments of Level Payment Loans.

We can start by setting out the annual debt service payments for a $100,000 Level Payment Loan at an interest rate of 7% at various terms ranging from 5 to 30 years. This is shown below on Table 9-B.

Table 9-B

A	B	C
5 years	$24,389	N/A
10 "	14,238	42%
15 "	10,979	23%
20 "	9,439	14%
25 "	8,581	9%
30 "	8,059	6%

Please direct your attention to Column C, which indicates the percentage difference in annual debt service payment for the respective terms. This is where the impact of term on annual debt service payment really begins to show up.

In the first place, the numbers in Column C read from the bottom up. In other words, referring to the last two lines at the bottom of Column

C, you will note that the difference between the annual debt service for a twenty-five-year loan and a thirty-year loan is 6%. This means that the annual debt service on the thirty-year loan is 6% less than the annual debt service payment on the twenty-five year loan.

(Please note that the numbers in Column C are not cumulative. You cannot find the difference in annual debt service payments between a fifteen-year loan and a five-year loan by adding the percentages in Column C. But, if needed, you can find the difference in annual debt service payments between any two loans by dividing the later loan by the earlier loan, and then subtracting the result from the number 1. For example, the difference in annual debt service between the fifteen-year loan and the five-year loan is 10,979 / 24,389 = .45; then 1 - .45 = .55. So, in this case the annual debt service payment on the fifteen-year loan is .55, or 55%, lower than the annual debt service payment on the five-year loan. And so on.

Please also note that we can characterize these numbers differently simply by reversing the way we consider them. For example, we can say that the annual debt service payment declines by 42% when the term of the loan is extended from five to ten years. Using the exact same numbers, but reversing the emphasis, we could also say that annual debt service payments increase by 71% when the term of the loan is reduced from ten to five years. Our overall purpose, however, is to seek out ways of lowering the annual debt service payments which the ratepayers have to make; so it is appropriate for us to emphasize the lowering of rates. Therefore, instead of saying how much rates increase with shorter terms, we will emphasize how much they decrease over the longer terms.)

These differences in annual debt service payments are the most important point to be made about the impact of different loan terms. Just look how great the impact really is!

The difference between a ten- and a thirty-year term means a difference of 43% in annual debt service payments which your ratepayers have to make. (8,059 / 14,238 = .57; the 1 - .57 = .43, or 43%) If you can lower their annual debt service payments by choosing a financial program which permits thirty-year terms over another program which permits only ten-year terms, they should be very happy ratepayers, indeed.

In the same vein, please note that the difference between just a twenty- and a twenty-five-year term means an annual debt service payment reduction of 9%. This is noteworthy because many financial advisors and investment bankers simply recommend twenty-year terms out of habit. This is because there is little distinction in the bond market between a twenty-year bond and a twenty-five-year bond. But by just making the suggestion that you would prefer a twenty-five-year term, you may be able to effect a substantial reduction in your ratepayers' annual debt service payments.

Finally, please note as well, that the difference between the annual debt service payments on a loan with a fifteen-year term versus a twenty-five-year term is 22%. (8,581 / 10,979 = .78; then 1 - .78 = .22, or 22%) This is noteworthy because fifteen years is often the maximum term which a bank will allow. But by going to the bond market, instead, you may be able to reduce your system's annual debt service payment by 23%. As you will see in the final section of this chapter, below, even if the banker offers you a lower interest rate, you may still achieve the lowest annual debt service payment by going to the bond market. This is because the lower interest rate may not result in an annual debt service payment which is lower than that on the bond with the higher interest rate but longer term. In short, despite a slightly higher interest rate, you might still wind up with a lower annual debt service payment, which is the ultimate goal for your ratepayers.

Now that you have seen how extending the term of a loan reduces its annual debt service payment, you might think that this is the last word the subject. Unfortunately, it is not.

Table 9-B showed you the impact of term on rate; but it did so <u>only at one rate of interest.</u> Unfortunately, the impact of term changes, depending on what rate of interest is used.

To illustrate this point, please look at Table 9-C, below, which presents almost the same data as in Table 9-B. The only difference is that the annual debt service payment on the $100,000 loan is shown at three rates of interest: 1%, 5%, and 10%, in Columns A, C, and E, respectively.

Table 9-C

Years	A	B	C	D	E	F
	1%		5%		10%	
5	$20,604	N/A	$23,099	N/A	$26,380	N/A
10	10,558	49%	12,950	44%	16,275	38%
15	7,212	32%	9,634	26%	13,147	19%
20	5,542	23%	8,024	17%	11,746	11%
25	4,541	18%	7,095	12%	11,017	6%
30	3,875	15%	6,505	8%	10,608	4%

Again as in Table 9-B, above, Columns B, D, and F are to be read from the bottom up, and they are not cumulative. Hence, the annual debt service payment on a loan with an interest rate of 1% is 15% lower if a thirty-year term is used than if a twenty-five-year term is used.

Now, please compare the numbers in Columns B, D, and F. Note that the higher the interest rate, the smaller are the differences in annual debt service payments.

There are two important points here. First, the impact of term on annual debt service payment changes depending on the rate of interest. And, second, the lower the interest rate, the greater the impact of lengthening term on lowering annual debt service payments.

Now, let us look at the impact of longer terms on the other type of loans: Level Principal Payment Loans.

Level Principal Payment Loans.

Table 9-D, below, sets forth the various annual debt service payments on a Level Principal Payment Loan of $100,000, at an interest rate of 7%, with terms ranging from five to thirty years at 5-year intervals.

There is one problem, however. Level Principal Payment Loans, as you know, have different annual debt service payments each year. For our purposes, we can, for the moment, use the first, or initial, annual debt service payment as a basis of comparison. We will deal more fully with the problems this creates at the end of this chapter. For now, we must just remember that the annual debt service payment with which we are dealing is the initial annual debt service payment.

Table 9-D

	A	B	C
	5 years	$27,000	N/A
	10 "	17,000	37%
	15 "	13,667	20%
	20 "	12,000	12%
	25 "	11,000	8%
	30 "	10,333	6%

Again, as in Table 9-B, above, the numbers in Column C read from the bottom up; and they are not cumulative. For example, a thirty-year term results in an initial annual debt service payment which is 6% lower than that of a loan with a twenty-five-year term. And, a twenty-five-year term results in an initial annual debt service payment which is 8% lower than that of a loan with a twenty-year term.

Now, as was the case with Level Payment Loans, as illustrated in Table 9-C, above, the impact of term on Level Principal Payment Loans also changes depending on the rate of interest used.

Table 9-E, below, demonstrates the impact of longer terms on initial annual debt service payment, using the same loan data as in Table 9-D, above, but at rates of interest of 1%, 5%, and 10%.

Table 9-E

Years	A	B	C	D	E	F
	1%		5%		10%	
5	$21,000	N/A	$25,000	N/A	$30,000	N/A
10	11,000	48%	15,000	40%	20,000	33%
15	7,667	30%	11,667	22%	16,667	17%
20	6,000	22%	10,000	14%	15,000	10%
25	5,000	17%	9,000	10%	14,000	7%
30	4,333	13%	8,333	7%	13,333	5%

As you can see by comparing Table 9-E with Table 9-C, the lower the interest rate, the greater the impact of lengthening term on lowering initial annual debt service payments. As you will also note, the impact appears to be slightly higher for Level Principal Payment Loans than it is for Level Payment Loans. We use the word "appears," because, as you recall, with the Level Principal Payment Loan we have only been looking at <u>initial</u> annual debt service payments, not all of the annual debt service payments.

Let us see if this phenomenon persists beyond the initial annual debt service payment. Let us see if lower interest rates mean a greater impact of term on annual debt service payment.

Table 9-F, presents the same data as in Table 9-E, above; but instead of looking at the initial, or first, annual debt service payment, we are now looking at the <u>fifth</u> annual debt service payment.

Table 9-F

Years	A	B	C	D	E	F
	1%		5%		10%	
5	$20,200	N/A	$21,000	N/A	$22,000	N/A
10	10,600	48%	13,000	38%	16,000	27%
15	7,400	30%	10,333	21%	14,000	12%
20	5,800	22%	9,000	13%	13,000	7%
25	4,840	17%	8,200	9%	12,400	5%
30	4,200	13%	7,667	6%	12,000	3%

Please compare the last two lines of Column F in both Table 9-F and Table 9-E, above.

As you can see in Table 9-F, two annual debt service payments at the bottom of Column F represent decreases of 5% and 3%, respectively. In Table 9-E, the same two data points are 7% and 5%, respectively. Table 9-E represents the <u>first</u> annual debt service payment. Table 9-F represents the <u>fifth</u> annual debt service payment.

From these two tables we can deduce that the effect of interest rates on the impact of term on annual debt service diminishes over time for Level Principal Payment Loans.

We can also go back and compare the data with the comparable data in Table 9-C, above. The two pertinent data points in Table 9-C are 6% and 4%, respectively. Thus, we can see that the Level Payment Loan falls between the first and fifth payment of the Level Principal Payment Loan. So, although the impact of interest rates starts out being greater for Level Principal Payment Loans, the effect diminishes over time such that the overall impact of interest rates on the impact of term on annual debt service payment is greater for Level Payment Loans than for Level Principal Payment Loans.

Here are some additional tables which confirm our conclusions. Table 9-G presents the same data as appear in Table 9-F, except that the tenth annual debt service payment is used instead of the fifth annual debt service payment, as the basis of comparison.

Table 9-G

Years	A	B	C	D	E	F
	1%		5%		10%	
5	N/A	N/A	N/A	N/A	N/A	N/A
10	$10,100	N/A	$10,500	N/A	$11,000	N/A
15	7,067	30%	8,667	17%	10,667	3%
20	5,550	21%	7,750	11%	10,500	2%
25	4,640	16%	7,200	7%	10,400	1%
30	4,033	13%	6,833	5%	10,333	1%

Table 9-H, below, presents the same data for the fifteenth annual debt service payment.

Table 9-H

Years	A	B	C	D	E	F
	1%		5%		10%	
5	N/A	N/A	N/A	N/A	N/A	N/A
10	N/A	N/A	N/A	N/A	N/A	N/A
15	$ 6,733	N/A	$ 7,000	N/A	$ 7,333	N/A
20	5,300	21%	6,500	7%	8,000	-9%
25	4,440	16%	6,200	5%	8,400	-5%
30	3,867	13%	6,000	3%	8,667	-3%

As you can now see from the last two lines in Column F, instead of annual debt service payments decreasing as the loan term is extended, they are now actually <u>increasing</u>. In specific, and contrary to everything we have seen so far, in <u>year fifteen</u>, the annual debt service payment on a <u>twenty-five</u>-year loan is actually <u>greater</u> than the annual debt service payment on a <u>twenty</u>-year loan!

Table 9-I, below, presents the same data for the <u>twentieth</u> annual debt service payment.

Table 9-I

Years	A	B	C	D	E	F
	1%		5%		10%	
5	N/A	N/A	N/A	N/A	N/A	N/A
10	N/A	N/A	N/A	N/A	N/A	N/A
15	N/A	N/A	N/A	N/A	N/A	N/A
20	$ 5,050	N/A	$ 5,250	N/A	$ 5,500	N/A
25	4,240	16%	5,200	1%	6,400	-16%
30	3,700	13%	5,167	1%	7,000	-9%

Lastly, Table 9-J, presents the same data for the <u>twenty-fifth</u> annual debt service payment.

Table 9-J

Years	A	B	C	D	E	F
	1%		5%		10%	
5	N/A	N/A	N/A	N/A	N/A	N/A
10	N/A	N/A	N/A	N/A	N/A	N/A
15	N/A	N/A	N/A	N/A	N/A	N/A
20	N/A	N/A	N/A	N/A	N/A	N/A
25	$ 4,040	N/A	$ 4,200	N/A	$ 4,400	N/A
30	3,533	13%	4,333	-3%	5,333	-21%

As you can see, the reverse phenomenon, where the annual debt service payments are increasing with the longer terms, is now becoming much more pronounced. It has now reached Column D, meaning that even at an interest rate of only 5%, the twenty-fifth annual payment of a thirty-year Level Principal Payment Loan is actually greater than the twenty-fifth annual debt service payment of a twenty-five-year Level Principal Payment Loan!

*   *   *   *   *

Now, what conclusions can be drawn from all of this discussion about term?   What is important about loan term, as far as the ratepayers of a water or wastewater district are concerned?

Just this: the longer the term, the lower the annual debt service payment. And, the lower the interest rate, the more so this is true.

Most project financings for water and wastewater systems will involve the use of Level Payment Loans. As such, the longer the term, the lower will always be the annual debt service payments which the ratepayers have to make. In addition, as interest rates decline, this factor becomes more important. At lower rates of interest, extending the term of a loan means even lower annual debt service payments.

Chapter 10

Rates

When it comes to borrowing money, most people think they have made a great deal if they get a low rate. Sometimes they can be very wrong. Here is an example.

You have just bought a new home and you are returning from the closing. You run into a new neighbor who has just bought the house next door, which is relatively comparable to yours. You strike up a conversation about your respective experiences in getting a mortgage. Your new neighbor is pretty happy with himself. He tells you he just got a great deal on his mortgage loan: the bank gave it to him for an interest rate of 8%.

Suddenly, your heart sinks. You know he is going to ask you how you did with your bank. You are now feeling pretty bad because you got a 9% rate on your mortgage.

Now all of a sudden you remember reading Chapter 10 of this book. Your confidence starts to return. You realize that your house and your new neighbor's house are pretty similar so they must have cost about the same amount. He doesn't look like a millionaire, so you guess that the two mortgages are probably for about the same amount, as well.

Now, with renewed confidence -- and before he can ask you what rate you got -- you beat him to the punch. You say: "That's great, what is your annual payment?"

Your new neighbor proudly tells you that his annual payment is "only" $11,683. You then just mention, with barely concealed triumph: "Gee, mine's only $9,734." Touche.

Your new neighbor has an 8% mortgage. You have a 9% mortgage. But your annual payments are $1,949 <u>less</u> than his. How so?

This is an example of how lower rates do not always translate into lower cost. And, it is absolutely critical to remember, whether you are a new home buyer or the chairman of the board of a major water district, that <u>annual</u> <u>cost</u>, <u>not rate</u>, is the crucial number. Annual cost is what comes out of your pocket and the pockets of your ratepayers.

It is neither fair nor accurate to say that rate is unimportant. It is important. It is, in fact, a concept which contributes greatly to the amount you actually wind up paying. But it does not absolutely control the amount you will pay. It is a factor, but it is only one factor. In the next few paragraphs we will see just how this one factor affects the annual payments which you and your ratepayers actually pay.

There are four questions which need to be addressed concerning the concept of Rates.

The first is where rates come from; how rates are set.

The second is legerdemain. Sleight of hand. These are the games -- usually political games -- which people sometimes play with the concept of Rate.

Third is the interplay between principal and interest; how these concepts combine to make annual payments.

Last is the relationship between rate and time. What the "yield curve" is, how it works and how it affects annual debt service payments.

                    *   *   *   *   *

How Rates Are Set.

There are literally bookshelves in major business libraries filled with erudite tomes discussing the precise mechanism of how rates are set. I commend these volumes to you for your reading pleasure. However, in teaching a class, which deals with loans and lending, every year at an institute at the University of Oklahoma, I tell them this:

"In the City of Zurich, Switzerland, there is a bridge near the train station over the River Limmat. Under the bridge there lives a Gnome who has a telephone which only accepts incoming calls. Every morning before the stock exchanges open in New York, the telephone rings. It is the Chairman of the Board of the Federal Reserve System of the

United States of America. He asks what the federal discount rate is. The Gnome tells him. The Chairman hangs up. He then tells all the banks and financial markets. Based on whatever the federal discount rate is, the banks and markets then, in turn, set their own rates.

"Now, as you go through life, you might hear other explanations as to how rates are set. You may even think that one of these other explanations is closer to the truth. But one thing is certain about my explanation: you will never hear anything more accurate."

I stand by the above explanation. It is as accurate as any other explanation you will hear. Here is another one that is only slightly different:

Rates -- whether the prime rate at your local bank or the rate passed on to the Chairman of the Fed each morning by the Gnome in Zurich -- have three basic components: inflation, cost and profit.

Rather than taking an abstract example out of some textbook, let us build one from the bottom up with common sense. Let us say you are going to lend me $100 for one year and you are considering what rate to charge.

First is inflation. You read in Chapter 3 about the time value of money. $100 today will not buy you the same amount of groceries in one year. Let us say -- based on past experience --that you think that you will need $104 a year from now to buy the same amount of groceries that your $100 will buy today. This means that I would have to pay you at least $104 just to keep your buying power even. This means a minimum rate of 4%. This is the inflation factor.

Next is cost. Let us say that it costs you $2 -- in pen, paper, stamps, whatnot -- to make the loan. You will certainly want to recover these costs.

What about risk? Covering your risk may be an intangible cost; but it certainly is a real cost. If you loaned $100 to fifty different people, you might feel that one of them will not pay you back. In other words, on 50 loans of $100 each, or $5,000, you might have one total loss of $100, or two partial losses of $50 each, for a total loss of 2% on your loan portfolio. In such case, you would estimate your losses at 2%. This means that your total cost, including estimated losses, is $4, or 4%. When we add in the 4%, or $4 for inflation, we are now up to an 8% rate.

Finally, after covering your costs and after charging enough to keep inflation from reducing your spending power, you decide you would like to make a 2% profit. To do so, you must charge me 10%. You will lend me $100 now, if I agree to repay you $110 in one year.

This is exactly how it works. But there is a bit more to it.

Remember the Chairman of the Fed and the Gnome of Zurich. Well, the Fed Chairman knows what the inflation rate was last year; but only the Gnome knows what it will be next year. The Fed Chairman knows what his costs are; but only the Gnome knows how much profit to make.

How does this really work in the real world? Something more like this.

You have $100. You have the option of investing this money or lending it to me. You look at what the bank or other investments will pay you. Then you look at what rate you think the bank would charge me for the same loan. You decide to charge me more than you can get from your other investments but just less than the bank would charge me.

If you followed the procedure exactly as it was just described above, you would be very wise. You would just have accurately measured supply and demand. This is precisely what the Gnome does every morning just before the Fed Chairman calls.

Inflation. Cost. Profit. Those are the three elements from which rates are derived. Each, however, contains unknowns. Measuring supply and demand -- which is what other borrowers and lenders, including the Gnome, think about these unknowns -- is the best way to arrive at appropriate rates at which to either borrow or lend.

Now, having established how rates are made, a few words about the dark side of rates.

Sleight of Hand, by the Numbers.

Mark Twain once said: "There are liars, damned liars and statistics." He was right. There is virtually no limit to the mischief a political interloper can make, given the right set of numbers. Managers and directors of public water and wastewater systems must not only be on guard as to the effect of their fiscal policies, but they must also be on guard as to how those policies can be characterized by political opponents, axe grinders and other general misanthropes.

For example, no one buying a $3.95 widget at a local hardware store will be deterred from making the purchase because the sales tax is 5%, versus 4%. At a 5% rate, the total cost of the purchase is $4.15. At a 4% rate, the total cost would be $4.11. The difference, after all, is only four cents.

But, as any politician who has ever voted to raise any tax will tell you: the difference between a 4% sales tax and a 5% sales tax is not a 1% increase, it is a 25% increase!

In the hands of a political opponent, that increase from 4% to 5% is a 25% increase. An increase from 4% to 6% is not a 2% increase, it is a 50% increase. God help the politician who votes to raise a tax from 1% to 2%. That poor hapless soul is -- as far as the opposition is concerned -- voting for a 100% tax increase! Nowadays, politicians who vote for double digit -- much less triple digit -- tax increases can measure their government careers in the days left till the next election.

It is an axiom of politics that an increase from 4% to 5% is a 1% increase to the incumbent; but it is a 25% increase to his challenger.

Where does the truth lie? As a matter of fact, politicians are always talking about "half-truths." This is a good example of what they are talking about; it is a half-truth. Each interpretation is half true.

The increase in the sales tax causes only a 1% increase in the cost of the $3.95 purchase. The extra four cents, caused by the tax increase, exactly equals 1/100th of the original purchase price. But it does increase the tax paid from $0.16 to $0.20. An increase of four cents on top of the already existing $0.16 tax is, in fact, a 25% increase. There is no denying either statement. It is both a 1% increase and a 25% increase. Mark Twain would just smile. This is exactly what he had in mind. Numerical legerdemain. Sleight of hand by the numbers.

Where does the truth really lie? In point of fact, the correct characterization should really be that it actually is a 25% increase in the tax, but only a 1% increase in the cost.

This brings us back to our overriding concern: cost.

Principal, Interest and Annual Debt Service Payments.

Apart from the obvious political concern, which all water and wastewater system managers and directors should rightly have about possibly adverse characterizations of their actions, the major concern is the impact of rates -- in this case interest rates -- on the total cost of service to system users.

The first point to remember here is that interest rates are only one component of annual debt service payments; and, annual debt service payments are only one component of the annual charge for water or sewer service. Interest rates are important; but their importance must be gauged in light of total service charges.

Having said this, we should now make the point that a change in interest rates from 4% to 5% is, in fact, a 25% increase in the <u>interest rate</u>; but it does not mean a corresponding 25% increase in <u>annual debt service payments</u>. An increase in the interest rate on a 20-year loan from 4% to 5% does <u>not</u> increase the annual debt service payment by 25%. The annual debt service payment on a 20-year loan of $100,000 at an interest rate of 4% is $7,358. The annual debt service payment on a 20-year loan of $100,000 at 5% is only $8,024. The increase is $666, which is only a 9% increase.

The reason for this is the interrelationship between principal and interest in calculating annual debt service. The relationship is a dynamic one where each of the elements changes over time.

What is important to remember is that **the percentage change in interest rates is less than the percentage change in annual debt service payments.** So, you must remember, if you are ever in a situation where you must agree to a higher interest rate on a project loan, that the annual debt service payments which your ratepayers make will increase, but by a lesser amount.

One final note before going on to the infamous "yield curve."

A few paragraphs above we pointed out that rate was only one component on annual debt service; and that annual debt service was only one component of the total system user fee. Let us restate that concept in an example.

Recall, from the discussion above, our $3.95 purchase in the hardware store. We established that the 25% increase in the sales tax caused a 1% increase in the purchase price. Of the total purchase price of $4.15, 1%, or $0.04, was attributable to the increase in the sales tax.

Now, let us say that this same customer made two purchases that day, one subject to sales tax, and the other not subject to sales tax. Let us say that the purchase not subject to sales tax was also for $3.95.

Now, the customer's total bill is $4.15 plus $3.95, or $8.10.

Out of the new total bill of $8.10, the $0.04 attributable to the increase in the sales tax is now only 0.5%.

The point here is that board members are often confronted with having to vote today for a project loan proposal with a higher rate of interest than the same proposal would have carried, say, three months ago. As such, they may have to face some political heat because of the increase. If this is ever the case, those board members must prepare their rebuttal with a few points in mind.

First, it is not how much the rate increased that really matters; rather, it is the annual debt service payment. Increases in rate do not translate into equivalent increases in annual debt service payments. Second, annual debt service is only one (and it may not be a large one, at that) component of the total system user charge.

So the directors' rebuttal should, of course, state the reason why it was imprudent to proceed three months ago, and then go on to point out that the allegedly huge increase in interest rates -- that their opponents complained of -- actually resulted in only a few pennies, well spent, on the monthly bill.

Now, for the last, but certainly not least, matter to consider concerning rates.

The Yield Curve.

The "yield curve" is one of those phrases that always comes up during project financings. It is frequently used by financial advisors and investment bankers. It seems shrouded in mystery. Especially when they call it an "inverted yield curve." It is, in fact, one of those phrases which make people think that basic finance is a form of black magic.

The first thing to do is to take away the quotation marks. It will no longer be called the "yield curve," but, rather, the yield curve. The next is to state that it is not a curve at all. At least not in the mathematical sense of a curve. Last, there is no equation or formula for the yield curve. All of the mathematicians in the world working together, simultaneously, forever, could not derive a yield curve.

Everyone who walks by the window of a bank now and then and looks at the signs in the window knows that you can earn different rates of interest depending on how long you are willing to leave your money in the bank. If you sat down and made a graph -- with the different rates of interest along the vertical, or y-axis, and the different time periods along the horizontal, or x-axis -- and put a dot for each different interest rate at each time period, and then connected all the dots with a line; you would have just created the yield curve.

As you can see, it is really not a curve at all. It only becomes a curve if you want to display the information graphically. Of course, most financial publications, such as The Wall Street Journal or The Bond Buyer, do present the data in graphic form. Hence, it has come to be called -- by people in the finance business -- the yield curve.

Figure 10-A, below, is an example of what a yield curve might look like on any given day. Please note that Figure 10-A presents the yield curve for each of three separate bond issues sold at approximately the same date.

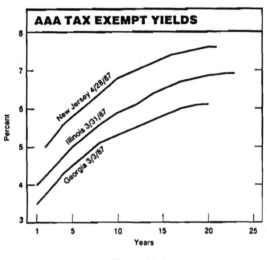

Figure 10-A

As you can see in the Figure above, interest rates are increasing over time. Rapidly at first; then more slowly. Between years 15 and 20, the rate barely rises at all. The curve is virtually flat.

There are three brief observations to be made about the yield curve.

The first is the yield curve is a useful device if you are going to finance a project in the bond market or with a financial institution whose rates vary over time. In such case, you would be well advised to obtain not necessarily a yield curve per se, but, rather, a listing of the rates for various loan or bond maturities. Your financial advisor or banker should be able to supply you with one. This should be extremely useful to you in making up your mind as to the most desirable maturity for your loan or bond.

For example, if the yield curve is flat between years 15 and 20, and you have the option of either maturity, you might well opt for the longer maturity which will, in such case, always mean lower annual debt service payments for your system's users. (See Chapter 9, Term.)

The second observation relates to the shape of the yield curve.

You will almost always see yield curves shaped like the one in Figure 10-A, above. That is, it will be low on the left end -- corresponding to the shorter terms -- and will rise to the right, reflecting higher interest rates for longer terms.

Once in a great while, you may see an "inverted" yield curve. In an inverted yield curve, the higher interest rates are at the shorter maturities and the lower interest rates are at the longer maturities. In other words, the curve is high on the left end, and low on the right end. The most notable time this occurred was in 1981, when the prime rate in the United States was 21%. At that time, the long-term rates were actually lower than the short-term rates!

How so? What determines the slope of the yield curve? This leads us to our final observation about rates, which is why rates -- and, hence, the yield curve -- change over time.

The answer harkens back to Chapter 3 of this book and the discussions there concerning the time value of money. The value of money changes over time because of inflation.

In the example early in this chapter about the Gnome of Zurich, we said that rates were determined based on three factors: inflation, cost, and profit. There is no reason why, in setting a given rate, that either cost or profit should escalate over time. (They do increase, of course, in terms of actual dollars and cents. But they should not increase as percentages; and rates are percentages.) Inflation, however, is a different matter.

As we previously discussed, the only clue we have to tomorrow's inflation rate is yesterday. But what about next year, or ten years from now?

When we say that yesterday's inflation rate is a clue to tomorrow's rate, what we are really saying is that we are familiar with the relative strengths and weaknesses of the various forces at work today and in the recent past. We then look forward to see if there are any events -- such as wildly excessive government spending -- which would upset the current dynamic equilibrium of economic forces. Seeing none, we simply assume that these economic forces will continue to work in the same way, producing inflation for tomorrow and the foreseeable future at just about the same rate.

So, in other words, for the short term, yesterday's inflation rate will be tomorrow's inflation rate.

But, as we look beyond just tomorrow or next year, we begin to lose confidence in this fragile assumption. What about another oil embargo? What if, God forbid, there is another war in the Middle East? What effect will the forthcoming nuclear disarmament have on our economy? We may think we know the answers to these questions now; but what about five years from now?

That is what is happening to the yield curve and to the long-term rates it describes. They are reflecting uncertainty. The uncertainty of the future and what it holds for the economy.

When faced with uncertainty, what do people do? They hedge their bets. The fact that interest rates increase over time is the direct result of borrowers and lenders hedging their bets about the future.

Why did the yield curve invert in 1981? The answer is that the people (including some economists) felt that the economic troubles the country was going through at that time were acute. That they were not going to last more than a few months or a few years at best. So, at that time, short-term rates were very high, and long-term rates were much more modest. That, however, was an anomaly. Usually, the shortest term rates are the lowest. Rates usually rise relatively rapidly over the next few years. Then they plateau. The rise reflects the growing effect of uncertainty over time. And the plateau reflects the exhaustion of the uncertainty process; the fact that it is unreasonable to assume that inflation will inexorably increase forever.

That is what interest rates are, where they come from, how they work, and, most importantly, how they affect the eventual costs which the users of water and wastewater systems have to pay.

Chapter 11

Financing Costs

In the preceding two chapters of this section, we have dealt with financial factors which affect the annual debt service payments, which ratepayers of water and wastewater systems ultimately wind up paying. In specific, we looked at the effect of rate on annual debt service payment in Chapter 9, and we looked at the impact of term on annual debt service payments in Chapter 10. Rate and term are two of the three mathematical variables (The third is principal.), which comprise the actual equation used to calculate annual debt service payments.

In this chapter we are not going to deal with mathematical variables. Rate and term are abstractions. They are very important, even critical, to the financing process; but they are abstractions. On the contrary, in this chapter, we are going to deal with factors which increase cost in plain dollars and cents. The costs in this chapter are not the kind which can be looked up on a table, or which can be derived with a formula; rather, they are the type of costs which are paid by check.

Let us begin with a definition.

Financing costs are those expenses directly related to the obtaining of funds for capital improvement projects. They are the costs of obtaining the funds, themselves, or of making the funding possible in the first place.

Financing costs do not include interest payments.

Below are the eight most common forms of financing costs.

- Commitment fees or points.
- Counsel fees.

- Financial advisory fees.
- Servicing fees.
- Investment banking fees.
- Credit enhancement fees.
- Rating agency fees.
- Printing and miscellaneous costs.

As you can see, by their very nature, with the exception of the commitment fees and the underwriter's discount, the above costs are paid to third parties which facilitate the loan or bond transaction. The commitment fee and the underwriter's discount are paid directly to the bank or the initial bond purchaser (the underwriter), as a concession or inducement for them to make the loan or purchase the bonds, respectively.

Almost all financing costs are paid at closing. The exception to this rule is the commitment fee. Commitment fees are generally paid upon acceptance of the commitment to finance. This means that they are paid before the closing, although they may be recouped at the closing.

We will now take a close look at each of these types of financing costs.

Commitment fees or points.

Commitment fees or points are commonly associated with bank loans as opposed to the issuance of municipal bonds.

Although the nomenclature is by no means clear or universal, commitment fees are generally 1% or 2% of the loan principal amount and are paid upon acceptance of the bank's commitment to lend. The term "points," on the other hand, most often refers to fees which are also 1% or 2% of loan principal and which are paid at the closing out of loan proceeds. On any given loan, there may be neither, either, or both; although the total of such fees seldom exceeds 2% of loan principal.

As you have seen in several previous chapters, "points" are simply an additional payment to the lender which has the dual effect of increasing the banker's yield on the loan and also of increasing the borrower's annual debt service payments and his True Interest Cost (TIC). They simply constitute additional compensation to the lender, and must be calculated into the real cost of the loan whether by the Annual Payment Method or the Total Payment Method.

Commitment fees, on the other hand, are ostensibly a form of compensation paid to the bank or lending institution for reserving funds from the time of commitment to the time of disbursement. In addition, they are also supposed to protect the bank against the possiblity that you will not close your loan, i.e., their already reserved funds will not ultimately be used in a timely manner.

We say "ostensibly" because, notwithstanding what they appear to be, or how anyone might choose to characterize them, commitment fees are simply a means of increasing the yield of the lender. They, too, are additional compensation to the bank. Furthermore, if their true purpose were simply to protect the bank against the borrower's failure to close, then the commitment fee would be refundable to the borrower at closing, wouldn't it? But, of course, it isn't.

In addition, if commitment fees were actually related to the cost of reserving funds they would have two characteristics. First, the amount of the fee would relate directly to the bank's cost of funds. This means that if we were in a 12% interest rate environment, the commitment fee would be 1% a month. If the going rate were 6% a year, then the commitment fee would be 1/2% a month. As a matter of fact, most commitment fees are between 1% and 2%, regardless of the interest rate environment. The period for which the funds are reserved are generally from one to six months, again regardless of the interest rate environment.

The second characteristic would be that if the fee were actually related to the time period between the commitment and the closing, then the fee should relate to the volatility of interest rates during such period. That is, the fee should directly relate to the possibility that between the time of commitment and the time of closing, rates would rise so much so that at the time of disbursement, the bank would actually be making a submarket rate loan. In point of fact, this is also not true. Banks charge commitment fees when rates are flat. They even charge commitment fees when rates are falling. The only thing which can be said is that in the early 1980's when interest rates were rising abruptly, the banks did actually behave like their commitment fees were related to rate volatility. At that time, they actually shortened their commitment periods.

None of this is to say that there is anything wrong with commitment fees, or points. They should simply be recognized for what they are and calculated into the total borrowing cost by either the Annual Payment Method or the Total Payment Method.

Counsel fees.

Counsel fees are quite straightforward. They are fees paid to the lawyers who close the transaction, and, in many cases, assist in negotiations.

In addition to the fees paid to various attorneys, each of them will expect to be paid for their actual and necessary expenses involved in the transaction, including travel costs to and from meetings.

Unfortunately, when it comes to loans for water or wastewater systems, there is usually more than one and often several lawyers to pay. Let us review the cast of all possible attorneys on a given transaction.

House counsel.

The house counsel, or in-house counsel, is a member of the staff of the water or wastewater system, if, of course, the system is large enough to warrant a full-time lawyer on staff. Most are not. As such, many smaller systems have a part-time counsel, who is otherwise free to practice law privately. It is as if the system is just another, albeit important, client to such lawyer.

If the attorney is a full-time staff member, then there are no additional direct payments to that attorney for assisting in the project financing. If the attorney is part-time, his work on a capital improvement project may be covered by his monthly retainer. It may be on an hourly basis which may be partially covered by the monthly retainer. Such work may also be on an hourly basis over and above the monthly retainer.

There are two points to remember here. The first is that project financings can get very complicated, and the areas of the law on which they impinge can be quite esoteric. Not all lawyers will be able to competently handle project financings. This is certainly not because such lawyers aren't smart enough; but rather it is because project financings happen so infrequently that many lawyers simply aren't up to date in all of the areas of law involved. So, the first point is to ascertain whether your in-house counsel or your part-time counsel has the experience necessary to properly advise you on the transaction.

Unless you do a project financing at least every other year, you should not expect your in-house counsel to have requisite experience. The same is true of your part-time counsel.

The point, here, about experience is that if a lawyer does not have certain expertise at hand, he can certainly do the readings and take the other necessary steps to inform himself in the areas of law in which he may not be up to date. This process, although worthy, is expensive. Bringing any lawyer up a learning curve can be very costly. Always remember, you are paying for the lawyer's time. The more time it takes for him to familiarize himself with project financings, the more it will cost your ratepayers.

This leads us to the second point which is how to handle the cost. Always negotiate the cost in advance. This is true of all of the lawyers involved in a project financing.

The best basis for the negotiation is an hourly fee against a maximum. As such, you will be able to contain the learning curve costs. And, by using the maximum in your financial projections, you will have a firm basis for your estimates.

Outside counsel.

Many systems engage outside counsel to assist on the financings of capital improvement projects. The term "outside counsel" is a casual phrase. What is really meant here is a law firm with special expertise in project financing. Their "outsidedness," if you will, is not what is important; it is their special expertise.

Here again, a few rules to follow. Get estimates. Make sure the estimates are stated in hourly terms with a maximum. Use the maximum as the basis for your project cost estimate.

Lender's counsel.

As anyone who has ever purchased a home knows, when dealing with a bank, not only do you pay your own lawyer, but you also pay the bank's lawyer. The same is true for water or wastewater systems, or for any other type of business, for that matter.

If your correspondence with the bank does not include any reference to the bank's counsel fees, don't be shy. Ask. Do not think for a moment that because such reference may not be there, that you will not have to pay these fees. You will undoubtedly be presented the bill at the closing. And, it is far better not to be surprised.

In addition, not only is it appropriate to ask about the counsel fees, but it is also appropriate to ask how much they will be, and to get that information in writing. Here, again, the bank's counsel is not likely to give you a specific number; but they should be willing to give you an estimate. Under such circumstances, you should have your counsel insert language in all relevant correspondence with the bank that your system agrees to pay the bank's counsel fee "an amount not to exceed x dollars." You should then hold them to that agreement.

There is one exception: unforeseen circumstances. As much as we all might like to insist on a pre-agreed upon fee regardless of any unforeseen circumstances, you may not be able to get either the bank or its counsel to assent to such an unforgiving formula. After all, you, yourself, might create the unforeseen circumstance. It would not be realistic to expect anyone to work for free to solve a problem which you, yourself, created.

Under such circumstances, you should be willing to agree to pay the bank's counsel "an amount not to exceed x dollars, unless otherwise agreed to, in advance, and in writing." Thus, if some unforeseen event does arise which adds complications to the transaction, the lender and his counsel know they have to come back to you to get a written increase. If they don't, you don't owe them an additional penny.

This is a very fair formula for dealing with unforeseen events. You should insist upon it, in writing. And, then, you should use this amount to calculate the true project cost to your ratepayers.

Where you have a loan, you have a lender, and a lender's counsel. Where you have a bond, you generally have an underwriter, who purchases the bond, and an underwriter's counsel.

Underwriter's counsel.

The role of the underwriter's counsel is similar to that of the lender's counsel. And you should deal with the matter of the underwriter's counsel fee in the same manner as you would deal with the bank's counsel fee.

In other words, you should insist on an estimate in advance and in writing. Your agreement with the underwriter should include a paragraph to the effect that you agree to pay the underwriter's counsel fee "in an amount not to exceed x dollars unless otherwise agreed to in advance and in writing."

You should be aware that lawyers who represent banks tend to do a great deal of legal work for the bank. Bank clients are usually big clients. Lawyers who represent underwriters do not necessarily do great amounts of work for the underwriter. Bankers, therefore, tend to have better control over their lawyers than do investment bankers.

The point here is that an underwriter's counsel is more likely to balk at your request for a firm estimate.

Do not be deterred. If the lawyers are testy and obdurate on this point, stick to your guns. There are other lawyers in the world. And, ultimately, there are other underwriters in the world, as well. Insist on the estimate - in writing. And, insist that the language suggested above, or some reasonable variation thereof, be included in your written contract with the underwriter. You are entitled to this. Just remember, one of the most unpleasant experiences in the world is to be handed a large, unexpected legal bill at the end of a bond closing.

And.....while we are on the subject of being handed large legal bills at the end of a bond closing, we might as well move on to the bond counsel.

Bond counsel.

In addition to paying for the lawyers on both sides of the transaction, whenever you finance a capital improvement project with a municipal bond, you will find there is one more lawyer whom you must pay, the "bond counsel." A bond attorney is a specialist in the several areas of law which involve municipal bonds.

Literally no one will purchase a municipal bond which does not have the written opinion of a bond counsel both as to the validity of the bond and as to the fact that the interest on the bond is exempt from federal income taxation, as well as state income taxes, if applicable.

There are two important things to remember about bond counsels. They are expensive and, even though the water or wastewater system, itself, is the client, they tend to be very independent-minded.

Bond counsel seem expensive because, unlike the other lawyers in the transaction, you are paying for their opinion, not their time. It should come as no surprise in this age of electronic information that bond counsels have many sets of documents tucked neatly into the folds of their word processors. The impressive tomes which they produce owe more to the quality of laser-jet printer than to the fact that some young lawyer sat up all night handcrafting every clause. No, bond counsel are not paid for their time.

As said previously, no one will buy a bond without a bond counsel's opinion. That is what you are paying for. And it is expensive. If for some technical reason your bonds have not been validly issued under state law, or if the interest thereon is not, for any reason, exempt from federal income taxation, then the bondholders who bought those bonds will undoubtedly sue the bond counsel on whose opinion they relied. So, as you can see, the bond counsel has a great deal of responsibility and a great deal of liability. Some wags say that the majority of the fee goes to pay the premium on the bond counsel's malpractice insurance. Whether there is any truth to such sayings or not, the bond counsel is, nonetheless, expensive.

Bond counsels tend also to be very independent-minded. Because it is their opinion which is valued, bond counsel value their integrity over their clients. (This, by the way, is as it should be.) But because of this tendency, they often get tetchy when you try to negotiate with them, especially on their fees.

On most moderate-sized financings, the bond counsel will get a small percentage of the par amount of the bonds. One half of one percent (0.5%) is a typical fee. With smaller transactions, there may be a flat fee. Thus, a bond counsel might say that his firm usually gets a half a percent or a minimum of $3,500.

But, by the same token, a bond counsel who expects a minimum fee, should also be held to a maximum fee, especially for large, relatively uncomplicated transactions. Once your system's projects start getting close to $10,000,000, you should definitely start putting a maximum in the bond counsel retainer agreement. For a $10,000,000 bond, a 0.5% bond counsel fee would be $50,000. At this level you should certainly be able to negotiate. After all, this is a very nice sized fee. There aren't a lot of bond counsels, but there are enough. You can be certain that unless your transaction involves bizarre or unusual circumstances, a reputable bond firm should be willing to do the work for $50,000.

Let us leave the section on attorneys by saying that of all of them, the bond counsel is, by far, the most important. On many projects, it is the bond counsel who drives all of the parties and who actually gets the deal done. And, having said that bond counsel is important, we now turn to the single most important source of expertise on project financings for water and wastewater systems: the financial advisor, or FA.

Financial Advisory Fees.

A good financial advisor is the most important member of the team of experts who assist your water or wastewater system in financing its projects. As a matter of fact, a good financial advisor can advise you on what all of the other experts on the team should be doing. The FA can even put the team together for you and provide the leadership to get the transaction closed.

Your FA should be your closest advisor on the project financing. If you select the right FA, he will wind up advising you on far more than just narrow questions of finance. In most cases he will actually wind up driving the deal, organizing all of the steps necessary to close. The FA should be chosen first and his advice should be looked to in choosing all of the other members of the team.

When choosing your FA, take your time and be sure that you find someone with whom you and your colleagues at the water or wastewater system are comfortable. Oh, of course, be sure the FA has all the right credentials on paper. That is important. And, be especially sure that the FA has done many similar (i.e., water or wastewater) projects in your state. This will insure that the FA knows all of the state laws and state procedures for financings of this type. Although the FA is not a lawyer, he will be able to tell you how other water and wastewater systems deal with certain state law requirements on a day to day basis. This type of advice can be invaluable.

The FA will help you choose your outside counsel. He should be familiar with firms throughout the state which regularly do the type of transactions with which your system is involved. He should also have a good idea what these firms charge for their services.

The FA should, of course, be able to advise you on all of the financial questions involved with your project. What is the best term? What rates are available at that term? Assuming you have more than one financing option, what are the additional financing costs associated with each? And the two most important questions: Which will provide the lowest annual cost to system ratepayers? And, which will provide the lowest total cost to system ratepayers?

Needless to say the FA will be able to do all of the calculations necessary for board meetings and public hearings. The FA can even do runs of "what ifs" to give you and your colleagues a sense of what might happen under various cost scenarios.

FA's used to charge a fee based on a percentage of the transaction. Nowadays, however, FA's seem to be going to a split fee basis. They charge an hourly rate and get a bonus, if you will, in terms of a small percentage of the final transaction. The hourly rates are likely to be in the range of $75 to $150. The "bonus" is likely to be in the range of 0.25%.

FA's are generally amenable to a cap or maximum on their hourly fees. Such payment clauses as "$100 an hour up to a maximum of 50 hours are not uncommon. As far as unforeseen events are concerned, this should not be difficult. The FA should spend a large part of his billable time in your office. You should be able to casually keep track of the hours spent. And, if the transaction gets complicated, or if many more hours than planned are required of his time, you should not be surprised at a request to increase the agreed upon number of hours.

The FA is your consultant. And, as is the case with the attorneys involved in the transaction, the FA will expect to be reimbursed for his actual and necessary expenses, including travel costs to and from meetings.

Servicing Fees.

Servicing fees occur in both loan and bond transactions. They cover the administrative costs of collecting the system's loan or bond payments and crediting them to the proper accounts.

In the case of a bond transaction, the servicing fees are generally called trustee fees, and they are larger than the servicing fees involved in a loan transaction. This is because in a bond transaction, the trustee must not only collect your payments and credit them to the proper account, but he must also pay each of the bondholders their proportionate share of each payment when due.

On a bond transaction, whether it is publicly sold or privately placed, the investment banker is seldom the ultimate owner of the bonds. On a private placement, the investment banker never owns the bonds at all. He actually arranges for the sale of the system's bonds to a third party prior to the closing. On a public sale, the investment banker, or underwriter, actually purchases the bonds at closing; but then he usually sells the bonds to his clients as soon as possible. So, on bond transactions the ultimate owner is rarely the investment banker with whom you are dealing.

This is why a trustee is necessary. The trustee actually acts on behalf of the bondholders, whoever they may be. He collects your payments and sees to it that each bondholder gets his proportionate share.

In addition, should a system become delinquent or actually default on any of its payments, the trustee acts as the legal representative of the bondholders in their efforts to secure their rights against the defaulting system.

Because the trustees duties go far beyond those of a simple servicer (i.e., the trustee does far more than just collect funds and has a much greater legal liability to the bondholders), his fees will be commensurately higher.

In the past, it was common that servicing fees or trustee fees were expressed as a percentage of the outstanding principal balance of the loan. This was a crazy system, especially for just servicing. It obviously takes the same amount of effort for a bank to collect and credit a $1,000,000 payment as it does a $1,000 payment. Thus, there is no reason at all why a servicing fee should be based on the amount of the loan.

Servicing fees should be flat fees; and they should clearly be a part of the system's negotiations with its lending bank. The system's FA should be able to advise as to what servicing fees are appropriate.

Both servicing fees and trustee fees should involve a set-up charge, which is payable at closing, plus an annual fee thereinafter.

As previously stated, trustee fees on a bond transaction should be higher; and there is even some justification for basing them, at least in part, on the outstanding balance of the bond issue. They should be higher in general because trustees have more extensive duties. The justification for basing them on the outstanding principal balance is because, if the trustee errs, he may become liable to the bondholders for the whole debt. This is a serious responsibility; and trustees expect to be compensated for taking on this responsibility.

Nonetheless, even in the case of a trustee, most of the duties are administrative and bear no relation to the size of the bond issue. Furthermore, should there be a problem with a system's payments or with securing a system's compliance with the terms and conditions of the bond indenture, the trustee can be compensated separately for any extraordinary actions in this regard.

The point to remember here is to seek the advice of the FA; and then to shop, wherever possible, for the most reasonable trustee fees. When you are borrowing from a bank, you have no say at all about the servicer; in virtually all cases, it will be the bank itself. But when you have a bond issue, the choice of trustee is up to you. The system's FA will be able to furnish a list of banks, including regional and national banks, who are willing to serve as trustee on water and wastewater system bonds. By shopping these banks under the guidance of its FA, a system should be able to settle on reasonable trustee fees for its transaction.

Investment Banking Fees.

Investment banking fees are the compensation which an investment banker receives for either buying or arranging for the sale of bonds. Such fees are, of course, limited to bond transactions.

There are three types of bond transactions. The first is public sale. The second is negotiated sale. And, the third is called private placement.

Whether a water or wastewater system's bonds are sold publicly, by negotiated sale, or by private placement is first and foremost a matter of state law. Your attorney and your FA should be able to tell you immediately whether your state requires public sale. If there is no state law requirement, the choice is generally up to the system. I say it is generally up to the system, because with very small bond issues, say less than $250,000, a public sale may just not be economically feasible.

Public Sale.

With a public sale, the bonds are sold by bid. Under such circumstances, when all of the other requisites of the transaction are complete and the legal documentation is in order, the system's FA will arrange for notices of sale to be published in authorized newspapers. He will also generally directly contact investment banks and banks in the region whom he knows to be regular purchasers of water and wastewater system bonds. On the appointed day, bids will be submitted and opened in public; and the bonds will be sold to the lowest responsible bidder.

The banker who wins the bid will be compensated through the bid structure. For example, the winning bid might be 98 1/2 at 7%. This means that the stated interest rate on all the bonds will be 7%, and that the system will receive 98.5% of the proceeds. Under normal circumstances, the winning banker, or underwriter, will reoffer the bonds to the public at par, or 100%. In this case, the underwriter's compensation is 1.5% of the bond issue. (Please see Chapter 5 to determine the True Interest Cost (TIC) for such transaction.)

This type of fee is called the underwriter's discount. The winning investment banker doesn't necessarily keep all of this money. He keeps a part of it for managing the transaction on behalf of all bond purchasers. But, especially on larger issues, the underwriter will share the larger part of this compensation with other investment bankers who assist in selling the issue. The other investment bankers technically purchase the bonds from the underwriter, but at a lower discount.

Negotiated Sale.

For negotiated sales, there is no public bidding procedure. Rather, under the guidance of the system's FA, the credentials and experience of several underwriters are reviewed to determine, based on past performance, which firm is likely to offer the system the lowest cost, including the underwriter's discount, on a TIC basis. Once this determination is made, a contract is signed with the underwriter. When the transaction is ready to be closed, the underwriter then actually prices the system's bonds. And, at closing, the underwriter purchases the bonds at that interest rate.

Negotiated bids may look questionable, but they really are not. If, based on the advice of the FA, the system does not like the price quoted by the underwriter, it is perfectly free to reject the price. If this happens on more than one occasion, the system may actually have grounds to rescind the contract with the underwriter. This, however, rarely happens. Underwriters make their living selling bonds. They do not make their living by deceiving bond issuers. If they did so, they would not last long in the highly competetive bond market. Furthermore, the FA would certainly know if an investment banker had a reputation for lowballing estimates, and would advise the system simply to avoid doing business with such an unscrupulous firm.

Private Placement.

The third basis for selling bonds is by private placement. In a private placement, the investment banker does not purchase, or underwrite, the bonds himself. Rather, he arranges for the sale of the bonds directly to one or more, usually, large institutional investors. For this reason, private placements are more common with smaller bond issues.

Under these circumstances, since the investment banker does not actually purchase the bonds, he is not called an underwriter; rather he is called a private placement agent. Notice the use of the word "agent."

With a private placement, the investment banker actually serves as an agent of the water or wastewater system to effect the sale of its bonds, much like a real estate agent serves a homeowner to effect the sale of a house.

Here again, under the expert advice of the FA, a system would request proposals to serve as private placement agent from competent firms in the region. The system would review these with the FA and probably interview several of the best candidates. Once a system was satisfied with a particular firm, it would then execute a private placement agency agreement with that firm. The agreement would require the agent to place the bonds, on a best efforts basis, at the best all-in cost to the system. When the transaction was ready to close, the placement agent would shop the bonds in the market to obtain the best possible price for the system, and would report that price to the system. Here again, the system is free to accept or reject the price.

The difference between underwriting and private placement is the difference in the legal obligation of the investment banker to the system. In an underwriting, the banker must purchase the bonds at the agreed upon price. With a private placement, he is only obligated to use his best efforts to sell the bonds at the agreed upon price. In other words, if, between the date upon which the price was agreed to and the date of closing, the market went up like a rocket, the underwriter would still be obligated to buy the bonds at the stated price. But, if for any reason, the placement agent's client refused to accept the bonds at the agreed upon price, the placement agent would be under no obligation at all either to buy the bonds himself or to find a new buyer at the old price.

Now, all of this may seem somewhat precarious, but in real life it is not. Certainly, in sharply rising markets, deals do fall apart once in a while. But this only happens once in a great while. As you might imagine, investment bankers protect themselves by lining up purchasers in advance whom they know will honor their obligations. The old saying that "whatever goes around, comes around" is also true in the bond market. No one would do business for long with people who reneged on their deals. Furthermore, the next time, the market might go the other way in which case a reneging purchaser might find himself out in the cold.

As you can see from the above, the liability and responsibility of underwriters are different from those of placement agents. You might guess from your reading about the different liabilities of trustees and servicers, that there might be a difference in compensation level, as well. If you made that assumption, you would be right. But, unfortunately, the question of compensation is not only more complicated than that, but it also far more important than just a simple matter of compensation. Here is why.

The absolute best rate obtainable on a private placement will never be as low as the best rate on an underwriting. The reason for this has to do with the size of each market. Furthermore, because the markets are of different sizes, selling in one requires more work than selling in the other. Hence, the compensation levels differ, as well.

The market for private placements is relatively small. Bonds sold through private placement generally are split into very large denominations, the only purchasers for which are institutions and the occasional wealthy individual. Because this market is relatively small and the denominations of private placement bonds are so large, relatively speaking, it does not take investment bankers a great deal of effort to sell them. On the other hand, institutions and wealthy individuals tend to have very professional staffs who know the market inside out, and who are very aggressive about pricing. For that reason, and also because large denominations of small bond issues tend to be less liquid (i.e. They are not easily resold.), buyers of private placements want higher interest rates.

There is an obvious tradeoff here. The higher interest rate will, at least to some extent, be offset by a lower fee to the private placement agent.

The public market, on the other hand, is composed of literally millions of individuals, like you and me, who do not spend every workday buying and selling water and wastewater system bonds. The public market will, therefore, accept lower interest rates. The other side of the coin, however, is that to deal in the public market requires bonds of relatively small denominations; and it requires a great deal of effort to sell many small pieces as opposed to a few large pieces. For this reason, publicly traded bonds (including those which are issued in a negotiated sale), carry lower rates of interest, but require higher investment banking fees. They also command higher investment banking fees to compensate the underwriter for the risk he takes that rates may rise and he may be stuck with the bonds. Here, again, is a tradeoff.

Which is better? Public sale? Negotiated sale? Or, private placement?

As stated above, there may be legal or economic reasons why you do not have all three options open to you. But no matter whether you have three options or no options, your FA will be able to inform you what your options are, in any given set of circumstances, and to assist you in calculating the investment banking fees for the transaction and converting such fees into True Interest Cost (TIC).

Credit Enhancement Fees.

The expense of credit enhancement fees applies only to a transaction which, one, is a bond transaction, and, two, is one in which the issuing system elects to use credit enhancement, assuming it is eligible. In other words, credit enhancement is by no means an expense which one automatically adds to the bill.

There are two types of credit enhancement: municipal bond insurance and bank letters of credit. The phrase "credit enhancement" is itself a sticking point in the finance industry. Those companies which provide credit enhancement want the world to know that the words mean just what they say. Such companies are willing to "enhance" the credit of a water or wastewater system, but they are not willing to substitute their credit for the ststem's credit. In other words, they will make a good credit better; but they will not make a bad credit good.

This means that the first question is eligibility. To be eligible for credit enhancement, a water or wastewater system must have at least one of the four highest credit ratings from one of the four nationally recognized credit rating agencies. The four major rating agencies are: Standard & Poor's, Moody's, Fitch Investors Service and Duff & Phelps. The four highest rating categories (using Standard & Poor's nomenclature) are: AAA, AA, A, and BBB.

The four highest ratings are collectively known as "investment grade" ratings. An investment grade rating means that the bond is eligible to be purchased by trusts, pension funds, and the like; which have fiduciary duties to invest their clients' funds in only the most sound investments. Credits which do not fall into the four highest categories are regarded as speculative investments. They are called, simply, non-investment grade, which is abbreviated to "NIG." Occasionally an NIG rated bond can be credit enhanced, but rarely so. The credit enhancers are penalized by insurance and bank regulators for guarantying NIG credits.

So, if your system does not have an investment grade rating, or cannot get one, it will most likely be unable to avail itself of credit enhancement. The only exceptions relate to strong credits where an investment grade rating is unavailable for technical reasons.

The purpose of credit enhancement is to lower TIC. In short, the only time a water or wastewater system would want to purchase credit enhancement is to lower its interest rate such that, even after paying the credit enhancement fee, it had a net savings in TIC. In other words, the savings must more than offest the cost of the credit enhancement. Unfortunately, not all credit enhancers lower interest rates by the same amount. Therefore, to calculate savings accurately, you must not only know the precise cost of the credit enhancement, but also the precise amount by which the bond interest rate is likely to decline if your system elects to purchase the enhancement.

To cut through some of this complication, we will first review the two forms of credit enhancement and then we will run through a couple of illustrations of how to calculate the savings.

Municipal Bond Insurance.

The most widely used form of credit enhancement is municipal bond insurance. There are five companies in the United States which offer municipal bond insurance. All have at least one AAA rating from a major rating agency. But, unfortunately, all AAA ratings do not trade at identical rates.

Bond insurance premiums are calculated on total debt service. This means the actual amount (not the discounted amount) of all annual debt service payments.

Let us use a Level Payment Bond of $1,000,000 for a term of twenty years at a rate of 7% as an example. The annual debt service payment on such a bond would be $94,393. The total of all annual debt service payments would be the annual debt service payment times the number of years in the term, or 20 x $94,000, for a total of $1,887,859. This is the number upon which the bond insurance premium is based.

The premium rate is based on the existing credit rating of the water or wastewater system. If your system enjoys a AA rating, your premium could be as low as 0.35%. If yours is the rare, insurable NIG rated system, your bond insurance premium rate could be as high as 1.4% of total debt service.

To calculate the premium, you simply multiply total debt service by the premium rate. For the AA credit in our example, the bond insurance premium would be $1,887,859 x .0035, or $6,608. For the NIG credit in our example, the premium would be $1,887,859 x .014, or $26,430. As you can see, the difference can be a great deal of money. The AA premium is just over one-half of one percent of the bond principal amount. The NIG premium, on the other hand, is over two and one-half percent. A difference of almost 500%!

A final note about bond insurance: beware of minimum premiums. Certain bond insurance companies require mimimum premiums. If your bond is below several million dollars, ask your FA to inquire about minimum premiums.

Bank Letters of Credit (LOC).

The other type of credit enhancement is the bank letter of credit (LOC).   LOC's are similar to bond insurance except in three very important ways. First, letter of credit fees are paid <u>annually</u>, whereas bond insurance premiums are paid once, upfront. Second, letters of credit are rarely issued for the full term of the bond. This can present some real difficulties. Third, few banks still have the AAA ratings, which the bond insurors have, so it is more difficult to calculate true savings.

Let us say your system obtains a seven year LOC on its twenty-year bond issue. In such case, the bond documents will be changed to reflect the fact that if the bank does not renew the LOC, or if you are unable to obtain a substitute credit enhancement of comparable quality, then at the end of that year <u>all of the outstanding principal on the bonds will be due and payable</u>.

Now, the documentation is structured such that if your system is unable to substitute credit enhancers or come up with the funds(!), then the LOC bank must pay the bondholders. This means the bondholders are fully protected (which they must be or otherwise the LOC would truly be useless). But it then means that suddenly your system owes the bank all of the then outstanding principal balance.

As you can imagine, this situation could get very messy. No one likes to be in the position of having to issue new bonds or being otherwise forced to borrow a huge sum of money in a short period of time.

The second difference is that LOC fees are paid annually. In addition, LOC fees are calculated on the <u>outstanding principal balance</u>, not total debt service.

Most LOC fees for water and wastewater systems are in the range of 0.3% - 0.7% of outstanding principal a year. Furthermore, rather than being based on the underlying credit quality of your system, they are based on the legal and financial structure of the system. If the system is part of a unit of local government with general taxing powers, and if general tax revenues are pledged to the bond payments, then the LOC fee is likely to be closer to 0.3%. If the only revenues which support the bond payments are the water or sewer revenues of the system, then the LOC fee is more likely to be in the 0.7% range.

N.B. This does not mean, if a system has the optimum legal and financial structure, but a shoddy financial history, that a bank will issue an LOC, anyway. It will not. No bank will cover a shoddy credit for any price. The point is just that if a bank does agree to issue an LOC for a system, it will base its LOC fee on structure.

Calculating the total cost of LOC's is different depending whether the bond is a Level Payment Bond or a Level Principal Payment Bond.

If the bond is a Level Payment Bond, the cost of the LOC can be determined by the TIC method. If the bond is a Level Principal Payment Bond, then you must multiply the LOC by the outstanding principal balance in each year, and then discount all of the annual fees back to obtain their total Present Value. Once you have ascertained the Present Value of the LOC fees, you can compare them directly to bond insurance premiums (since they will then both be denominated in Present Value dollars), or determine the LOC's impact on your rate by converting the Present Value dollar amount to TIC.

Calculating Savings on Credit Enhancements.

To calculate the savings on credit enhancements, both bond premiums and LOC fees must first be converted to TIC. For example, using the Level Payment Bond above, the TIC of the AA bond insurance premium would be about 0.09%. In other words, the cost of the bond insurance premium would add about 0.09% to TIC. Now, let us say that your system's FA determines that your bond would sell at a rate of 7.25% without the credit enhancement and a rate of 7.1% with the credit enhancement. Without considering the premium, the savings to your system would be the 0.15% difference in the rates. Since, however, you must pay the fee, you must deduct the increase in TIC due to the enhancement fee from the difference in the two rates. In our example, the difference between the two rates is 0.15% and the increase in TIC caused by the credit enhancement fee is 0.09%. Deducting the increase in TIC from the difference in rates indicates that the real savings to your system is 0.06% in TIC. By reversing the TIC method, you can calculate the exact amount in Present Value dollars. The exact savings would be $5,024 on the $1,000,000 bond.

Now, as you can see, whatever savings there may be depends greatly on the rate at which the credit enhancement trades. For example, based on our example, we saw that at least one AAA enhancer was trading at 7.1%. At the same time, another AAA insuror might be trading at 7.2%. If the second enhancer charged the same premium, there would be no savings at all.

This is the point. There is a fairly wide trading range among all of the credit enhancers, both banks and bond insurors. Your FA should know the trading range of each, or he can easily find out the trading range of each by asking the investment bankers who regularly trade in credit enhanced bonds. Once a particular credit enhancer's trading range is known, you can then ascertain any savings by converting the premium to TIC and subtracting this cost from the difference between the enhanced rate and the rate on unenhanced bonds of your system.

This may sound complicated, but it is not. A good FA can boil the problem down to a few straightforward conclusions such as: Unenhanced rate, 7.25%. Enhanced rate, 7.1%. Potential savings, 0.15%. Cost of enhancement, 0.09%. True savings, 0.06%.

Rating Agency Fees.

This is the simplest cost element to understand. If your system intends to issue a bond, and your FA believes the system has the credit strength to obtain an investment grade rating on such bonds, then he will undoubtedly recommend that the system apply for one. The fee is the cost of obtaining the rating.

This fee is paid directly to the rating agency at the closing. The rating agency fee is a flat fee and it is based on the complexity of the review which the rating agency staff conducts of your systems financial statements and operating history. Fees run from a few thousand dollars for a smaller system with simple accounting procedures and management history, to several tens of thousands of dollars for larger, more complicated systems.

Rating agencies will agree to the amount of fees in advance. It is part of the FA's job to contact the rating agencies on your system's behalf and to negotiate the rating fee. The FA will also be able to guide the system's financial staff through the rating process.

The rating agency staffs are very professional and their reviews are quite rigorous. It is a fair thing to say that, if you apply for a rating, the fee charged by the rating agency will be the least of your worries.

Printing and Miscellaneous Costs.

Managers of water and wastewater systems are well advised to keep cost sheets on each financial alternative available. During each meeting he should list the types and amounts of third party costs involved with each proposal. At the end of each discussion of each proposal, the manager should recite the list of costs to the FA and the other participants and ask directly whether there are any additional costs involved with the particular proposal.

There generally will be some miscellaneous costs. But they should not involve large amounts of money. They should easily be able to be identified in advance. Among the more common types of miscellaneous costs are the following:

* Cost of printing bond documents. The document which legally offers the bonds for sale is called the Official Statement, which is abbreviated to "OS." Investment bankers generally have the OS's printed by a professional printer. For simple bonds, this should not run more than $1,000 - $2,000.

* Surveys. Bankers generally require an updated survey of property which is to be mortgaged. The cost of a survey, which is done by a professional surveyor, can run from a few hundred dollars for a small parcel of property to several thousand dollars for large parcels.

* Title insurance. Title insurance is commonly required where a mortgage on land and buildings is involved. The cost of the premium is generally only a few hundred dollars

* Appraisals. Again, when real property is involved, an appraisal is often required. If the property has been recently acquired, then only a confirming appraisal is usually required. Not every appraiser will be acceptable. There is a professional designation within the appraisal industry called "M.A.I.", which means Member of the Appraisal Institute. This designation involves a

course of professional study plus continuing education requirements. It is an indication that the appraiser is a professional and that he is aware of recent developments in appraisal standards. As such, most lenders will require an M.A.I. appraisal.

Depending on the size of the property, the appraisal will run several thousand dollars. A confirming appraisal will be about 25% to 50% of the cost of a full appraisal.

* * * * *

The above eight categories comprise the vast majority of financing costs. Each has been only briefly described. The purpose here is to give water and wastewater system managers just a sketch of what each category is to serve as guidance when discussing the project with the FA and the various other advisors.

As has been mentioned several times, the key to controlling and estimating financing costs is the system's FA. He will be able to identify each financing cost you will encounter on any given financing. He will be able to explain them thoroughly. He will assist in negotiating firm, written estimates. And, he will be able to calculate their impact both on annual cost and on total cost.

But before leaving this topic, it might be well to recap some of the categories by providing below, on Table 11-A, a theoretical but thoroughly realistic estimate of the financing costs involved in the issuance of a $1,000,000 bond to fund a project for a small, but experienced and financially sound system with an A credit rating.

Table 11-A

Financial Advisor Fee	$10,000
Commitment Fee	N/A
Trustee Fee*	2,500
Underwriter's Counsel	2,500
Rating Agency Fee	3,500
Bond Counsel	6,500
Underwriter's Discount	20,000
Bond Insurance Premium	9,500
Printing & Miscellaneous	2,000
	$56,500

* Please note the Trustee Fee is $2,500 for setup plus $500 a year
thereinafter.

   As you can see the financing costs equal $56,5000 for a $1,000,000
bond, or 5.65% of bond principal. By using the TIC tables in Appendix
C, you can see that if the interest rate on the bond is 7% and the term
is 20 years, the financing costs will increase TIC from 7% to 7.72%.

Chapter 12

Delay

Every chapter in this book has at least mentioned the concept of Present Value or Future Value. Chapter 3 has an extensive section on the "Time/Value Theory" of money. Throughout this book, the point has been made , over and over again, that value changes over time.

This fact is perhaps the single most important concept in finance. But, despite its paramount importance, it is still a major problem for many people. The problem, I believe, is with the word "concept." People have a tendency to think that "concepts" somehow aren't real. They are right. Concepts are not real.

Charity is a concept. But what about writing a check to the United Way? The act of writing the check to the United Way is called an act of charity. It is, if you will, a physical manifestation of the concept. What, then, are the physical manifestations of the concept of the Time/Value Theory of money? What physical evidence is there that the amount of money you have to pay today for something will be different from the amount of money you have to pay tomorrow?

The best physical evidence lies in your checkbook. How much did you pay for health insurance last year? How much are you paying now? If you rent, how much was your rent last year? How much is your rent this year? If you pay tuition for a child, how much was the tuition last year? How much is the tuition this year? If you add up all of the checks you wrote last year for household expenses, are they more or less than what you are paying this year?

If you are not paying more now, write a book and tell the world how you do it! It will surely be a best seller; because literally everyone pays more from one year to the next.

The cause of this phenomenon is inflation.

When you read about inflation in newspapers or listen to commentators discuss it on radio or television, it seems like a remote abstraction - a concept.

Inflation is a concept, just like charity is a concept. Just like the Time/Value Theory of money is a concept. And, just like charity, it has its own physical manifestations. Unfortunately for us all, the physical manifestation of inflation is that we must pay more and more for the goods and services we need.

As far as water and wastewater systems are concerned, the all-too-real physical manifestation of inflation is that a length of pipe and an hour of labor cost more today than they did last year.

This brings us to the point of this brief chapter. A budget which does **not take the effects of inflation into account is a bad budget.** And, more specifically, a project estimate which does not take into account the effects of inflation is a bad project estimate.

Now, the question arises. How does inflation creep into a project estimate. The answer is: Delay.

Let us say, as an example, that a water or wastewater system's consulting engineer estimates that a particular project will cost $1,000,000. The date of the engineer's estimate is <u>January 1</u>.

By the time the staff has an opportunity to review the project and discuss it with the board, a month elapses. It is now <u>February 1</u>.

The board decides to proceed. But the board decides the project is too costly to be covered by current revenues and is disinclined to consider a one-time surcharge or levy on the ratepayers. They feel the project should be financed over time. So, the board authorizes the staff to retain the services of a professional Financial Advisor (FA).

The staff prepares a Request for Proposal (RFP). They send the RFP to the system's outside counsel for review. The counsel reviews the RFP and returns it for board action. The board approves the RFP and it is published in the appropriate newspapers. It is now <u>March 1</u>.

The staff reviews the proposals submitted by the various FA's and makes its recommendation to the board. The board approves retaining the recommended FA. It is now <u>April 1</u>.

The FA familiarizes himself with the system itself, its management, its financial history, and, most importantly, the project and its costs. He makes a preliminary recommendation that the project should be financed with a municipal bond to be issued through the state bond bank. The board approves the preliminary recommendation and authorizes the FA to discuss the project with the state bond bank. It is now May 1.

The FA and the system staff meet with the staff of the state bond bank who encourage the system to proceed with the financing through the bond bank. They tell the system to prepare an official application. The FA and the system staff report the results of their discussions with the bond bank to the board. The board approves the preparation of a formal application to the bond bank. It is now June 1.

The FA, the system staff and the outside counsel prepare the official application to the state bond bank, and the board approves it. The application is submitted. It is now July 1.

The state bond bank notifies the system that their preliminary review is favorable but that an Environmental Impact Statement (EIS) will be required. The board authorizes and approves an RFP for an independent consulting firm to prepare the EIS. It is now August 1.

The staff reviews the proposals from the independent consulting firms and reports their recommendation to the board. The board approves the recommendation and authorizes the hiring of the independent consultant to prepare the EIS. It will take six months to prepare. It is now September 1.

The holidays come and go. The old year ends, and a new year begins. As far as the project is concerned, it is now Year Two.

The independent consultant submits his report. The board reviews and approves the EIS, and authorizes its submission to the state bond bank. It is now March 1 of Year Two.

It takes the state bond bank sixty days for their staff to review all of the project submissions and for their board to approve the issuance of bonds on behalf of the project. It is now May 1 of Year Two.

Under state law, the issuance of bonds requires a public hearing. Thirty days legal notice is required to hold a public hearing.

The notice is prepared by outside counsel and published in the appropriate newspapers. The public hearing is held. It proves to be a blessedly uneventful occasion. But, it is now July 1 of Year Two.

The state bond bank issues bonds only three times a year. It is too late to include the project in the August 1 bond issue, so the bond bank schedules the project for its next bond issue in December.

In December, the bonds are issued and the project funds are placed in the system's bank account. It is now December 1 of Year Two.

The holidays come and go. The new year arrives. It has now been two full years since the original cost estimate for the project.

Assuming the rate of inflation for the intervening two years has been only 4% in each year. The project cost has now climbed to $1,081,600.

Now, if the system did not take into account the effects of inflation on the project cost, then the system will be $81,600 short. Imagine what the ratepayers will say at the second public hearing, which the board and the management must then call to explain why the project did not come in on budget, and why, therefore, the ratepayers must suffer an $81,600 surcharge in order to complete the project! These are the types of situations which all managers and directors seek to avoid.

On the other hand, assume that the cost increase was taken into account and that the size of the bond issue was increased commensurately to cover the increase in project cost. No nasty public hearing will have to be conducted. No public explanations will have to be given. Nonetheless, the fact remains: for no reason other than delay - in this case, perhaps, perfectly legitimate delay - the cost of the project inflated by 8.16%. The project cost the ratepayers $81,600 more than it would have at the time of the original cost estimate.

As you can see, delay costs money -- even unavoidable delay.

Now, we could go over the above example and conduct a lively debate over whether any of the delays were avoidable. This is not the point.

Now, there are two points to be made here. The first point is: avoid delay. Wherever possible. Delay costs money.

The second point is that, since delay is in many cases utterly unavoidable, be absolutely certain that the delays are estimated as accurately as possible and, most importantly, that the cost of such delays is fairly estimated, as well. This is not only the most prudent course of action to follow, but it will also obviate confrontations with angry ratepayers.

It is an unfortunate fact of life that even unavoidable delays look like mismanagement to the general public. And there are few stories which sell more local newspapers, than "exposes" of "mismanagement" by public officials.

Can delays be accurately estimated? Certainly.

Here's how. Ask the system staff for a schedule of the internal review and approval process. Ask the FA for a schedule of the external review and approval processes. Then get the staff and the FA to combine and integrate these schedules. Next, add some generous leeway, at least sixty to ninety days. Finally, and most importantly, after conferring with your staff, your FA, your engineers, your lawyers and all of the other participants in the project, set a target date for

<u>commencement of the project and let everyone know that you will hold</u> <u>them to their schedules</u>. You might also assign the FA the responsibility for overseeing everyone's schedules and reporting to the board immediately if anyone gets behind.

In addition to all of the above, water and wastewater system management must also be aware that different financial alternatives have different delay factors. One financial alternative will always take more or less time than another to implement.

For example, let us say that in the above example, the FA estimated that the system could issue its own municipal bond. Let us say that by doing so, a full year's delay could be avoided. In such case, our system would have been able to save $41,600.

Perhaps.

But what if the system were able to save itself $50,000 of financing costs by going through the state bond bank (which is often the case)? Then, it would be wiser to proceed with the financing through the state bond bank.

The point here is that delays can be estimated. The cost of such delays can also be estimated. In addition, once the costs of various financial alternatives have been identified, they can be compared and a prudent decision, based on sound financial principles, can be made. Furthermore, without considering the cost of such financial alternatives, any decision made will be, by definition, defective.

Chapter 13

Imposed Costs

Imposed costs are those costs which are extrinsic to the water or wastewater project, per se, but which are required as a condition of financing the project by a particular lender or a particular financial program.

In the preceding chapter, there is an example which describes a project funding through a state bond bank. The example recites the fact that the state bond bank requires an Environmental Impact Statement (EIS) as a condition of funding. This a perfect example of an imposed cost.

Chapter 12 was concerned with the delay caused by having to obtain the EIS. This chapter will concern itself with the cost of having to obtain the EIS, as an imposed cost. In other words, as far as this EIS is concerned, there are two types of costs associated with it. The first is the cost of the delay caused by postponing the financing of the project until the EIS is completed and approved. And the second is the cost of paying for the EIS itself. This chapter will concern itself with this latter type of cost, as well as the other types of imposed costs which are present in virtually all alternative financial programs.

Unfortunately, trying to list all of the possible imposed costs in various alternative financing programs is like trying to list all of the possible things which might go wrong with a project. The list is not infinite; but there are so many variables that trying to catalogue them is unproductive.

Rather, this chapter will concentrate on identifying the major catagories of imposed costs as well as identifying where they are most likely to crop up in various financing programs.

We can begin this discussion by creating a new axiom of finance, as follows:

The more government regulations are involved in any financing program, the more that program will contain imposed costs.

In saying this, we create a spectrum of all financial alternatives. At one end of the spectrum are the least regulated programs, which contain the fewest imposed costs. At the other end of the program are the most regulated government programs, which contain the largest number of imposed costs.

Let us begin at the end of the spectrum with the fewest imposed costs. As we move on, however, please note that we will not always be moving from Program A to Program B, and so on. Our purpose here, in this chapter, is not to rank financing programs in terms of the number of imposed costs. We are not going to score the various alternative financing programs. Rather, we will start with the simplest types of imposed costs and work through the more complex and subtle forms of imposed cost, regardless of whether any particular imposed costs falls within any one program or another. In addition, the fact that a certain type of imposed cost is mentioned in conjunction with only one financing program is not meant to imply that it is absent from all other programs.

Having said that, the most common imposed costs, which water and wastewater systems are likely to face, are found in conventional loan transactions with commercial banks, as well as just about all other financing programs. They are:

Most Common Types of Imposed Costs.

- A survey of any land which is to serve as collateral for the loan.

- The appraisal of real property which is to serve as collateral for the loan.

- Title insurance which guaranties to the lender that the owner's legal interest in a property is valid and enforceable.

There is nothing new about any of these costs. What is new, perhaps, is the consideration of such costs as imposed costs. But, in fact, that is what they are.

These costs are extrinsic to the project. In and of themselves, they have absolutely nothing to do with providing fresh drinking water or disposing of wastewater. Rather, they are costs which a water or wastewater system must incur to prove to the lender or provider of funding that it owns all the property it says it does (the survey), that its property is worth what it is supposed to be (the appraisal), and, that its ownership interest is legal and valid (title insurance). In other words the system must incur these costs because the lender or provider of the funding says so. The lender imposes the requirements of survey, appraisal and title insurance on the borrower, and the borrower must pay the costs of meeting these requirements. Yet, they have nothing to do with water or wastewater.

These particular imposed costs - survey, appraisal and title insurance - are now so common that they should almost be considered financing costs. In fact, as you may have noticed in the last section of Chapter 11, entitled "Printing and Miscellaneous Costs," they were, in fact, listed as financing costs.

The point here is not what name or label we attach to these costs. Rather, it is that they exist. They must be paid. If you would rather think of them as financing costs, go ahead.

Adverse Refinancings.

Another more subtle type of imposed cost involves the adverse refinancing of an existing debt. Here is how this type of imposed cost arises.

Let us say a water or wastewater system already has a $1,000,000 mortgage on its real property as a result of the financing of a previous project. Let us also say that this same system now has another $1,000,000 project which it must also finance. The problem arises from the fact that both lenders want first mortgages. Neither will subordinate. Neither will permit a parity mortgage. What can be done?

In such case, the system must refinance the first mortgage. This means that the system will increase the amount of the new project financing to include the outstanding principal balance of the old mortgage, plus any interest due, plus any prepayment penalties due. At closing, the system will pay off its old mortgage, so that the new lender (assuming, of course, that he agrees to it) will be the sole first mortgagee, albeit for a much greater dollar amount.

This type of forced refinancing can result in additional imposed costs in two ways. First, there may be prepayment penalties. Second, the terms and conditions of the refinancing may be less favorable and more expensive than the existing mortgage.

Either eventuality results in higher costs to the water or wastewater system. And the higher cost can be considered an imposed cost because the event which created the additional cost, i.e., the refusal to subordinate, was one of the terms and conditions imposed by the new lender.

Audits.

Probably the next most common form of imposed cost is the requirement for audited financial statements. Now, for many water or wastewater systems, especially the larger ones, the cost of an audit is moot since they normally have their financial statements audited, anyway. Yet, many systems do not have audited financial statements. They may manage to get along famously without audited statements and they may have maintained excellent financial records. All of this good work will come to no avail when it comes time to finance a capital improvement project. There are a few government financing programs - notably the Farmers Home Administration's Water and Waste Disposal Program - which exempt the smallest systems from the audit requirement. But virtually all other financing programs and all other lenders require the borrower to provide audited financial statements.

(N.B. Lenders may waive an audit requirement in the first year only for systems which have never had audits before. But they will require that such systems obtain audits in the future.)

Imposed Costs Inherent in Municipal Bonds.

The next area to review is municipal bonds. As you know, the interest on municipal bonds issued by water or wastewater systems, which are, or are a part of, a unit of local government, is exempt from income taxation by the federal government and by almost all state governments. As such, as you might guess, the issuance of such bonds is regulated, and heavily regulated, by the federal government through the Internal Revenue Code (IRC).

You already know from Chapter 11 that the issuance of municipal bonds requires both a bond counsel and a trustee. Since a municipal bond cannot, under most circumstances, be issued without a bond counsel or a trustee, some would argue that the requirement to have both a bond counsel and a trustee should be classified as imposed costs. Perhaps so. But, again, it is not the nomenclature or the category which is important. Rather, it is the identification of the cost as a cost that is important, as well as the proper inclusion of such costs in the total project cost estimate.

Most of the rules in the IRC governing the issuance of municipal bonds relate to the question of whether the interest on such bonds is, in fact, entitled to the exemption. This in turn begets questions about such matters as the issuer's legal status (as a unit of local government), the purpose of the bonds, how the bond proceeds will be used, when the bond proceeds will be disbursed, as well as how any repayments, reserve funds or undisbursed bond proceeds will be invested. Many of these rules concern the infamous concept of "arbitrage." Arbitrage, in general, refers to the happy circumstance of investing the proceeds of a loan, such that you are earning more interest on your investment than you are paying in interest on the loan. Since the low interest rates on municipal bonds are a direct result of the exemption from taxation, the federal government is zealous to guard against abuse of the exemption by arbitrage. The mechanism they have devised for doing so is the promulgation of voluminous regulations to cover virtually all possible permutations of events surrounding the issuance of a tax-exempt bond.

Some of these regulations, as you might imagine, result in imposed costs on the systems which issue the bonds. Although there is an exemption for very small and infrequent bond issuers, most water and wastewater systems, for example, will be required to set up elaborate accounting mechanisms to track the interest earnings on all unexpended bond proceeds in order to comply with the IRC regulations against arbitrage. These tracking systems are often managed by the trustee. In addition, the results of such accounting will have to be reviewed by the bond counsel every year to assure compliance. Regardless of how perfunctory these reviews may become, they still require time and effort on the part of the trustee and of the bond counsel. And, as you know, the time and effort of any professional cost money. But regardless of who performs these services, they still constitute an additional cost, extrinsic to the project, which is imposed by the IRC.

Another such cost, is the cost involved in holding a public hearing, if one is required.

In an effort to guard against the privilege of the tax-exemption's being extended to non-governmental entities, the IRC provides that if more than 5% of the proceeds of a bond issue goes to benefit any private business concern, then the bond issuing system must hold a public hearing on the question. This rule affects many, many water and wastewater systems. This is because many such system's private business customers provide more than 5% of the system's revenues. In such case, the system is required to hold a public hearing. And, although the holding of a public hearing does not usually involve a great deal of money, in and of itself, it also involves a delay of about sixty days.

So, between the cost of the delay and imposed costs of the notice publications and the hearing itself, the imposition of this requirement is no small matter.

Finally, as we leave tax-exempt municipal bonds, we now proceed across the spectrum into the realm of the financing programs with greater numbers of imposed costs. This is the arena of government programs, where imposed costs truly abound.

Of all of these, the two most infamous imposed costs which appear in government project finance programs are the so-called Davis-Bacon Act requirements and the previously mentioned requirement for an Environmental Impact Statement (EIS).

We have already discussed the EIS requirement both here and in Chapter 11. We have seen that it involves both the cost of delay and the imposed cost of the preparation of the EIS itself. We have seen that the preparation of the EIS can take six months. At the prevailing rate of inflation, this delay adds about 2% to the cost of a project, or $20,000 to a $1,000,000 project cost. This alone is significant. But when the imposed cost of preparing the study, which for even a small project can easily run between $5,000 to $25,000, is added to the cost of the delay; it is clear that the EIS requirement can constitute a major component of total project cost.

The Davis-Bacon Act requirement does not cause delay; but it is perhaps the most significant of all the imposed costs.

The Davis-Bacon Act is a statute which was enacted by Congress to provide that all federal projects paid the prevailing <u>union wage</u> to all construction workers on the project. In addition to purely federal projects, i.e., the public works projects of the federal government itself, the act also applies to several types of project finance programs which the federal government administers or sponsors. As you will see in

Chapter 20, one such program containing the Davis-Bacon Act requirement is the State Revolving Fund (SRF) program, sponsored by the U.S. Environmental Protection Agency.

The problem with the Davis-Bacon Act is its impact on small, rural water and wastewater systems. The problem arises from the fact that union scale wages are higher than non-union scale wages. Furthermore, most union labor is clustered in cities, where the cost of labor, in general, is higher than it is in the countryside. So, the effect of all this is that the cost of union labor is higher than the cost of urban non-union labor, which, in turn, is higher than the cost of rural labor.

Thus, for a rural water or wastewater system to comply with the provisions of the Davis-Bacon Act often requires it to pay wages which are significantly higher than the prevailing wage in the local community. These added wages constitute an imposed cost in the truest sense of the word.

To give you an idea of just how much of an increase the imposition of the Davis-Bacon Act requirements can add to the cost of a project, a regional coordinator for EPA estimated in 1990 that for his region, the imposed cost of having to pay Davis-Bacon wage scales added between **5% to 20% to total project cost!**

Consider this for a moment. The gentleman did not say that the cost of labor increased by 5% to 20%. Rather, he said that the total project cost increased by 5% to 20%. This is an astonishing number. For a $1,000,000 project, this means an imposed cost increase of between $50,000 and $200,000.

The lesson of this story is clear. If the labor component is a major part of the total cost of your system's project, then the imposed costs of the Davis-Bacon Act requirements may so burden the project that other, alternative non-governmental programs, or even governmental programs without the Davis-Bacon Act requirements, become far more cost effective.

As you can see, imposed costs can have subtle forms and effects. At first blush, who would think that an EIS requirement could add from 2.5% to 5% to total project cost? What manager of a water or wastewater system is there, who, after reading the legal boilerplate about having to pay wage scales consistent with the provisions of the Davis-Bacon Act, would ever dream that these few simple words could add as much as $200,000 to a $1,000,000 project cost?

Before concluding this chapter, let me say that if the little stories about the EIS and the Davis-Bacon Act have not given you pause, consider this: The Davis-Bacon Act requirements are one of the mandates of the USEPA's SRF program. There are sixteen other such mandates.

The conclusion here is that imposed costs are not only real costs, they can be extremely significant costs, as well. In order to avoid being caught out on a limb by any imposed cost on any program, be certain to ask the question: what requirements must our system agree to in order to obtain the funding through this program? Ask this question of yourself as you read through information on any financing program. Ask the question of the FA. And, ask the question of the staff of each financing program.

Once you have ascertained that you will have to agree to do something, your staff and the system's advisors can estimate the cost of doing it and, if applicable, any delays which will ensue.

Once all of these costs have been identified and measured, they can be factored into the total cost of alternative financing programs. The alternative financing programs can then be compared, and a decision can be made based on sound financial principles.

Chapter 14

Ineligibility

Ineligibility refers to those costs which simply cannot be financed, either as a matter of law or a matter of policy, through a particular financing program, whether public or private.

In the last chapter dealing with imposed costs, we mentioned that such costs could be quite subtle. We noted that an imposed cost could easily be concealed among the many terms and conditions of a particular financing program.

This is doubly true of ineligible costs.

In fact, ineligibility is a far more insidious cost category because, in most cases, the only items which are listed in any program description are the eligible costs. There is no list of ineligible costs. Therefore, in order to determine which of the cost items of a given project, if any, are ineligible, you would literally have to sit down and match each cost category involved in the project with the list of eligible costs for any particular financing program.

The point here is that when you see the concept "Eligible Project Costs" in literature describing a financing program, you must be aware that it has a hidden counterpart called "Ineligible Project Costs." These ineligible costs still have to be paid for, even if they cannot be financed through a particular program.

In addition, every program, whether public or private, has ineligible costs. There are two broad categories of such ineligible costs. They relate to quantity or quality.

Quantitative ineligibility means that a percentage of a given cost category cannot be financed.

Qualitative ineligibility means that the entire cost category, itself, cannot be financed.

Let us begin with a look at quantitative ineligibility.

Quantitative Ineligibility.

Anyone who owns a home is familiar with quantitative ineligibility. It is the amount of the down payment.

Banks will not (or are, at least, not supposed to) finance 100% of the purchase price of a home. Bank regulations generally prohibit the financing of more than 80% of the value of a home, or more than 95% of the value of the home, if the owner purchases Private Mortgage Insurance. (There are some exceptions for veterans, first time buyers and others.) In other words, between 5% and 20% of the cost of the home is quantitatively ineligible under conventional bank mortgage lending programs.

Did you ever think of it this way? Did you ever think of a downpayment as an ineligible cost?

But, as you can see, the downpayment truly does represent an ineligible cost.

Is this really any different than if banks put signs outside their doors which said: No Kitchens Financed? Aren't both of these prohibitions just variants on the same point of principle: that you cannot finance all of the house?

No. In fact, there is a difference between only financing 80% of a house and not financing kitchens. The former is an example of quantitative ineligibility, and the latter is an example of qualitative ineligibility.

There is a very good reason why the bank will not finance 100% of a house. The reason is liquidity. If you defaulted on your mortgage, the bank would want to foreclose on the house and then sell it as fast as possible. If the bank had loaned you 100% of the purchase price and you defaulted on your mortgage after the real estate market declined 10%, or even 5%, then the bank would have a difficult time recovering the full amount of its loan in a timely manner. In other words, the bank might have to wait quite a while to get all its money back. Its loan would be illiquid.

On the other hand, if the bank only loaned 80% on the house and you defaulted under the same conditions, the bank would, theoretically at least, have no difficulty recovering the full amount of its loan right away.

Thus, as you can see, in the case of a home mortgage, an 80% loan is more liquid than a 100% loan.

This is, in fact, a general rule: <u>the lower the loan-to-value ratio, the more liquid the loan</u>. This means that the less a bank lends against collateral of a certain value, the greater is the likelihood that the bank will make a full and timely recovery on that loan.

As you might imagine, this rule is responsible for a great deal of quantitative ineligibility.

Some examples of quantitative ineligibility which relate to water and wastewater systems are the following loan-to-value ratios for various types of assets. Please note that loan-to-value ratios always express the positive, i.e., eligible amounts. The ineligible amounts are the reciprocals of these values. So, if 80% of a cost is eligible, then 20% is not.

- Raw land: 50%.
- Buildings, machinery and equipment: 75%-80%.
- Inventory or other disposible assets: 60%-70%.

Qualitative Ineligibility.

Qualitative ineligibility, as you will see, means that the particular financial program you have chosen to finance your system's project will not permit certain categories of costs to be included. The implications of this, at first, can be quite subtle; but they can sometimes be quite dire.

Say, for example, that two years ago your system was offered an extensive piece of property, at an extremely attractive price, on which to drill a series of new wells. In addition to the land that was absolutely necessary, the system's board decided to buy a considerable amount of surrounding acreage. They felt this was the most prudent course of action. On the one hand, the additional acreage might be needed for future wells, if the system's needs expanded. But, on the other hand, even if the system didn't need any further wells, the additional acreage would assure good wellhead protection for the original well field. So the board went ahead and bought the property and financed it with a short-term bank loan. The bank would only lend 50% of the cost of the land, but let us say that the owner of the property would accept a subordinated note, or second mortgage loan, for the balance.

The board's intention was to roll the acquisition of the land into a much larger project which would involve drilling the new wells, adding new storage capacity, as well as connecting the new well field to the system itself.

Let us now assume that they have completed all the planning and the engineering studies as well as the design of the new improvements. They are now ready to finance the entire project.

Unfortunately, the board now learns that most of the land can't be financed through the program they have chosen. The program they have chosen allows the system to pay for the project over a twenty-year period, but it will only permit land costs to be included in the project to the extent of the minimum acreage required for the wells themselves. In other words, no credit will be given for wellhead protection, and no credit will be given for future expansion.

At first this does not sound so drastic. After all, they already have interim bank financing as well as financing from the previous owner. So what if the majority of the land cannot financed through the program the board had chosen. How hard can it be to get additional financing for the extra acreage?

Well, consider this. The bank and the previous owner might have been very happy indeed, to give the system an interest-only bridge loan for a couple of years on the land. They did so in anticipation that the loan would repaid in its entirety in a year or so when the full project was financed. Now both the bank and the previous owner learn that most of the land will not be financed with the project. Now they are being asked to finance it. And, what is worse, they won't even have mortgages on the full property, since part of the mortgage must be split off as security for the part of the land that is being financed through the financing program.

The bank decides it won't give the system a loan for more than five years and at a very substantial rate, on the excess acreage. The previous owner will only offer the same terms.

The point here is that the ineligibility of a project cost, such as the excess acreage, means that that particular cost must be paid for by some other means. Perhaps, such costs can be paid for by another financial program. Perhaps, no one will finance them. If no one will finance these costs, then they must be absorbed directly by the ratepayers in the form of a one time assessment.

The effects of an ineligible project cost on a water or wastewater system may be quite mild, especially if the cost is small and there is a reasonable alternative readily available. But, on the other hand, if the ineligible cost is large and there is no alternative available, you might well have some angry ratepayers who are forced to come up with a large sum of money on a one-time basis. In any event, as we shall now see, ineligibility will result in adverse consequences to system ratepayers.

Cost of Ineligibility.

Let us assume, as an example, that out of a total project cost of $1,000,000 to be financed through a 20 year loan at an interest rate of 7%, only $900,000 is eligible. The balance of $100,000 represents the cost of the excess land. Let us further assume that the remaining $100,000 will be financed by a bank and by the previous owner at a 10% rate, but that it must be paid off in five years.

Now, from our tables in Appendix C, we can determine that the annual debt service payment on the $900,000 loan at a 7% rate for a 20 year term would be $84,954. The annual debt service payment on the $100,000 loan at a 10% rate for a 5 year term would be $26,380. The system's combined annual debt service payment would, therefore, be $111,334.

If the system had been able to finance all $1,000,000 through the financing program, the annual debt service payment would be only $94,393. Therefore, because the excess land is ineligible for the program, the cost to the system's ratepayers increases from $94,393 to $111,334 - an increase of 18%!

Please note that the above example illustrates <u>qualitative ineligibility</u>. As you recall, the under the financing program chosen by the system's board, excess land is an ineligible cost category. The result of this exclusion, however, is that $100,000, or 10%, of the total project cost cannot be financed through the program.

Question. How does the above example differ from a financing program which will only fund 90% of a project? Where 10% of project costs are <u>quantitatively ineligible</u>.

The answer is: it doesn't. There is no difference. The effects of qualitative ineligibility and quantitative ineligibility are the same. They both result in higher cost to the ratepayers.

The reason for creating a distinction between qualitative and quantitative ineligibility is simply to emphasize the fact that these hidden costs lurk in two areas. On one hand, when a program description says "80% financing", then system management must realize that they must get the remaining 20% from another source. On the other hand, when a program description says "no excess land," or more probably, "only land necessary to maintain wellhead," then the system must find another source of funds to pay for the balance of the land it wants to purchase.

Furthermore, when you read the words "another source," you should be thinking "another (expensive) source." Financing small pieces of projects can be very costly. If you decide to pursue a 90% financing program, for example, for your water or wastewater project, do not kid yourself into thinking than your work is 90% done. It has only just begun. Putting the last 10% of the funding together will prove to be far more arduous than arranging the first 90%. And more expensive.

The reason that ineligible costs are so expensive to finance is that, as you saw in the above example about the excess land, they relate to things which are, most often, only part of a greater whole, or a fraction of something. In some cases, the ineligible portion might not even be a real thing in its own right - such as 20% of a house. What can you do with 20% of a house?

The point here is that the part of the project cost which is ineligible in any given program will, by its very nature, be difficult to finance. When something is difficult to finance, it will also be expensive to finance. Second mortgages carry higher rates than first mortgages. Second liens are more expensive than first liens.

In our example about the excess land, the parcel would first have to be divided into the eligible and ineligible parts. New surveys would have to be done. New deeds would have to be created. In addition, since one of the principal reasons for acquiring the excess acreage was wellhead protection, it is most likely that the deed to the ineligible acres would contain restrictions against uses inconsistent with wellhead protection. The more restrictions there are in a deed the less valuable is the land. The less valuable the land, the more difficult and expensive it will be to finance. In short, the ineligible portion of a project will be both more difficult and more expensive to finance.

So, to conclude this discussion about the effects of ineligible costs, there are two points to consider.

First, be sure to identify all of the elements of the cost of your system's project and then match them to the eligibility criteria of each financing program you are considering. Be sure to look for both quantitative and qualitative eligibility criteria.

Second, before you commit to any financing program, be sure to identify alternate sources of funding for all of the ineligible costs. Be sure to do this in advance. One of the worst things that can happen is to commit to a particular financing program, only to learn later that the remaining ineligible costs cannot be financed at all.

Chapter 15

Coverage Ratios

The two preceding chapters of this book note that both imposed costs and ineligible costs often take subtle forms. They are often concealed or difficult to identify.

The cost of "coverage" is the most subtle of all. It is so because, although it is a cost, it is not an expense.

This may seem a cryptic and tiresome remark; but it is true. The ratepayers have to pay the cost of coverage to the system; but the system does not use the money to make debt service payments.

Let us begin to explain this phenomenon with a definition of "coverage." Coverage is the amount by which a system's net income or cash available for debt service exceeds the system's annual debt service payments.

(The term "cash available for debt service" has a very precise meaning. Please see Chapter 1, where there is a discussion of the precise categories of both income and expense which comprise the term "cash available for debt service.")

Coverage is always expressed as either a ratio or a percentage. It is also commonly referred to as the "debt service coverage ratio."

So, for example, if a water or wastewater system has, in any given year, $125,000 of cash available for debt service and annual debt service payments of $100,000, then it has coverage of 125%, or a debt service coverage ratio of 1.25 (i.e., 1.25:1).

Now that we know what coverage is, let us look to why it is such a subtle and invidious form of cost.

Let us assume, in our little example above, that the operation and maintenance expenses for this obviously smaller system comprise 50% of the system's budget. And, let us also assume that annual debt service payments comprise the remaining 50% of the system's budget. In such a circumstance, the annual budget would be composed of $100,000 of annual debt service payments, as indicated above, and $100,000 of operations and maintenance expenses, for a total of $200,000.

But, let us also say that the system's lender requires a debt service coverage ratio of 1.25. In such case, the system would have to collect another $25,000 from its ratepayers.

This means that the total charge to ratepayers would be $225,000.

A Financial Statement for such a system would look something like Figure 15-A, below:

Figure 15-A

Regular Income	$225,000
Operations & Maintenance (i.e. cash) Expense	-100,000
Cash Available for Debt Service	$125,000
Annual Debt Service Payment	-100,000
Net Income (Loss)	$25,000

Please note the line "Net Income (Loss)," in the above example. The amount of funds listed on this line can be interpreted as the amount of revenues which are collected from the system's ratepayers, but which are neither used for operations & maintenance nor annual debt service payments. As such, the question then arises as to what these funds are used for?

The answer is both nothing and anything. In other words, such funds may be used for any lawful purpose. Or, they may also simply be put into a bank account, and not used at all.

The point here is twofold: first, the funds are excess funds; and, second, the amount of the funds is mandated. In other words, the funds are not absolutely necessary to run the system. This does not mean that such funds will not be used for good and proper purposes. For example, they might be used for preventive maintenance. They may be saved as a reserve fund for replacement and repair of equipment. They may also be saved in an account and used to prepay loans or other debts of the system.

So, the point here is not that these extra funds are squandered or wasted. Rather, it is just that they are not absolutely necessary to run the system.

What if, since the funds are not absolutely necessary, the board of directors decided to reduce rates by $25,000? This leads us to the second point.

The board cannot, realistically, reduce rates by $25,000. The reason for this is that the additional funds must be collected pursuant to the agreement with the system's lender. In other words, the board could reduce rates. It is within their legal authority to do so. But, if they did so, and the debt service coverage ratio fell below the amount required by the lender (in this case, assume the lender requires a debt service coverage ration of 1.25.), the system would be in technical default default of its loan agreement.

(A "default" generally occurs on a loan when the borrower fails to make a full and timely payment under the loan agreement. A "technical default" occurs when the borrower fails to comply with any of the other terms and conditions of the loan agreement.)

So, if a board reduced rates to the point where the debt service ratio fell below the amount agreed to in the loan agreement, the lender could go to court and force the board to increase rates to maintain the required debt service coverage ratio.

This seldom happens. Responsible boards do not willfully violate signed agreements. But what does happen -- and with some frequency -- is that from year to year the cost of operation & maintenance will rise with the rate of inflation; and boards do not correspondingly raise their rates. After two or three years, these systems find themselves seriously in violation of the debt service coverage covenants of their loan agreements. The bankers will always be watching the system's annual financial reports. Once they become alarmed that expenses are increasing and revenues are not rising apace, they will realize that their debt service coverage ratios are eroding and their lawyers will soon be on the system's doorstep.

Another problem which arises with some frequency is the unforeseen maintenance expense. If the system has no reserve account out of which it can pay for such extraordinary expenses, then it will have to pay for the unplanned cost out of current income. This will show up on the annual financial statement as a reduction in Net Income, which in turn will mean a reduced debt service coverage ratio. If this happens a couple of years in a row, or in two out of three years, the bankers will soon be at the door wanting to know why, and wanting the system to raise its rates to make certain that the debt service coverage ratios are adequately maintained.

So, the point we made a few paragraphs above is not so precious, after all. Coverage really is a cost, but it really is not an expense. It is a cost because the funds to provide the coverage actually come out of the pockets of the ratepayers. But it is not an expense because these funds are not spent on the project. They may not even be spent at all; but they still have to be collected under the terms of the loan agreement. So, the ratepayers must pay for coverage.

Now, how often does the question of coverage come up? The answer is: with growing frequency.

For example, let us say that a water or wastewater system originally financed itself with a loan through the Water and Waste Disposal Program of the Farmers Home Administration. Let us say that the system now wants to do a major upgrade of the entire system. This means that the assets used to collateralize the original loan will have to be used to collateralize the second loan, as well. Under these circumstances, the system will be surprised to learn that one of the major terms and conditions of the second loan will be that the system must now maintain a 1.2x debt service coverage on <u>both</u> loans.

In like manner, take the case of a system which has grown steadily over the years from a very small to a very solid, mid-sized entity. At the same time, this system has enjoyed superb management which has operated the system in a sound and fiscally conservative manner. During its early years it could only get credit through government programs. Whenever it had to go to a bank to finance vehicles or other light equipment, it had to pay very high rates. Now, after years of this travail, it has finally achieved fiscal independence. It is now going to receive a credit rating of "A" from Standard & Poor's. Now, it will be able to borrow funds virtually anywhere. Now it will no longer be a slave to government programs or to the local banking community.

What a shock it will be to the system's directors when they learn that after their years of toil and diligence, that in order for their bonds to be awarded the coveted "A" rating, they will have to maintain a 1.25x debt service coverage ratio. The system's ratepayers might actually wind up paying higher rates because of their system's success.

The point here is that coverage ratios cost ratepayers money. The other point is that requirements for various coverage ratios come from a variety of financial programs. They are not limited to private sector financial programs; government programs may require them, as well. In addition, as can be seen from the above example of the system which just received its "A" rating, coverage requirements are not limited to poor or small or even new water or wastewater systems. They can be imposed, and are imposed, on systems with varying financial health and histories.

In summary, coverage -- or more properly, excess coverage -- is not a project cost per se, nor is it a financing cost per se. Rather, it is a legal requirement imposed on a borrower by a lender. Its causes may vary widely from government regulation to simple bank policy. Nonetheless, regardless of the cause, the effects of excess coverage requirements are to increase costs to ratepayers of water and wastewater systems across the country.

Section III

Chapter 16

Public and Private
Funding Programs

Public and Private Funding Programs.

The first section of this book explained what annual debt service payments actually are. It also described how to convert dollars into rates and rate increases into dollar increases so that sense could be made out of the competing claims of various financing programs.

The second section of the book presented and described seven of the most important factors which can inflate project cost.

Section Three will now present the four most commonly used programs to finance water and wastewater projects.

It is important to note at the onset that the following four chapters will not attempt an exhaustive description of each of the four financing programs. Rather, each program will only be described in brief terms.

You may recall from the Introduction to this book that its purpose is to serve as a handbook - a reference book which can sit in the bottom desk drawer for months or even years until a project needs to be financed. Furthermore, between the time I write these words and the time you have a real, live project and really need to read them, at least one major change will have likely occurred in most of the programs, which would have the effect of rendering the information in this book obsolete.

So, the real thrust of Section Three is not simply to present these financing programs, but rather to present them in terms of the seven critical cost factors described in Section Two of this book.

In other words, as far as the government financing programs are concerned, you can pick up the telephone and request exhaustive amounts of information on all aspects of such programs. What information you do not get in the mail, your lawyer can look up in the county law library.

For the non-formal financing programs, such as conventional bank loans and municipal bonds, as much background information as you will ever want to see will be available through your system's Financial Advisor.

What you will not get from any government agency or any library is a breakdown of each financing program into the seven critical cost factors described in the preceeding seven chapters. Nor will they show you how to convert dollar costs to rate increases, and vice versa. And, most importantly, they will not show you how to identify and compare the real costs involved in alternative financing programs.

It should also be emphasized that the analytical methodology contained in Section One of this book and the seven critical cost factors described in Section Two of this book are not confined merely to the four financing programs described in Section Three of this book. On the contrary, both the methodology and the cost factors apply to all financing programs, both public and private.

So, if there is a statewide program or a local financing program for water or wastewater projects in your area, both the methodology and the cost factors in Sections One and Two of this book will absolutely apply to those programs and any others, as well.

Therefore, the next four chapters will each cover one financing program; but they will do so by discussing which of the seven cost factors apply to that program and to what degree.

The financing programs to be discussed are:

- Municipal Bonds.
- Conventional Bank Loans.
- The Water and Waste Disposal Program (WWDP) of the Farmers Home Administration (FmHA) of the U.S. Department of Agriculture (USDA).
- The State Revolving Fund Program (SRF) of the U.S. Environmental Protection Agency (USEPA).

The first two programs, as you know, are part of the private sector. The last two are government programs.

Government funding programs involve public money. Tax dollars cannot be left to the professionalism of individual bankers nor to the integrity of the capital marketplace. Public funds must have safeguards. And the safeguards must be built into the program.

In like manner, although the reason the Congress created a particular funding program undoubtedly fulfilled a single social purpose, the Congress never does things in a singular manner. Whenever they create one piece of policy, they want it to jibe with their other major pieces of social policy as well.

As such, you should not be surprised to find that the government funding programs are replete with rules and regulations which are largely absent from the private funding programs.

(It should be noted that although Municipal Bonds are funded from the private market, they are, because of their federal income tax exemption, heavily regulated by the United States Treasury.)

In other words, if you want to borrow money from the government, you must obey their rules about civil rights, environmental protection, historical preservation, and so forth. Some of these policies result in nothing more that a requirement to fill out another form. Some, however, not only cost money but, as you will see, they cost real money.

Chapter 17

Municipal Bonds

The issuance of municipal bonds enables water and wastewater systems to raise the funds necessary for capital improvement projects directly from the investing public. In the most literal sense, by issuing municipal bonds, a water or wastewater system can borrow the funds it needs directly from tens or hundreds, and, for larger bonds, even thousands, of small investors.

This transaction is made possible by the presence of the investment banker. It is the investment banker who serves as the intermediary between the many investors who provide the funds and the systems which use them.

A system which wishes to issue municipal bonds must negotiate the terms and conditions of the bonds directly with the investment banker who acts on behalf of the investors who purchase the bonds.

There are two reasons why investment bankers serve well in this capacity. The first is that investment banking is not an unsupervised activity. Uncle is watching. Investment bankers are regulated both by the United States Securities and Exchange Commission (SEC) and by either one of the major stock exchanges or the National Association of Securities Dealers (NASD). The second reason is that if they do not know what they are doing, they will go broke. In other words, an investment banker must be familiar with the general terms and conditions which the investing public will accept. If he makes a mistake and permits unacceptable conditions to be included in the transaction, then there is a strong likelihood that when he goes to sell the bonds to the public, they will not purchase them. The investment bank will be stuck with unsaleable bonds. The nice young investment banker who puts his firm in this unenviable position is due for an abrupt career change.

Why should water and wastewater systems be interested in issuing municipal bonds? The answer is: low rates of interest. The reason for this is that the interest on the bonds is exempt from federal income taxation and is exempt from income taxation in most states. Because investors do not have to pay income taxes on their interest earnings, they are willing to accept a lower rate of interest. This means that water and wastewater system will wind up paying less interest on their borrowings.

There is one other point which should be made about the municipal bond market. When we speak of the "investing public," we are actually talking about two distinct markets. The first market really is the public. It is you and me and Aunt Millie.

Investors like us commonly invest a part of our savings in municipal bonds because, in general, they are good, safe investments. And, as you will see in a few paragraphs, they generally carry quite favorable rates of interest.

The other distinct market is what is known in the public finance world as the "institutional market." The institutional market consists of large corporations and funds which desire the tax exempt income. The institutional market is distinct from the public market (which is also called the "retail market") because the goals and objectives of the institutional market are distinct from those of the vast majority of individual investors. We will discuss this a little further below when we talk about Rate and Term.

Suffice it to say that the most attractive features of issuing municipal bonds are the low interest rates, and, as we will see, the relative ease with which they can be issued.

Now, however, we will look into the seven cost factors as they apply to tax-exempt municipal bonds.

1.) **Term.** Municipal bonds have by far the most flexible terms. A water or wastewater system can issue bonds with terms ranging from less than a year to forty years. Service life of assets is, however, a factor in determining term. The term of a bond is generally limited to the service life of the asset it finances.

In addition, the market itself has a bearing on term. By this we mean that there are large numbers of both individual investors and institutional investors who will generally buy bonds with terms of up to twenty years. Beyond twenty years, the individual investors tend to disappear from the market. After twenty-five years, there are almost no individual investors left in the market, at all. What is left are the institutions. The institutions know that the market for bonds with over 25 year terms is very thin. They therefore demand higher rates.

For bonds with terms of 35-40 years, there are literally only a handful of major national institutions left who will purchase such paper. They all know each other. They are all bloodless professionals. So, they will charge to the hilt.

The difference in the annual debt service payments between a twenty-year bond and a forty-year bond is about 22%. But the premium in higher interest rates, which the handful of forty-year buyers will want, may well eat up a good portion of that savings. When comparing twenty-year bonds with lower rates and forty-year bonds with higher rates, be sure to use annual debt service payments as the basis of comparison. By doing so, you will be able to precisely identify the financial advantages of going either way.

The credit rating of the bond also effects term. Whereas the market for investment grade rated bonds truly does extend out to forty years, the market for unrated bonds carries far shorter terms. All but the most sophisticated institutional investors (who will charge exorbitant rates) are reluctant to leave their money in an unrated investment, where the issuer is not well known, for more than twenty years. For this reason, unrated bonds seldom have terms of more than 15 years and frequently have terms of ten years or less.

2.) **Rates.** We must always begin any discussion of the interest rates on tax-exempt municipal bonds by saying that they are generally low, and that they are always lower than commercial rates.

This is very important. When people get involved in a project and they start comparing tax-exempt bonds, they start talking about how much higher one is than the other. At this point, it is good for everyone to take a break and remember that all tax-exempt bonds have lower rates than their taxable counterparts.

Now, in the tax-exempt municipal bond markets as is the case with all of the major capital markets, rates are set by supply and demand as well as investor expectation of future interest rates. Up-to-the-minute rates are collected and published electronically as national municipal bond indices. These indices cite the rates for higher quality, or investment grade rated, bonds.

Since the indices reflect a mix of investment grades, bonds with the highest ratings (AAA) will enjoy rates slightly lower than the national index. On the other hand, bonds which do not have a rating at all, or whose rating is not investment grade, will carry rates considerably above the national index. Furthermore, the spread between AAA rated bonds and unrated bonds may vary greatly from state to state, and even in some instances within states. It should also be noted that the market for unrated bonds is very thin and, because of this, rate spreads are very volatile.

The graphic representation of the national indices at the various terms is called the yield curve. Interest rates are depicted along the vertical, or y-axis, and time is depicted along the horizontal, or x-axis.

As we discussed in Chapter 10, the yield curve has for almost all of the last decade trended from lower left to upper right. That means the lower the number of years, the lower the interest rate; and, conversely, the higher the number of years, the higher the interest rate. The reason for this is uncertainty. Investors want higher rates the longer they are willing to go out because of the greater uncertainty associated with longer terms.

So, this is the normal posture of the market. Rates rise over term. If you want a twenty-year term, you should expect a higher rate than for a ten-year term. In addition, from our previous discussion of term, you should also be aware that for terms over twenty years, the market thins out dramatically and as you go over thirty years rates begin to rise rather abruptly.

3.) **Financing Costs.** The financing costs of issuing bonds will amount to several percentage points of the project cost. They will include the fees of the Financial Advisor, the outside counsel, the underwriter's discount or bond placement agent's fee, the investment banker's counsel fee and the bond trustee's fee, as well as printing and miscellaneous costs. Depending on the nature of the bond, the financing costs might also include a rating agency fee and a bond insurance premium.

Table 11-A in Chapter 11 estimates the financing costs on a $1,000,000 tax-exempt bond at $56,500, or 5.65% of principal. These additional costs have the same effect on annual debt service payments as if the interest rate on the bonds had increased from 7.0% to 7.72%.

Financing costs will be higher for underwritings than for private placements. Such costs will also be higher for rated issues than for unrated issues. And, insuring a bond issue will, by definition, increase upfront costs.

For example, a bond insurance premium on an A-rated bond issue might cost 0.95% of bond principal. This would translate into 0.12% of TIC on an annual basis. But the annual savings in interest rate generated by the insurance might be 0.25%. In such case, the water or wastewater system would have a net savings of 0.13% a year. Thus, financing costs alone should not dictate the use of one or another financing program. Rather, the net cost on a TIC basis must be determined and then compared against the other cost factors affecting all other financing programs.

4.) **Delay.** Delay is only a modest factor in bond financing. In order for a bond issue to obtain its tax-exempt status, several formalities must be observed under federal law and often under state law as well. These formalities, such as official notice and meeting requirements for system boards of directors or sometimes even public hearings, generally add from 60 to 120 days from the date on which the board decides to issue bonds until the bonds are actually issued.

5.) **Imposed Costs.** The issuance of tax-exempt bonds does not involve any imposed costs. Other than the additional financing costs involved in bond issuance (which, if we were really splitting hairs, could be considered a form of imposed cost), there are no other additional costs or charges as a result of issuing tax-exempt municipal bonds.

6.) **Eligibility.** When it comes to financing the capital improvements of water and wastewater systems, virtually 100% of the normal project costs will be eligible for funding with the proceeds of a tax-exempt municipal bond. This includes such "soft" costs as engineering studies, and the like.

There are a small number of items, such as labor training costs, which cannot be financed. But since they are not tangible assets with extended service lives, they are unsuitable for financing with any type of long-term debt.

7.) **Coverage.** The issuance of tax-exempt bonds does not, in and of itself, require that the revenues of a water or wastewater system be any higher than are necessary to make annual debt service payments. If the bonds are rated, however, by a national rating agency, that agency will, under current rating standards, require that the system's cash available for debt service exceed by 25% the system's annual debt service payments. In other words, they will require a debt service coverage ratio of 1.25 times.

In addition, on unrated issues the investment bankers themselves may impose coverage ratios. These coverages obviously create a cushion or margin of error against default. Investment bankers will require the additional coverage as a comfort to assure their investor clients that the possibility of default is truly minimal.

Chapter 18

Commercial Bank Loans

Commercial banks were once a major source of funding for public works projects of all kinds. However, since the advent of "money market" funds in the early 1970's, this is no longer true.

Before the 1970's, Americans had passbook savings accounts paying steady and seldom-changing rates of interest. Once people began making regular deposits, they seldom made withdrawals. The only reasons why a depositor would make a withdrawal would be either for an emergency or for a truly milestone event, such as the birth of a baby, the purchase of a new home or the tuition for a college education.

Once money market funds were created, however, people finally began to realize that their bi-weekly deposits were, in fact, investments.

Once they began to look upon their bank accounts and their money market accounts as real investments, the next major awakening was that people should compare these different types of investments to see which offered the best return.

It was no great surprise that the bi-weekly deposit in the local bank ran a poor second to the yields obtainable from the money market accounts. For several years thereafter, literally billions of dollars a week poured out of the banking system and into a growing multiplicity of new money market accounts. To ward off disaster and stanch the flow of withdrawals, the banks finally created and began offering their own money market accounts. In addition, the banks also began offering, for the first time, certificates of deposit in small denominations.

The principal casualty in this struggle for the American savings dollar was the old fashioned passbook savings account at the local bank.

The most significant aspect of this truly watershed event in the financial history of the United States was that, for the first time in history, local banks could not count on huge sums of money remaining static in passbook accounts for twenty, thirty or even forty years.

Once these long-term deposits disappeared, so too did the banks' ability to make long-term, fixed-rate loans. Without having long-term deposits, such as the passbook accounts, on which they paid a fixed-rate of interest and seldom suffered withdrawals, the banks were simply unable to deal with clients who needed long-term, fixed-rate loans.

This cataclysmic change in the American banking industry, the ramifications of which are still going on, sets the stage for the banking policy which water and wastewater systems face when looking for long-term loans to fund their capital improvement projects.

With this in mind, we will now examine the seven cost factors as they apply to commercial bank loans under the new constraints facing the banking industry.

1.) **Term.** The most radical difference wrought by the changes in the banking industry described above is that banks literally cannot offer long-term loans. For these reasons, the longest terms which banks will customarily offer in today's environment for loans to fund the capital improvement projects of water and wastewater systems are in the five- to ten-year range.

Only seldom will banks offer terms of up to fifteen years. And under such circumstances, such loans would never carry fixed rates. On the contrary, these loans will either have floating rates, or the rates will be adjustable annually or every few years, at the very most.

2.) **Rates.** As indicated above, the radical changes in the banking industry have put an end to fixed-rate loans. The reason for this is that money market and other such accounts pay a different rate of interest every day. If the banks are required to pay their depositors a different rate of interest every day on their money market accounts, then they certainly cannot risk fixing the rates which their borrowers pay to them.

In the past, fixed-rate loans were common. Today, they are so scarce as to be utterly non-existent. If willing to give a fixed-rate commitment, the rate is seldom fixed for more than a few years, at most, at which time the rate is reset according to the then current rate environment.

If state law permits a loan form which qualifies as a tax-exempt borrowing under the provisions of the United States Internal Revenue Code, then, depending on the degree to which the interest on the loan is exempt from federal and state income taxes, the bank will generally offer a rate which is substantially below the bank's normal lending rates.

If the tax-exempt loan is to be a floating rate loan, then it will be considerably below prime; and, it will generally be expressed as a percentage of prime. Rates between 60% and 90% of prime are common.

If the bank offers a fixed rate, then the determination of the interest rate will be more complicated. The exact rate offered will depend on three factors. The first factor is what rates the national indices are indicating for tax-exempt debt of similar maturity. The second factor is the bank's estimation of the credit worthiness of the borrowing water or wastewater system. The third factor is competition. If the bank knows that it is one of several quoting on a loan to a financially sound and well established system, then it will set a rate close to the national index rate. On the other hand, if the bank feels it has the field to itself, it may quote a scandalously high rate of interest for the same credit.

If the loan does not qualify for the federal income tax exemption, then a different interest rate regime can be expected.

If the bank offers a floating rate, it will be at prime or as much as 2.5% or 3% above prime. If the bank offers a fixed rate, then the rate will be determined according to the three factors cited above for determining the interest rate on a tax-exempt bond, with one exception.

The exception is that the various U.S. Treasury Bond rates will be used as a base rate instead of the national index for tax-exempt bonds. However, since the Treasury rates reflect the highest credit rating of all (the United States Government), and since the bond indices reflect only a blend of rates, the rate, which the bank will offer, will be significantly above the rate on the commensurate Treasury Bond.

3.) **Financing Costs.** The financing costs involved with commercial bank loans tend to be quite reasonable.

Borrowers are generally required to pay only for their own counsel as well as for the bank's counsel. This is, of course, in addition to paying for the system's Financial Advisor, who must be paid regardless of the financing program used.

Banks will also generally charge a commitment fee or points, which are paid upon commitment or at closing, respectively. The will also charge an annual servicing fee.

4.) **Delay.** Water and wastewater systems will face no delays in obtaining commercial bank loans.

5.) **Imposed Costs.** The imposed costs involved in commercial bank loans are quite modest. They involve such things as fees for additional land surveys or fees for appraisals, or confirming appraisals, of property which will be mortgaged as part of the financing program.

6.) **Eligibility.** When it comes to commercial bank loans for water and wastewater systems, the concept of ineligibility results from the banks' limits on the amount they will lend against certain classes of collateral. In other words, quantitative ineligibility. For example, a bank may simply not lend more than 50% of the cost of raw land or 80% of the cost of a new structure.

This can create serious problems for water and wastewater systems since the balance of the amount required is virtually unfinanceable. In almost all cases, the unfinanced amount must be paid in cash.

As such, if a system has no reserve funds, it must levy an assessment on the ratepayers. Such an assessment will invariably be expensive, highly unpopular, and may cause delays which will add even futher to total project cost.

7.) **Coverage.** Banks will generally require a certain amount of excess coverage. For older water and wastewater systems with outstanding financial histories and ample reserve funds, a bank may not insist on excess coverage. But for the average system, which has tried to set its rates at just about the income level it needs to pay its bills, banks will not be comfortable without at least some excess income.

The coverage ratios required, however, in most instances, are fairly reasonable. Required ratios of 1.1x, 1.15x and 1.2x are common. Ratios in excess of 1.2x are very uncommon.

As you can see, bank loans are quite straightforward and free of red tape. Their drawbacks are that they may not cover total project costs and they may be at higher, taxable rates.

Chapter 19

The Water and Waste Disposal Program
of the Farmers Home Administration
of the U.S. Department of Agriculture

The Water and Waste Disposal Program (WWDP) of the Farmers Home Administration (FmHA) of the U.S. Department of Agriculture (USDA) is one of the oldest and most successful financing programs in the country. Founded under the Consolidated Farm and Rural Development Act in 1940, the WWDP was part of the New Deal's effort to alleviate some of the financial burdens of rural America.

For the purposes of the WWDP, an area is currently defined as rural if it has fewer than 10,000 inhabitants, according to the last census. Eligible applicants for financial assistance through the WWDP are municipalities, special purpose authorities or districts and not-for-profit associations (cooperatives), which issue taxable or tax-exempt municipal bonds which are purchased by FmHA.

The WWDP is funded each year by appropriation. At the beginning of the 1980's, the appropriation for the WWDP was $1 billion. During the eighties, the appropriation dropped to $350 million. By the early nineties, it had bounced back to the $600 million level.

Since it began in 1940, FmHA has approved over 30,000 loans totalling over $10 billion.

The WWDP is actually both a grant and loan program. In the early 1990's, approximately $200 million of grant funds are being appropriated for the program in addition to the loan funds. Grant funds are used to meet the needs of communities which cannot afford badly needed water or wastewater projects.

For example, let us say that a state health department orders a sewer district to build a new $3 million treatment plant, which would require

211

annual debt service payments of about $300,000 a year.   FmHA determines, based on formulae which take into account population, income level and the sewer rates in adjacent districts, that the ratepayers in the sewer district can only afford to pay $200,000 a year. In such case, FmHA would probably fund the project with a $2 million loan, which would carry an annual debt service payment of about $200,000, and a $1 million grant. Since the grant would require no annual debt service payments, all the system would have to pay would be the $200,000 a year on the FmHA loan.

As indicated above, FmHA's financial criteria for eligibility are based on two major factors. The first is affordability, which is reckoned on the percentage of median household income which constitutes the annual water or sewer charge. The second is how the water or sewer charges will compare to the water or sewer charges in adjacent districts after the rates are increased to pay for the new project.

Obtaining an FmHA loan is not easy. The entire process is quite rigorous. First, a system files a preapplication with the district office of the FmHA. The preapplication includes a thorough project description and financial statements for the last three years. Once the preapplication is filed it is then subject to intergovernmental review within the state. Here local and regional planning bodies are invited to comment on the project. The project must also be vetted by the state historic preservation officer. In addition to other certifications and acknowledgements, the applicant is required to furnish a certificate from an "authorized individual" that private sector financing is unavailable for the project. This is because, under federal law, the FmHA is supposed to be the lender of last resort. The FmHA is not supposed to compete with banks and investment banks. The "authorized individual" can be a sympathetic banker, investment banker, or even, in most cases, the system's own Financial Advisor.

Next, FmHA district office personnel visit and inspect the system and forward their comments and recommendations to the state office. Once a preapplication has been approved by the state office, the applicant is then invited to submit a formal application. FmHA solicits the formal application only when it has funds available for the project.

When the applicant submits its formal application, FmHA will also want to see updated financials as well as copies of the water or sewer rate ordinances establishing service charges at levels sufficient to meet debt service.   FmHA will also require any Water Use Agreements, Connection Ordinances, Service Agreements, or Water Purchase Agreements. In addition, a variety of compliance (Civil Rights, etc.)

documentation is required as are all of the official resolutions and ordinances normally attendant to any municipal bond issuance.

Once FmHA has approved all of the above, the system is notified and the USDA's regional attorney begins working with the system's bond counsel to effect a closing.

As indicated above, each FmHA loan is a free standing municipal bond, similar in all other respects to the municipal bonds issued by any other water or wastewater authority or district.

Finally, it is worthy of note that FmHA usually requires quarterly payments on its bonds, and that the actual servicing of the loan is done by the FmHA district office.

Now, let us look at the seven factors which inflate total project cost and see what bearing they have on the WWDP.

1.) **Term.** The FmHA Water and Waste Disposal Program offers the longest term of any financing program available in the country.

Terms of up to <u>forty years</u> are available. FmHA may insist upon shorter terms if the service life of the assets being financed are significantly shorter than forty years. But in the main, FmHA will allow forty year terms for most major projects.

2.) **Rates.** FmHA lends money to water and wastewater systems at three different rates. They are called the Market Rate, the Poverty Rate and the Intermediate Rate.

The Market Rate is determined quarterly with reference to one of the major national indices of rates for tax-exempt bonds. On a daily basis <u>The Bond Buyer</u>, which is the bible of the entire public finance industry, publishes what is known as the "Twenty Bond Index." This is a composite of the rates of interest at which twenty, large, selected, investment grade-rated bonds traded on the preceding business day. The FmHA Market Rate is based on the midrange of the "Twenty Bond Index" for the preceding quarter.

The Market Rate is offered to eligible systems where the median household income of residents within the system's service area is above the state median household income level.

The Poverty Rate is 5%. It is available to eligible systems where the median household income of the residents of the district is below 85% of the state median income level.

The Intermediate Rate is available to eligible systems where the median household income is between 85% and 100% of the state median household income level. The Intermediate Rate is the arithmetic average of the Market Rate and the Poverty Rate.

3.) Financing Cost. The financing costs associated with the WWDP are limited to bond counsel fees. The FmHA requires what is known as a "red book" bond counsel. A "red book" bond counsel just means a "nationally recognized" bond counsel. Nationally recognized bond counsel are listed in The Bond Buyer's Guide to Municipal Finance. The cover of this book has been, traditionally, bright red. Hence, "red book" bond counsel.

The point here is that the red book bond counsel tend to be larger firms in larger cities; and they tend to be expensive. The reasons for this as well as an explanation of the role of the bond counsel is set forth in Chapter 11.

A word about the use of Financial Advisors is in order. FmHA does not require that systems use Financial Advisors to assist them. However, since the WWDP is only one of several financing options open to water and wastewater systems, it is still advisable for systems to retain FA's. Even once the financing options have been narrowed down to the WWDP, a Financial Advisor will generally prove to be well worth the cost.

4.) Delay. Getting an FmHA loan can take up to three years. The reasons for this delay are not all straightforward.

There is one cause of delay which is straightforward. That is the time it takes to do the Environmental Impact Statement (EIS). The time it takes to do the EIS is directly related to the environmental complexity of the project. If, for example, your system is just replacing a forty year old standpipe, then the environmental impact should be nil, and the EIS should be able to be prepared with alacrity. If, on the other hand, your system wants to lay a couple thousand yards of sewer main through a game preserve located on an ecologically sensitive wetland......well, I certainly would not hold my breath waiting for that EIS.

The other cause of delay is harder to describe. And it certainly does not show up in the program literature or the regulations. It is not caused by bureaucracy. Although the USDA is a mammoth organization, there is little bureaucratic delay involved in the FmHA program. Nor is there anything sinister about such delays.

The delays seem to emanate from the state FmHA staff's attempts to match the funds available with the preapplications in various stages of completion. Need is a factor. How long the preapplication has been pending is a factor. The cost of the project in terms of the number of people to be served may also be a factor. Finally, the sheer size of the project in terms of the funds allocated to the state for that year is also a factor.

The point is that projects may face considerable delay in obtaining funding for a variety of reasons beyond the control of the water or wastewater system's management. Since that is the case, system managements are well advised to sit down with their FA and the FmHA state and district staff to work out a realistic assessment of when a particular project is likely to be funded.

5.) **Imposed Costs.** FmHA does require "red book" bond counsel, as described above, which is expensive. The EIS may also be expensive, as described above, depending on how much the project impinges on environmentally sensitive areas.

But other than these two requirements, the WWDP does not impose any other costs on applicant water and wastewater systems.

6.) **Eligibility.** Happily, all standard project costs are eligible for funding under the WWDP. As previously noted, the WWDP loan, in most cases, takes the form of a tax-exempt municipal bond. In such case, you may wish to consult the discussion of Eligibility in Chapter 17, which covers municipal bonds.

7.) **Coverage.** For the first borrowing which a water or wastewater system makes through the WWDP, The FmHA does not require rates to be set any higher than is necessary to generate cash available for debt service equal to the system's annual debt service payments. FmHA does, however, require the funding of a modest debt service reserve over a ten year period. This means that cash available for debt service must be sufficient to include the annual contribution to this fund, as well as the annual debt service payment on the loan.

On subsequent borrowings which share the same system collateral with the WWDP loan, however, the FmHA requires a 1.2x debt service coverage ratio. This means that systems will have to set rates so that the cash available for the annual debt service payments on both loans exceeds the annual debt service payments on both loans by 20%.

This requirement, as previously stated, applies only to those subsequent loans which are on a parity with the collateral of the original FmHA loan. For subsequent loans whose collateral is subordinate to that of the original WWDP loan, the FmHA does not require this additional coverage.

*   *   *   *   *

In all, as you can see from the above, apart from the problem of possible extensive delays in funding, the WWDP is an extremely attractive program. Eligible systems, with populations of fewer than 10,000, are well advised to thoroughly review this financing option.

Chapter 20

The State Revolving Fund Program of the U.S. Environmental Protection Agency

The State Revolving Fund (SRF) Program is one of the newer funding programs available. Unfortunately, it is <u>only available for wastewater projects</u>. In addition, it is not available for privately owned wastewater systems. Only communities and public wastewater districts may apply.

The Clean Water Act of 1987 essentially replaced the old and widely known Title II grant program with a new revolving loan program established under Title VI.

Under the old Title II program, literally billions of dollars of outright federal grants had been made available to communities and wastewater systems for the construction or improvement of their water treatment facilities. Under the 1987 Act, some $18 billion was authorized for the grant program as well as for the new SRF Program for the years 1988-1994. Most of these funds, however, will not result in subsequent grants to local communities or wastewater systems. Rather the funds will be made available through the SRF Program in the form of loans.

Under the SRF Program, the U.S. Environmental Protection Agency makes grants to the states based on the population of the state. Under the terms of the grants, the states are constrained to organize loan programs to make the funds available to local communities and public wastewater districts. In addition, for every five dollars of federal grant funds, the states are required to contribute an additional one dollar. So, a federal grant of $25 million to a state will result in that state's having to contribute an additional $5 million, for a total SRF lending capacity of $30 million.

These funds are then loaned to communities and wastewater districts within the states for certain capital improvement projects. The beauty of this program is that when the loans are repaid by the communities and wastewater districts to the SRF, the SRF does not, in turn, repay the principal to the federal government. In short, the funds always remain within the state and available for further wastewater treatment projects.

The concept behind the SRF is twofold. First, from a public policy perspective, giving a community a loan, which must be repaid, as opposed to a grant, which does not have to be repaid, serves notice on the residents of that community that there is an actual dollar cost associated with every gallon of wastewater they produce.

Second, from a purely economic point of view, as a loan made to Community A is repaid, those funds become available for a subsequent loan to Community B. Assuming the states charge at least some rate of interest on their SRF loans, and that defaults and delinquencies are either minimal or nil, the pool of money in the individual SRFs will continue to grow steadily over the years. Hence, the fact that these federal grants to the states can be perpetually re-loaned is why the program is called a "revolving" loan fund.

Depending on the particular policies of the individaul states, the actual funding mechanism may either be a loan made to a community or wastewater system or a tax-exempt municipal bond issued by that community or system.

It is important to note that although most of the rules and regulations which cover the SRF program are federal, the administration of the program is done on a state by state basis. It is, therefore, most important that wastewater systems which are contemplating projects contact the state agency responsible for the program as early as possible in the planning process.

1.) **Term.** The maximum term of SRF loans may not exceed twenty years. Terms may be shorter if the service life of the assests being financed are significantly less than twenty years.

2.) **Rates.** The interest rates on individual loans are entirely within the discretion of the states. They may be as low as 0% or as high as the market rates for tax-exempt bonds. (They may actually be higher, but to date no state has attempted this.)

Please also note that even if the state sets the interest rate at 0%, the annual debt service payment on a twenty year loan will be 1/20th of the principal amount, or 5% of the amount borrowed.

3.) **Financing Cost.** The financing costs of the SRF depend on whether the individual states require local systems to issue their own tax-exempt municipal bonds.

If not, financing costs can be limited to the Financial Advisor's fees and the outside counsel's fees for reviewing the state's loan documentation.

If the local system must issue bonds, but the SRF is the purchaser of the bonds, then financing costs will be limited to the fees of the Financial Advisor, the outside counsel and the bond counsel.

If, on the other hand, the state program provides a subsidy for a tax-exempt bond issued by a local wastewater authority, and the bond is sold on the open market, then substantial financing costs may be involved. In addition to the fees of the Financial Advisor, the outside counsel, and the bond counsel, there will be underwriter's discounts or bond placement agent's fees, and investment banker's counsel fees as well as printing and miscellaneous costs. Furthermore, there may even be rating agency fees and bond insurance premiums. Bond insurance premiums alone may add from 1% to 6% to the cost of the project.

4.) **Delay.** Although the SRF program has supplanted the old grant program under Title II, in enacting the new SRF program under Title VI, the Congress incorporated many of the Title II requirements for eligibility. To comply with these standards, an extraordinarily long planning and design procedure is required. This adds at least one year to the period before which funding through the SRF can be obtained. In addition, until 1995, the federal grants which initially fund the SRF are subject to annual appropriation by Congress. And, since the states generally use priority lists to allocate their limited funds, there may be considerable additional delays before funding. Furthermore, to the extent that a local system's need for building its new project is at variance with the priorities of the state agency which administers the SRF, a local wastewater system may face several years of delay.

5.) Imposed Costs. In addition to the costs occasioned by the delays caused by the Title II requirements, these mandates, in and of themselves, add substantial cost to the program.

For example, one major element of additional cost is the imposition of the so-called Davis-Bacon Act requirements. The Davis-Bacon Act is a federal law which requires that union scale wages be paid to craftsmen working on the project. The problem here is that, outside of major urban centers, there may be no local union scale for craftsmen's wages. This fact does not relieve the wastewater system from the onus of the law. Rather, the law may require that the union scale craftsmen's wages in a major metropolitan center hundreds of miles away be used as the wage scale on the project. This requirement alone can add literally hundreds of thousands of dollars to the cost of a large project.

Furthermore, there are <u>seventeen</u> such Title II requirements which apply to the SRF. Experts within EPA, who are familiar with both Title II and the SRF have estimated that these seventeen requirements <u>add between 5% and 20% to the total cost of the project.</u>

6.) Eligibility. In addition to the increases in project cost occasioned by the Title II requirements, there are some costs which are simply not able to be financed, at all, under the SRF program because they are ineligible. The most obvious and significant of these ineligible costs are land for plant sites as well as the easements and rights-of-way for collector systems. As discussed in Chapter 14, these costs, which are ineligible under the SRF program, can prove to be devilishly difficult to finance by other means. The ineligibility of such costs under the SRF program may well mean that systems electing to finance projects under this program may have to fund these costs through short-term commercial bank loans at taxable rates.

It is worthy of note that, if only 10% of the cost of a particular project were not eligible for funding under the SRF program, and if the the ineligible 10% had to be financed over a five year term with a commercial bank loan at taxable rates, the payments on the ineligible 10% alone would equal nearly one-third of total annual debt service payments on the project!

The alternative is even worse: no financing at all. In such case, the wastewater system undertaking the project would have to levy a large, highly unpopular, one-time assessment on the system's ratepayers.

7.) **Coverage.** If the state administrative agency opts to organize the SRF as a subsidy program involving the issuance of tax-exempt municipal bonds by individual wastewater systems, and if the state does not purchase these bonds directly, then either the investment banker who underwrites or places the bonds or a bond rating agency may impose excess coverage ratios as a condition of bond issuance, as well. These matters have nothing directly to do with the rules and regulations of the SRF program itself. Rather, they are simply an exigency of the issuance of any tax-exempt municipal bond. As such, you may well want to consult the appropriate sections of Chapters 15 and 17, which discuss Coverage and Municipal Bonds, respectively.

*   *   *   *   *

As you can see from the above illustrations of the seven factors which serve to inflate the costs of wastewater projects, the SRF program has numerous difficulties which can add considerable amounts of money to standard project costs. Nonetheless, it must be noted that the predecessor program of the SRF, the old Title II grant program, was one of the most widely used means of funding wastewater projects across the country.

The SRF program is intended to supplant the old grant program. As previously mentioned, its principal purpose is to begin instilling in the American public the true cost of disposing of every gallon of wastewater. The old grants came out of everyone's federal tax dollars. As far as the ratepayers of a system which received a grant were concerned, it was painless. It was so, because all of their fellow taxpayers in the country were footing the bill along with the individual ratepayers. But these days are now gone. Now, the SRF program requires ratepayers to recognize and pay for the costs of collecting and treating the wastewater they produce. As such it a valuable program; and it is one which every wastewater system should examine as a funding alternative for needed capital improvement projects.

The SRF program does have its difficulties; but the EPA staff and the states are working to get the Congress to ameliorate these problems. As such, each wastewater system contemplating a major project would be well advised to retain a Financial Advisor who is familiar with the SRF program, and to investigate thoroughly whether this financial program is suitable to fund, on a cost effective basis, the project at hand.

Appendix A

The following are the steps involved in calculating the Annual Debt Service Payment for a Level Payment Loan with the terms and conditions described below:

Principal Amount - $100,000

Interest Rate - 10%

Term - 5 Years

1.) Take the interest rate (10%) expressed as a decimal (.1) and add it to the number 1, as follows: .1 + 1 = 1.1.

2.) Take 1.1 and divide it <u>into</u> the number 1, as follows: 1 / 1.1 = .9091.

3.) Take .9091 and multiply it by itself as many times as there are years in the loan term (5). In other words, multiply .9091 by itself five times, as follows: .9091 x .9091 x .9091 x .9091 x .9091 = .6209.

4.) Subtract this number (.6209) from the number 1, as follows: 1 - .6209 = .3791.

5.) Take this number (.3791) and divide it <u>into</u> the interest rate expressed as a decimal (.1), as follows: .1 / .3791 = .2638.

6.) Finally, take this number (.2638) and multiply it by the Principal of the loan ($100,000), as follows: 100,000 x .2638 = 26,380. This number, in dollars ($26,380), is the Annual Payment.

At this point, you should be very proud of yourself because you have just successfully used the following formula:

$$AP = P \times \frac{i}{1 - \left[\frac{1}{1+i}\right]^n}$$

Appendix B

The following are the steps for calculating the Annual Debt Service Payment on a Hewlett-Packard 12C calculator for a Level Payment Loan with the terms and conditions described below:

Principal Amount - $100,000

Interest Rate - 10%

Term - 5 Years

1.) Key in the Principal Amount, as follows: 100000.

2.) Press the "Change Sign" key (CHS).

3.) Press the "Present Value" key (PV).

4.) Key in the number of periods, as follows: 5.

(N.B. Please recall that throughout this book we are assuming loans with annual principal payments. Thus the number of periods is equal to the number of years in the term.)

5.) Press the "Number of Periods" key (N).

6.) Key in the Interest Rate, in a non-decimal format, for the number of periods.

(N.B. A non-decimal format means that 10% is written 10, and not .1. In addition, the "interest rate for the number of periods" means that, assuming the interest rate is an <u>annual</u> interest rate (which it always is), then you must divide the interest rate by the number of periods in a year.

For example, if you are calculating the annual debt service payment on a loan with a 10% rate, but <u>monthly</u> payments, then you would divide the rate (10) by the number of periods in a year (12), or 10/12, which equals .83.)

7.) Press the "Interest Rate" key (i).

8.) Press the "Payment" key (PMT), and the calculator will calculate and display the Annual Debt Service Payment which is $26,380.

Appendix C

Annual Debt Service Payments on a $1000 Level Payment Loan at Various Rates and Terms

Rate	Term								
	5	10	15	20	25	30	35	40	Years
5.00%	231	130	96	80	71	65	61	58	
5.25	233	131	98	82	73	67	63	60	
5.50	234	133	100	84	75	69	65	62	
5.75	236	134	101	85	76	71	67	64	
6.00	237	136	103	87	78	73	69	66	
6.25	239	137	105	89	80	75	71	69	
6.50	241	139	106	91	82	77	73	71	
6.75	242	141	108	93	84	79	75	73	
7.00	244	142	110	94	86	81	77	75	
7.25	246	144	112	96	88	83	79	77	
7.50	247	146	113	98	90	85	81	79	
7.75	249	147	115	100	92	87	84	82	
8.00	250	149	117	102	94	89	86	84	
8.25	252	151	119	104	96	91	88	86	
8.50	254	152	120	106	98	93	90	88	
8.75	255	154	122	108	100	95	92	91	
9.00	257	156	124	110	102	97	95	93	
9.25	259	158	126	112	104	100	97	95	
9.50	260	159	128	113	106	102	99	98	
9.75	262	161	130	115	108	104	101	100	
10.00%	264	163	131	117	110	106	104	102	

Table D-1   Table of Increases in True Interest Cost (TIC)

Rate - 5%

Rate	Term								
	5	10	15	20	25	30	35	40	Years
1	.36	.21	.15	.12	.10	.09	.08	.07	
2	.72	.41	.29	.23	.20	.17	.16	.15	
3	1.08	.61	.44	.35	.30	.26	.24	.22	
4	1.44	.82	.59	.47	.40	.35	.31	.29	
5	1.79	1.02	.73	.58	.49	.43	.39	.36	
6	2.14	1.22	.87	.70	.59	.52	.47	.43	
7	2.50	1.42	1.02	.81	.69	.60	.55	.50	
8	2.85	1.62	1.16	.93	.78	.69	.62	.58	
9	3.20	1.81	1.30	1.04	.88	.77	.70	.65	
10	3.55	2.01	1.44	1.15	.98	.86	.78	.72	

Table D-2   Table of Increases in True Interest Cost (TIC)

Rate - 6%

Rate	5	10	15	20	25	30	35	40	Years
				Term					
1	.37	.21	.15	.12	.11	.09	.09	.08	
2	.73	.42	.30	.25	.21	.19	.17	.16	
3	1.10	.63	.46	.37	.31	.28	.25	.24	
4	1.46	.84	.61	.49	.42	.37	.34	.32	
5	1.82	1.04	.76	.61	.52	.46	.42	.39	
6	2.18	1.25	.91	.73	.62	.55	.51	.47	
7	2.54	1.45	1.05	.85	.73	.65	.59	.55	
8	2.89	1.66	1.20	.97	.83	.74	.67	.63	
9	3.25	1.86	1.35	1.09	.93	.83	.76	.70	
10	3.60	2.06	1.50	1.21	1.03	.92	.84	.78	

Table D-3   Table of Increases in True Interest Cost (TIC)

Rate - 7%

Rate	Term								
	5	10	15	20	25	30	35	40	Years
1	.37	.22	.16	.13	.11	.10	.09	.09	
2	.74	.43	.32	.26	.22	.20	.18	.17	
3	1.11	.64	.47	.38	.33	.30	.27	.26	
4	1.48	.86	.63	.51	.44	.40	.36	.34	
5	1.85	1.07	.78	.64	.55	.49	.45	.43	
6	2.21	1.28	.94	.76	.66	.59	.54	.51	
7	2.58	1.49	1.09	.89	.77	.69	.63	.60	
8	2.94	1.70	1.24	1.01	.89	.79	.72	.68	
9	3.30	1.91	1.40	1.14	.98	.88	.81	.77	
10	3.66	2.12	1.55	1.26	1.09	.98	.90	.85	

Table D-4   Table of Increases in True Interest Cost (TIC)

Rate - 8%

Rate	Term								
	5	10	15	20	25	30	35	40	Years
1	.38	.22	.16	.13	.12	.11	.10	.09	
2	.76	.44	.33	.27	.23	.21	.20	.19	
3	1.13	.66	.49	.40	.35	.32	.29	.28	
4	1.51	.88	.65	.53	.46	.42	.39	.37	
5	1.88	1.10	.81	.67	.58	.53	.49	.46	
6	2.25	1.31	.97	.80	.69	.63	.59	.56	
7	2.62	1.53	1.13	.93	.81	.73	.68	.65	
8	2.99	1.74	1.29	1.06	.92	.84	.78	.74	
9	3.36	1.96	1.45	1.19	1.04	.94	.88	.83	
10	3.72	2.17	1.60	1.32	1.15	1.04	.97	.92	

Table D-5   Table of Increases in True Interest Cost (TIC)

Rate - 9%

Rate	Term								
	5	10	15	20	25	30	35	40	Years
1	.38	.23	.17	.14	.12	.11	.11	.10	
2	.77	.45	.34	.28	.25	.22	.21	.20	
3	1.15	.68	.50	.42	.37	.34	.32	.30	
4	1.53	.90	.67	.56	.49	.45	.42	.40	
5	1.91	1.12	.84	.69	.61	.56	.53	.50	
6	2.29	1.35	1.00	.84	.69	.67	.63	.60	
7	2.66	1.57	1.17	.97	.85	.78	.73	.70	
8	3.04	1.79	1.33	1.11	.97	.89	.84	.80	
9	3.41	2.01	1.50	1.24	1.09	1.00	.94	.90	
10	3.78	2.22	1.66	1.38	1.21	1.11	1.04	1.00	

Table D-6   Table of Increases in True Interest Cost (TIC)

Rate - 10%

Rate	Term								
	5	10	15	20	25	30	35	40	Years
1	.39	.23	.17	.15	.13	.12	.11	.11	
2	.78	.46	.35	.29	.26	.24	.23	.22	
3	1.17	.69	.52	.44	.39	.36	.34	.33	
4	1.55	.92	.70	.58	.52	.48	.45	.43	
5	1.94	1.15	.87	.73	.65	.60	.56	.54	
6	2.32	1.38	1.04	.87	.77	.71	.68	.65	
7	2.70	1.61	1.21	1.01	.90	.83	.79	.76	
8	3.08	1.83	1.38	1.16	1.03	.95	.90	.87	
9	3.46	2.06	1.55	1.30	1.15	1.07	1.01	.97	
10	3.84	2.28	1.72	1.44	1.28	1.18	1.12	1.08	

Appendix E

The following are the steps necessary to calculate the <u>increase</u> in True Interest Cost (TIC) on a Hewlett-Packard 12-C calculator for the example below:

Project Cost - $100,000

Additions to Project Cost - $8,000

Interest Rate - 10%

Term - 5 Years

1.) Key in the Project Cost, as follows: 100000.

2.) Press the "Enter" key.

3.) Key in the Additions to Project Cost, as follows: 8000.

4.) Press the "Addition" key (+). At this point, the display should read 108,000.

5.) Press the 'Change Sign" key (CHS).

6.) Press the "Present Value" key (PV).

7.) Key in the number of periods, as follows: 5. (For explanation of "periods" see Appendix B.)

8.) Press the "Number of Periods" key (N).

9.) Key in the Interest Rate, in a non-decimal format, for the number of periods, as follows: 10.   (For an explanation of this procedure, please see Appendix B.)

10.) Press the "Interest Rate" key (i).

11.) Press the "Payment" key (PMT).

12.) Key in the Project Cost, again, as follows: 100000.

13.) Press the "Change Sign" key (CHS).

14.) Press the "Present Value" key (PV).

15.) Press the "Interest Rate" key (i).

16.) Key in the Interest Rate, again, in a non-decimal format, as follows: 10.

17.) Press the "Subtraction" key (-). At this point, the calculator will be displaying "3.08," which is the increase in the True Interest Cost caused by the Additions to Project Cost. In other words, the true cost of borrowing is 13.08%, not 10%.

Appendix F

Principal Component of Annual Debt Service Payment for $1,000 Level Payment Loans

Rate	Term								
	5	10	15	20	25	30	35	40	Years
5%	$181	80	46	30	21	15	11	8	
6	177	76	43	27	18	13	9	6	
7	174	72	40	24	16	11	7	5	
8	170	69	37	22	14	9	6	4	
9	167	66	34	20	12	7	5	3	
10%	$164	63	31	17	10	6	4	2	

Index

1   Orang Laut

People of the sea

The scope of this book

The Riau Archipelago, which lies to the south of Singapore and is now part of modern-day Indonesia, was at one time a Malay sultanate. It is still home to a scattered population of roving fishing communities known locally, in Malay-Indonesian phraseology, as 'Orang Suku Laut' or 'Orang Laut', which is literally translated as 'people of the sea'. The people have been called 'Sea Nomads' in English-language writings on account of their full-blown nomadic way of life. The Orang Laut consider themselves to be the *'orang asli'* (indigenous people) and *'asli Melayu'* (indigenous Malays) of a vast maritime world known as the *'Alam Melayu'* (Malay World) which includes Riau. For centuries, the sea and coastal areas have been their life and living spaces. Nonetheless, the Indonesian and Singaporean authorities, like other governments attempting to administer nomadic and migratory people – be they of the desert, jungle or sea or the urban homeless – persist in seeing the Orang Laut as lacking a fixed attachment to housing in permanent villages – that is, not 'sedentarized'. The developmental pressure thus generated is one of the issues discussed in this book, along with those of social assimilation as citizens, territorial rights, religious conversion and cultural identity.

At this point, it may be useful to say something about the organization of this book. A text of this sort could have been structured in a variety of ways. The story of the impact of strategies for development on a people's life cannot easily be divided into particular categories of experience or discrete dimensions of social and cultural change. Therefore, although united by a set of common themes, each of the chapters in this book is a self-contained essay, which may, if desired, be read and understood on its own. I have arranged them in an order that I think most effectively brings out the connections between them. Nevertheless, the order is somewhat arbitrary and the reader need not feel constrained to follow it.

Chapter 1 begins the story by outlining the aims of the book and introduces the Orang Laut by presenting some basic ethnographic information on their area of habitation, their demographic data and population changes, as well as their ethnolinguistic background.

Chapter 2 traces the Orang Laut's social organization from as early as the sixteenth century to today. It examines the changes in their social organization and how these have affected the way they identify themselves. Issues pertaining to the formation, expansion and break up of village communities, households and the nuclear family are discussed.

Chapter 3 focusses on the former cultural and political construction of the Orang Laut as honoured subjects of Malay sultans and the later re-interpretation of their roles by Western colonialists. This provides the background to understanding how and why the Orang Laut have come to be marginalized and excluded from the wider Indonesian and Malay societies.

Chapter 4 looks at how Orang Laut living and working spaces are based on a system of collective ownership of territory that is intertwined with their kinship structure. This chapter examines their system of tenure and how it has functioned to sustain their lives for centuries. Indigenous title deeds are also discussed.

Chapter 5 examines the Orang Laut fishing economy. Their houseboat functions as both living space and production site. Both men and women are skilful spearers of fish and other maritime species. This chapter not only presents ethnographic data on their material and intellectual technology, but also shows how fishing is so important to them that it forms a central part of their identity. For them, there is no distinction between religion and economy; both the spiritual and physical worlds are part of the same living process.

Chapter 6 explores how the Indonesian government sees the cultural refinement – in this case the Islamization – of the Orang Laut as necessary for nation building to be successful.

Chapter 7 turns to another aspect of the Indonesian government's efforts for nation building in this area: economic development. This chapter scrutinizes the implementation of the Growth Triangle in Riau and delves into the details of what has been determined as an economic programme of globalization, internationalization, and supranationalization on the space that has for centuries had a deep emotional and personal meaning for the Orang Laut. Changes that have come about, such as border manipulations to create what have since been referred to as borderless zones, are discussed.

Chapter 8 concludes by looking at how global demands for strand and sea resources confront the Orang Laut with new forms of territorial tenure which either threaten their livelihoods or exclude them from their habitats and resource zones. It also examines how they have responded to these challenges and the adjustments that lie ahead of them.

Terminologies

Debates have raged over the most appropriate English term of reference for the Orang Laut. Various translations have appeared, including 'Sea Gypsies' (Thompson 1851: 140, Foley 1983: 38), 'Sea Jakun' (Skeat and Blagden

1906: 87), 'Sea Folk' (Sopher 1977: 47), 'Sea Hunters and Gatherers' (Sopher 1977: 218), 'Sea People' (Bradley and Harlow 1994: 166), 'People of the Sea' (Sandbukt 1982) and *Suku Sampan* or 'Boat People' (by some Indonesians). However, all of these terms have proven to lack scientific and ethnographic merit as they present difficulties owing to their narrow connotations, mis-identifications or erroneous ethnological implications. Of these, 'Sea Nomads' has steadfastly remained the term preferred by the general public, researchers and officials alike. This is not to say that the term 'nomad' has not been disputed. Opponents to this appellation maintain that it is etymologically derived from the Greek expression *nomás*, which means 'moving around and living on pasturage', so the word is irrelevant for maritime communities. Proponents of the term nonetheless argue that contemporary anthropo-logical usage has broadened its meaning to include all groups of people who practice spatial mobility to enhance their well-being and survival. It is interesting to note that the meaning of 'Sea Nomads' is so broad that it is convenient enough to mean not only the Orang Laut of the Riau-Lingga Archipelago, but also the Bajau Laut of the Sulu Archipelago of the Philippines, eastern Borneo, Sulawesi and the islands of eastern Indonesia, as well as the Moken and related Moklen of the Mergui Archipelago of Burma and southwestern Thailand.

The use of the shorter Malay term 'Orang Laut' instead of the longer form, 'Orang Suku Laut' as a term of reference has also been an issue of some controversy. Although it has been argued that Orang Suku Laut is a more accurate term to refer to the sea nomads because in Malay, the term 'Orang Laut' could apart from referring to sea nomads also include 'ordinary' coastal-dwelling Malays who are engaged in maritime occupations (Lenhart 1997; Sather 1998: 21). I shall maintain usage of the shorter term in this book because, in my fieldwork experiences, my seafaring friends preferred the briefer term 'Orang Laut' in their daily conversations. In the company of people with whom they feel at ease, they often simply speak of themselves as 'Orang Laut'. It is also noticeable that they try to avoid this term of reference if and when they feel themselves to be in the company of people who associ-ate this term with the insulting connotations of being backward and unprogressive. Under such circumstances, they will make a situational switch and identify themselves as indigenous Malays of the region.

Areas of habitation

Today the Orang Laut are found scattered throughout the Riau-Lingga Archipelago of Indonesia, the east coast of Sumatra, the larger islands of Bangka and Belitung in south Sumatra and the southern coasts of the Malay Peninsula (Johor). In fact, until recently, the island of Singapore was also home to many Orang Laut.

The Riau-Lingga area where most Orang Laut live today is rapidly being transformed into an international economic zone for trade and industrial development. The Orang Laut are therefore in a congeries of local and global interactions. This area is situated close to the equator. A hot and wet climate prevails throughout the year. Temperatures during the day range from 28 to 38 degrees Celsius. Nights are slightly cooler at 25 to 30 degrees Celsius. The Orang Laut divide the year into four seasons, namely the seasons of North Winds (October through January), East Winds (February through May), South Winds (June through August) and West Winds (September). The monsoon season begins towards the end of the year and heavy rains continue to fall throughout January. The rest of the year is generally dry with intermittent showers. Temperatures soar in the hottest period of May through August.

Population

Riau has a complex geography and history. Geographically, Riau denotes both mainland and archipelago Riau. The mainland consists of the watersheds of four major river systems on eastern Sumatra (the Rokan, Kampar, Siak and Inderagiri and the immediate offshore islands). The archipelago comprises over three thousand islands stretching from the Strait of Malacca over the South China Sea to Borneo. The total population of Riau is 4,733,948 (<http://www.riau.go.id/penduduk/penduduk.php>). Around 725,865 people inhabit archipelago Riau, while the rest are to be found on Riau mainland (ibid.).

The ethnic composition of Riau includes – apart from the Orang Laut, who live in the sea and near river mouths – a mix of Malays (who mostly live in the capitals of former sultanates), tribal groups (such as Sakai, Talang Mamak, Bonai and Akit who collect forest products), the Bugis (who came from Sulawesi), the Banjarese migrants from Southeast Kalimantan (who settled on Sumatra's east coast in the late nineteenth and early twentieth centuries), the Javanese transmigrants (who have moved because of the state's transmigration programme), the Batak (who came from North Sumatra), the Minangkabau, Chinese, Flores and a whole host of other ethnic groups (Colombijn 2003: 342).

Only a few statistical data on the Orang Laut population are available, but they are insufficient in number and of such inconsistency as to render impossible even the crudest demographic analysis. Table 1.1 presents estimates from

Table 1.1 Orang Laut population figures, Riau

DBMT 1990	Kantor Sosial 1993	Edhie Djatmiko 1993	Kantor Sosial 1993 (verbally communicated to Chou)	DBMT 1994/ 95	Safian Hussain 1999	Lenhart 2002[1]
5,402	4,580	9,582	3,309	7,179	12,000	3,000–5,000

the *Direktorat Bina Masyarakat Terasing* (DBMT) (1990, 1994/95), which was a division of the Indonesian Republic's Social Department, *Kantor Sosial* (1993) (Office of the Department of Social Affairs) of the Riau Regency and researchers Edhie Djatmiko (1993), Safian Hussain (1999), Lioba Lenhart (2002) and myself (1993).

These figures are highly disputable for several reasons. The Orang Laut are constantly moving around and the lax system of registering births and deaths in the archipelago accounts largely for the imprecision of their population figures. In fact, not all areas inhabited by them are even known to government officials. Furthermore, government officials regard the Orang Laut as a backward people who are a source of embarrassment to a modern state, hence the deliberate lowering of their population figures. It therefore remains uncertain how many Orang Laut there are today.

Appearance

Indonesians often describe the Orang Laut as having a 'black' complexion. They are invariably spoken of as having a general appearance of being 'uncouth' and 'hideous' because of their 'kinky hair', 'scaly skins', and 'bow-leggedness'. This deformity is said to have resulted from extensive use of their toe to hold the wooden cross-bar while steering their boat.

Such subjective observations, which are undoubtedly very much influenced by attitudes of prejudice, are subject to much qualification. The Orang Laut are of a dark complexion. However, they are no darker than most Malays and it is almost impossible to distinguish one from the other. There is no evidence of the Orang Laut being bow-legged or scaly-skinned, and their hair varies from wavy to absolutely straight. They are generally of the same height and weight as the other island dwellers, averaging approximately 1.6 metres and 60 kilogrammes. They are also just as handsome. Some even had a mixture of Chinese-looking features due to intermarriage.[2]

Language

Orang Laut mobility brings them into contact with various ethnic communities. More often than not, in their interaction with these communities, they are multilingual. They speak a mixture of Riau Malay, Bahasa Orang Laut ('Orang Laut Language' and 'Language of the Sea Nomads'), subdialects pertaining to different locales and Teochew.

'Riau Malay', that is Malay with 'perhaps some local pecularities', as described by Voorhoeve (1955: 22), is widely spoken in Riau. A study by Foley in 1983 highlighted the problem that a full dialect survey of Sumatra Malay that includes Riau Malay has yet to be carried out and it is 'quite likely that several distinct languages [are] simply collapsed under the term, "Malay dialects"' (Foley 1983: 38). A more recent study by Safian Hussain (1999: 639) identifies two types of Riau Malay, namely, Archipelago Riau Malay

Dialect (consisting of 18 sub-dialects) and Mainland Riau Malay Dialect (comprising 13 sub-dialects).

Studies on the language of the Orang Laut are lacking because it has been difficult to locate their exact sites (Foley 1983).[3] The few studies that have been done are divided in their opinion about the relationship of the language of the Orang Laut to Riau Malay. Bradley and Harlow (1994: 166) studied the language of the Orang Laut, whom they call 'Para Malay', and determined that they speak 'divergent dialects' and 'local varieties' of Malay that are closely related to Riau Malay. However, Kähler (1960) who offers one of the most detailed analyses of the language of the Orang Laut, emphasizes that there is a considerable difference between their language and Riau Malay. Voorhoeve (1955: 23) is of a similar opinion. Although the language of the Orang Laut has been distinguished as 'Bahasa Orang Laut' by Esser (1951: 1), Foley (1983: 38) finds that it is impossible to even ascertain if the Orang Laut speak one or two languages due to the absence of adequate studies.

David Gil, a linguist who had been working in Riau on an aberrant speech variety that he called, 'Riau Indonesian' was invited by fellow anthropologists Geoffrey Benjamin, Vivienne Wee and myself to Orang Laut fieldwork sites to study the language of the Orang Laut. Upon arriving at Tanjung Berakit on the northeast tip of Bintan Island, he was astounded to hear the Orang Laut speaking something quite different from the speech variety that he had been studying. They were conversing in the normal Johor-Riau-Malay that formed the original basis for the standardized Indonesian language (Benjamin 2002: 64).

My Orang Laut informants are indeed proud to say that they possess their own Bahasa Orang Laut. Each *suku* ('group', a term that is discussed in greater detail in the next chapter) seems to have its own dialect, and it is the sum total of all of these dialects that forms the Bahasa Orang Laut. Differences between dialects may not be very great, but rivalling Orang Laut may pretend not to comprehend each other. Sometimes, indeed, when they want to disassociate themselves from each other, they stress the differences in their dialects by claiming that their speeches have nothing in common at all. The following is a quick illustration of the relatively regular variations between the Suku Mantang, Galang and Barok.

In addition to their dialects, many are also able to converse in *Bahasa Indonesia* (Indonesian) and *Bahasa Melayu* (Malay). The first is the official *lingua franca* of Indonesia, while the latter is spoken by Malays in Singapore, Malaysia and Brunei.

As the Orang Laut move among different territories, they are often also able to speak the 'sub-dialect' (Safian Hussain 1999: 639) of that particular area. Examples are the Orang Laut on the islands of Teluk Nipah and Nanga who are able to speak what they call '*Bahasa Galang*' (Galang language), which is in fact the sub-dialect of the Galang region.

As most middlemen in their trading activities are Teochew-speaking

Table 1.2 Orang Laut dialects

	rumah (house)	*durian (durian fruit)*	*berat (heavy)*
Mantang	Гumah	Nəɣian	Bəɣat
Galang	Umah	Dəyan	Bəat
Barok	Humah	Hian	Bəhat

	daging (meat)	*jaring (dragnet or seine)*	*bahu (shoulder)*
Mantang	Dagiŋ	Jaɣeŋ	Bau
Galang	Dəgiŋ	Jəaŋ	Bəuk
Barok	Dəgeŋ	Jaheŋ	əu

Source: Safian Hussain 1999: 657–658

Chinese, many Orang Laut are also fluent in Teochew. Some women who have worked for Chinese households have learnt to speak Teochew well. They were in fact very keen to let me know that they were multi-lingual as it reflected their vast exposure and knowledge in the maritime world. Therefore, from time to time, our conversations would be interspersed with a combination of different languages and dialects. In Dapur Enam, the Orang Laut even asked if I would teach them and their children English. They were model students and the women, in particular, were eager to learn.

Literacy level

The majority of the Orang Laut have never received any form of formal education. Only a few of them attended schools, and even so, many have not gone beyond a year at the most. This is partly because of their unpredictable fishing hours, but mostly because their children are bullied in schools by the Malay children. I was therefore curious to know how a few of them, in spite of never having attended school, were able to read and write. They explained that they learnt by copying letters from pieces of newspapers used by shop-keepers to wrap items and numbers from clocks. Contrary to the general view that the Orang Laut are not keen to attain a level of literacy, I discovered during the course of my fieldwork that they were indeed keen to learn to read and write from me.

Most of the government's attempts to get Orang Laut children to attend a formal school have failed. However, there was one literacy project which scored some success with the Tiang Wang Kang community. Here, Christian missionaries implemented a flexi-hour programme in which there were no fixed school hours. The Orang Laut were free to fish as and when they needed to. They could receive lessons whenever they were free. Consequently, there are more of them on this island who are able to read and write as compared to those on other islands.

Literature review

Chinese and Arab accounts made the earliest mention of the sea nomads in Southeast Asia before 1500, noting that they dominated the maritime world of the region and were very cruel pirates whom sailors dreaded (cf., Chau 1911 and Ferrand 1913). From the sixteenth to the eighteenth centuries, more information could be glimpsed from European (mainly Portuguese, Spanish and English) records of exploration and travel commentaries (cf., Barros 1777, Pires II 1944, Combés 1667, Valentijn 1724, Forest 1780 and 1972). These writings were highly exaggerated descriptions based on personal 'enquiries and conjectures' (Forrest 1780: Preface). The mariners were a novel ethnographic phenomenon and, although frequent mention was made of their pervasiveness in the waters of the region, their habitat remained unknown to the Europeans. These accounts were also affected by the political climate of the time. After the Dutch displaced the Portuguese from Malacca early in the seventeenth century, no further mention of the sea nomads could be found in Portuguese papers. From the early nineteenth century to the early twentieth century, the study of sea nomads was done mostly by colonial administrators. Two typical descriptions of the mariners are:

> The Sea Gypsies in their boats behaved like wild animals. Whenever they saw a crowd of people coming, if there was time they made off quickly in their boats: if there was not time they leapt into the sea and swam under water like fish, disappearing from view for about half an hour before coming to the surface as much as a thousand yards away from the place where they entered the water. Both men and women behaved like this. As for the children words fail me . . . the children remained wild, to such an extent they became ill with fright at the sight of people.
>
> (Hill 1973: 95–96)

and

> The condition of the lower class of Malays in these parts is wretched beyond what we should conceive to be the lot of humanity in an inter-tropical climate, almost the whole of their life being spent upon the water in a wretched little canoe in which they can scarce stretch themselves for repose. A man and his wife and one or two children are usually to be found in these miserable sampans; for subsistence they depend on their success in fishing.
>
> (Findlayson and Raffles 1826 in Gibson-Hill 1973: 122)

British colonial administrators such as Findlayson and Raffles (1826 in Gibson-Hill 1973: 122) described the sea nomads as a 'lower class of Malays'. There was also the compendium of Malayan ethnography by English ethnologists Skeat and Blagden in 1906. They introduced a classification of peoples based on their perceptions of the physical anthropological distinctions of the primitive tribes of Malaya. Notably, the Orang Laut were mistakenly classified as the primitive forest nomads of the interior of Batam Island, possessing no boats and fearing water. This misinformation seriously distorted the understanding of the boat-dwelling nomad culture and misled later researchers on Malaya, such as Heine-Geldern (1923). Skeat and Blagden (1906) were also writing with a political frame of mind that ignored mainland Johor as forming an ethnographic continuum with the Riau Islands and the east coast of Sumatra. This led them and others towards an ill-informed ethnographic understanding of the sea nomads whose culture-history does not necessarily correlate with constructed political borders and boundaries. Much of the early works on the sea nomads relied on observations from a distance and not much was done to map out and understand their migratory movements. It goes without saying that these early colonial writings were not inclined to focus on issues concerning indigenous conceptions of space and claims to territory, lest they overshadow colonial sovereignty in the area.

A long period of silence followed after the colonial administrative reports. It was almost half a century later, in 1954, that the sea nomads were thoroughly analyzed in a doctoral thesis by David Sopher. It was initially published in 1965 with the title *Sea Nomads: A Study Based on the Literature of the Maritime Boat People of Southeast Asia* and later reprinted with a postscript in 1977 under the slightly revised title of *The Sea Nomads: A Study of the Maritime Boat People of Southeast Asia*. This work has since become a classic in the study of sea nomads. For the first time, a comprehensive study of the boat-dwelling mariners covered over 2,000 miles. Sopher explored beyond the confines of political borders and boundaries to present an overview of their cultural geography, migratory movements and environmental relationships. However, this was a geographical-cum-historical study based on secondary material and it lacked information about the contemporary circumstances of the sea nomads.

Around the time of Sopher's work, two more academically sound publications emerged. They were Hans Kähler's (1960) linguistic study, *Ethnographische und linguistiche Studien über die Orang Darat, Orang Akit, Orang Laut und Orang Utan im Riau-Archipel und auf den Inseln an der Ostküste von Sumatra*, and Nicholas Tarling's (1963) historical study, *Piracy and Politics in the Malay World*. Unfortunately, these better informed studies did not inspire any steady research interest, and for the next decade, research on the sea nomads stagnated.

The 1970s ushered in a golden age for the study of sea nomads. Whether they were historical studies based on sound archival research or ethnographies derived from fieldwork and close contact with the subject of study, scores of books, monographs and articles appeared (see Chou 2006a: 1–11). An invaluable project to compile a comprehensive bibliography for works on the sea nomads was undertaken by the Centre of Southeast Asian Studies, Kyoto, and their findings were published in the first two issues of the *Sama Bajau Studies Newsletter* (1995 and 1996).

This recent body of work is of great value. Much of what has been documented pertains to issues concerning the socialization of time, space and identity. Through sound research techniques and fieldwork participation, there came great accumulation of more accurate and refined ethnographic detail. These works present fundamental theoretical revisions about the pre-history and indigenous culture history of the region. They have opened an understanding of the social organization and cultural premises that structure the sea nomads' daily life, and insights have been developed into questions of identity, social hierarchy and stigmatization at the local, transnational and supranational scales of their mobility.

Two recent works can serve to illustrate some research trends. The first is James Warren's 1998 monograph, *The Sulu Zone: The World Capitalist Economy and the Historical Imagination*. This work focusses on the relation-ship between social groups like the Sama Laut, Iranun, Balangini Samal and Taosug and their physical environment. It develops a critical understanding and discussion of historiographical methods and models in examining the eco-nomic and cultural border zones in a changing global–local context. It reframes and re-presents the ethnohistory of the Sulu zone in its own terms rather than taking its point of departure from the history of Western imperi-alism and expansion in the region. Inspired by historiographical formulations and debates of the neo-Marxist dependency theorists and linking them with Wallerstein's (1974) World System approach and the theories of Latin American economists and Annales historians, Warren (1998: 14–15) demon-strates the evolution of the areas and concept of 'periphery' as a 'series of intersections, encounters or "historical accidents usually due to contact with foreign formations" '. The second work is one of my own essays (2005), which looks at political and social tensions between fixed national boundaries-cum-inflexible state requirements of citizenship in contrast to the flexible require-ments of the Orang Laut. These two examples reflect increasing attempts in recent research to capture and place the dynamics of sea nomads within the wider social science of sovereignty, borders, citizenship and nation states.

Even though great strides have been made in the study of sea nomads in Southeast Asia, much more research is required. This exercise of revisiting older writings and reviewing recent works has opened potential directions that could lead to new ranges of knowledge in the field. Much of the older ethnographic record, which narrowly looked at the organization of sea nomads' travel routes, their techniques for spatial production of locality and the often humdrum preoccupations of small-scale communities, can be reread from other points of view. These issues and ethnographic details can be reconceived as dynamic and exciting interstices of transnational mobility and cultural belongingness.

Southeast Asia, which comprises several bordered nation states, is a space of deep emotional and personal meaning for the sea nomads. They claim ownership of and sovereignty over this entire space in spite of interference posed by current political borders. Their continuing seafaring activities, mobility and sense of multi-local belongingness combine to create an endemic sense of anxiety in the minds of state officials today, just as they did in earlier times for former colonial masters. Their persistent wide distribution throughout the region surely bears to very real, though different, indigenous perception and mapping of the region that have for them prevailed for centuries. However, research pertaining to sea nomads' (particularly the Orang Laut's) conceptions of mapping spaces and ownership of territory has sadly been overlooked. These issues demand attention today as competing global and national development-oriented programmes clash with escalating claims of ownership of traditional territories and resources by indigenous groups all over the world.

This book is then ethnography about Orang Laut perceptions of their inalienable gift of territory, an issue which is central to understanding the culture-history of Southeast Asia but has so far received little attention. It addresses the startling gaps of past and present writings that have failed to give priority to those aspects of sea nomads' culture that they themselves regard as most important. In this book, we see how the ideas of community, citizenship and legal territorial rights are reconfigured, reinterpreted and reconceptualized in the encounter between mobile populations and nation states.

This book also diverts from doing the classic anthropological village ethnography. The Orang Laut do not always live in any one fixed place. Some are now less nomadic than others. Yet even they sail to different places for varying periods of time. Some sail for a day, others for months. They are organized into several extended families scattered throughout the archipelago and operate in different networks of social, political and economic organization. To capture vital observations on inter-group relations, it has thus been necessary to employ new methods to study the interaction between different groups of Orang Laut and other peoples scattered throughout the archipelago. Fieldwork for this ethnographic account thus covers clusters of island communities in the Riau Archipelago as detailed in Map 1.1.

Significance of the Orang Laut

The Orang Laut are of special significance to anthropologists because they are the people who gained mastery of the seas, bridged the barrier between land and sea and exploited this cultural variety for their own advantages (Urry 1981: 7). They played a key role in the prehistory of the Indo-Malaysian region. Geographically, the region comprises thousands of islands. When the ocean levels altered at the end of the Pleistocene, the sea constituted a barrier to wider integration and marine innovation became necessary for integrating the region (Urry ibid.: 4). Despite geographical and cultural barriers, the Orang Laut were instrumental in controlling the seas and integrating communities and regions via a complex network of communication, trading and exchange relationships. So adept were they in this role that they became a centrally important factor in Malay political history. They were instrumental in the formation of the earliest historical states in the region because they

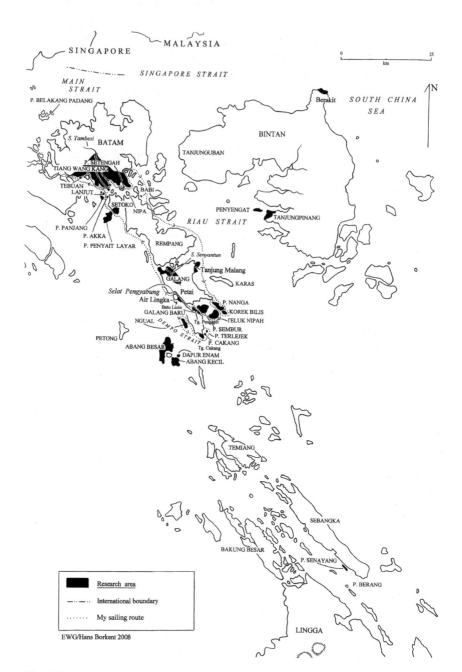

Map 1.1 Research area.

provided the naval might and communicative ties on which the hegemony of successive Malay states were based in an area of relatively sparse population. As a result, a mosaic of cultures occupying different environments developed. This structure generated growth of trade and exchange in addition to fostering socio-cultural diversity in Southeast Asia (Urry 1981: 7).

The Orang Laut are also fine examples of skilled mariners who move around to various places to appropriate resources within a relatively complex social and economic organization. This type of subsistence significantly contributed to the development of human societies. Island Southeast Asia comprises a combination of Asian and Australian terrestrial faunal zones. In terms of its marine fauna, it is of itself the world's only primary centre of diversity (Dunn and Dunn 1984: 252–253). As Dunn and Dunn (ibid.: 252) remarked, 'the tropical seas . . . of present day Southeast Asia are known to contain the greatest wealth of marine life'. This is because of the prevalence of two ecosystems unique to the tropics: coral reefs and mangrove swamps (ibid.: 252–253). The people of the Indo-Malaysian Archipelago have long depended heavily on the sea for their subsistence. In particular, the Orang Laut have responded to this richness by developing even more proficient fishing and strand-gathering technologies.

The issue of territorial rights[4]

Territorial rights, under which land rights is subsumed, is a complex issue in Indonesia. In pre-colonial time, disputes over land tenure were settled according to *adat*. The term '*adat*' means practices of social propriety, social norms, tradition and custom. Each community of people had their own set of *adat*, hence their own way of resolving territorial disputes. The expression '*adat* law' originated during Dutch colonial rule when the Dutch, in exercising indirect rule, consolidated local customary norms into *adat* law which then existed side by side with colonial laws. The purpose was manifold. First, this system of indirect rule was the best strategy to extract commodities from unwilling local producers. Second, *adat* law, which is often counter to Islamic law and colonial law, was perceived better suited for managing and solving the problems of everyday social life. Third, recognizing *adat* law made it easier for the Dutch to form alliances with traditional rulers to resist those rulers' adversaries, the Islamic leaders (Bowen 2003: 13). After Indonesia gained its independence, it continued with the system of legal pluralism so that both the civil law of the land and customary laws co-exist and supposedly have equal legal force.

Sukarno, the first president of the Republic of Indonesia, declared land to be 'an indispensable part of the Indonesian Revolution' (CCCIL 1988).[5] Development programmes carried out under his regime to reconstruct the nation provoked much hostility over land rights. Under the subsequent three-decade-long leadership of President Suharto, *Rencana Pembangunan Lima Tahun VI* or *Repilita VI* (Five-Year Development Plan), a chapter in Indonesia's twenty-five-year long-term development plan that included land rights issues, was affirmed and implemented to encompass the entire nation, in a consolidated, well-directed way involving uninterrupted and interconnected phases. The issue over land rights repeatedly surfaced and led to much animosity. In 2005, soon after President Susilo Bambang Yudhoyno took office, there was increased pressure throughout Indonesia for the implementation of decentralization and regional autonomy. Advocates of this legislation argued that it would enable the 450 districts in the country to gain more control over their natural resources and the local indigenous communities to participate in government. It was believed that this would lead to a return to local customary laws of governance and resource management that had been slighted in the recent past.

In spite of the positive outlook of a possible return to local customary laws, the International Work Group for Indigenous Affairs has cautioned that the new decentralization can adversely affect indigenous communities too. This is because local governments

> are now responsible for a larger portion of their operating budgets. There are thus pressures to raise new revenues that can often result in increased logging, mining and plantation [and coastal-tourism] expansion, which often disregard the land claims of indigenous communities.
>
> (Erni and Stidsen 2006: 303)

In this climate of change, several aspects of rural culture, in particular those people who have been categorized as the *suku-suku terasing* ('isolated people') by the Indonesian government, are severely criticized as a hindrance to progress. Rural communities have officially been classified in Indonesia into a hierarchy of a four-category scale. The categories are: *pra-desa* ('pre-villages'); *swadaya* ('traditional') villages; *swakarya* ('transitional') villages; and *swasembada* ('developed') villages. The first refers to groups of people, either of a clan or under a leader, but beyond government administration (*Pelita*, 26/12/1975 in Colchester 1986: 89). The categories correspond to ideas of 'the expected stages of development through which rural communities are to progress uniformly as they move toward true integration into an advanced and modern Indonesian nation' (Colchester 1986: 89). The Orang Laut are defined as the *suku-suku terasing* or *suku-suku terbelakang* ('isolated and backward people'), who form communities categorized as 'pre-villages' or 'traditional' villages.

Colchester's (1986: 90) translation of the explanatory memorandum which accompanied the 1979–1984 Five-Year Plan gives the identifying characteristics of the 'isolated people' as follows:

– Many of these people subsist partly by hunting, *fishing* and gathering forest products. Their *rudimentary economics* employ *extremely simple nomadic* farming *practices and equipment*. These farming techniques are *devastating the environment* and pose dangers to maintaining ecological equilibrium.

– Contrary to the State philosophy of *Pancasila*, the five principles, the first of which is belief in One Almighty God, these people have animistic religious practices

– Their *social systems are unstructured*, in which people live in *small scattered and dispersed groups*, isolated from the mainstream of religious, ideological, political, economic, social and cultural life. They *distrust anything coming from the outside.*

– These people, *depending* as they do *on the resources of their natural environment*, are nomadic and, therefore, it is impossible for the government to extend to them administrative and other services.

– These peoples are generally *minimally clothed*, covering only their vital parts.

– The *diet* of these peoples is *inadequate.*

– Their *dwellings* are merely places that provide shelter and a place to sleep. They are *far below the norms and requirements* that have been established for healthy, secure and pleasurable human dwellings.

– The *health conditions* of these peoples are far *below* generally accepted *norms* for healthy living.

– Formal *education* is *unknown*. Most of these people are illiterate.

– The art and culture of these people has merely achieved a *very primitive* level and their dances are still predominantly *magico-religious* in character.

– Their *economics* are centred on a system of *barter*. Monetary exchanges are still largely unknown.

– Most of these people remain *ignorant of* the *existence of the government* or of the *concepts of the Indonesian nation and state*. They have no sense of their duties as citizens.

– These peoples have no capacity or ability to withstand external and internal political threats. In the context of the state doctrine of total people's defense, these isolated communities constitute *weak groups* and regions within the total system of defense.

– These people are *not* yet in a position to enjoy the fruits of national development. Moreover they are *not contributing* anything to the *progress of Nation and State* (Colchester 1986: 90, *my emphasis*).

(The italicized terms are further discussed in this book.)

Characterized in this way, the Orang Laut are perceived as having no relationship with spaces that they occupy and, therefore, as possessing no concept of territory or permanent tenure. Their problems pertaining to territorial rights are compounded by several other factors. First, there is no official government policy in Indonesia concerning the rights of indigenous ethnic minorities. According to Erni and Stidsen (2006: 300), '[g]overnment officials argue that the concept of "indigenous" is not applicable in Indonesia, as almost all Indonesians (with the exception of the Chinese) are indigenous and thus entitled to the same rights'. Therefore, the government has refused to grant any privileges to groups who claim to be 'indigenous' people. For this reason, many communities have lost their land and have been unable to submit a legal claim for compensation.

Second, Indonesian land laws are supposedly enshrined in the Basic Agrarian Law of 1960 whereby customary law replaced formalized civil laws relating to land ownership. Under the traditional customary land rights system, land is an inalienable and common property of the community that cannot be bought, sold or leased.[6] However, this is simultaneously superseded by Article 5 of the 1960 Basic Agrarian Law which states that

> [t]he agrarian law which applies to the earth, water and air space is *adat* (customary) law, as long as it does not conflict with the national and State interests, based upon the concept of one nation, and with Indonesian socialism along with regulations that are issued according to this Law and with other legal regulations, all of the above with due regard to principles deriving from religious law.

In addition, Articles 3 and 14 of Basic Agrarian Law No. 5 weaken the position of indigenous peoples with respect to their land. Article 3 clearly states that *hak ulayat* ('traditional communal property rights') must have a form 'that is consistent with the national and State's determination, based upon the concept of one nation. Also, it cannot be contrary to other laws and regulations with higher force'. Article 14 adds to the complexity by stating that the government

> shall, within the framework of Indonesian socialism, make a general policy with regard to the supply, preparation and use of the earth, water and airspace along with the natural wealth they contain for the needs of the State, for the need of religion and other sacred uses, according to the principle of the One True God (*Ketuhanan Yang Maha Esa*), for the basic living needs of the people ... for the need to increase the production of agriculture, animal husbandry and fishing ... [and] for the use of developing industry, transmigration and mining.

Hence, land rights are under pressure to be relinquished in the face of competing claims by the state. The Basic Agrarian Law also only provides legal security to registered or titled property (Weinstock 1979: 99). This requirement overlooks the fact that documentary proof of title is irrelevant in traditional law.

Still another new legislation with respect to land, Presidential Decision No. 36/2005 on Land Acquisition for Development of Public Facilities, was passed in 2005. It permits the government to revoke people's property rights in the name of national development. This new law allows the government to evict people for the implementation of large-scale development projects. Even though there were widespread criticisms and protests against this law throughout the country and a call for it to be reviewed was heard in parliament, no alteration to this latest regulation has been attempted (Erni and Stidsen 2006: 304).

Most studies on the indigenous peoples of Southeast Asia are concerned with socio-cultural problems due to changes brought about by development programmes. Yet, as Benjamin (1989: 14) counsels,

> unarmed investigators have tended ... to take up implicitly (or even explicitly) an evolutionary-ladder approach to cultural differences ... Consequently differences that have in fact resulted from choice, such as following a foraging way of life instead of swiddening or an intensive agriculture one, have been seen instead as unavoidable steps in a progression from primitivity toward civilization.

Such characterizations of the indigenous peoples, as noted by Benjamin, have appeared often in textbooks, newspapers and other media. Attention must be paid to Benjamin's insightful advice. It is precisely the 'evolutionary-ladder approach' that relegates the Orang Laut to the state of being foragers moving around helter-skelter in the vast expanse of the sea in search of food. Land rights, in this case pertaining to sea and coastal territorial rights, are therefore denied to them because they are not recognized as producers, but as nomadic foragers. An oversimplified dichotomy has been created. On the one hand there is nomadism, with the unmonitored extraction or removal of resources from the environment for human use. On the other hand there is sedentarism, with responsible producers or developers who appropriate resources and convert them into objects of property that are properly managed for present and future needs. This mistaken view is also addressed in this book.

The case of the Orang Laut furthermore addresses an issue of great concern today: the question of being indigenous. Which groups regard themselves as indigenous? Which groups are allowed to regard themselves as indigenous peoples? Which groups actually succeed in being regarded by national and international governments as indigenous peoples? The category of 'indigenous peoples', which emerged as a retaliation to Western European colonialism, has also disturbed postcolonial governments which do not want to distinguish between indigenous and non-indigenous populations (Barnes 1995: 2). Indigenousness is a concept that has generated passionate debates. Common themes in these debates are the ambiguity of the term itself, ethnic identity, historical priority, self-determination, political, social and cultural freedom, ownership and control of space and resources, ecological exploitation and protection, relationship to the state and many others.

The challenges faced by the Orang Laut are often similar to those seen throughout Southeast Asia. The Maniq or Sakai in southern Thailand, for example, experience great difficulties because at least two conflicting development-oriented plans have been imposed upon them. In the early 1990s, Thai authorities turned their village in the forest into a major tourist site (Hamilton 2002: 87–88; Porath 2002: 112–115). Another Thai government initiative to transform the Maniq into civilized and modern Thais required replacing their bamboo and palm-leaf shelters with Thai-style tin-roofed houses and encouraged them to speak central Thai rather than their own language. Only very small bands of Maniq can now survive and those who insist on their own autonomy are denied official recognition and not classified for land-holding, health and welfare benefits (Hamilton 2002: 88). Even more obstacles confront the hill tribes in Thailand. They are denied Thai nationality and residence. It has even been reported that Thai armed forces have expelled long-settled communities into Burma at gunpoint (Colchester 1995; Tapp 1986; McKinnon and Bhrukasari 1986; Survival International 1987; Ekachai 1990). In Peninsular Malaysia, the Federal Constitution accords indigenous peoples a special status. They are given preferential treatment in education, employment, business opportunities and ownership or control of resources such as land (Mohammed Suffian 1972: 249 ff.). Nevertheless, in reality, areas reserved for their use are held by the state and may be reallocated and gazetted at any time for higher state priorities (Nicholas 1990; Colchester 1989, 1991). Over in Vietnam, a government plan from 1975 to the early 1980s for the spatial 'rationalization' of its people involved resettling many indigenous hill tribes of the Central Highlands (Evans 1995: 253). This was a strategy designed to dilute their concentration in the highlands in order to prevent them from staking claims for autonomous areas. However, with the state's colonization of the mountains and forest and the absence of the specific owners of these territories, long-term conservation of these areas has been badly affected (ibid.: 265).

Indeed, the matter of land and resource ownership has in recent years become one of the most pressing issues in the face of increasing needs of national governments to develop their regions and economies to profit from global demands. Indigenous communities in Southeast Asia in comparison to Australia, New Zealand, Canada and the United States are rarely accorded any rights to territories, and even when such rights are recognized they are seldom backed by legal sanctions (Hamilton 2002: 78–79). The people of the Philippine Cordillera are one of the first indigenous peoples in Southeast Asia to have staked a successful collective claim to ownership of traditional land and resources. As a group, they formed the Cordillera People's Alliance and fought against the construction of the Chico dams in 1981–1982 (Gray 1995).

These are but some of the emerging predicaments of indigenous peoples in the face of global development. Such predicaments have led to a renewed interest in ethnological, social and political comparisons of indigenous groups all over Southeast Asia and all over the world. The case of the Orang Laut is thus one in the grip of worldly encounter.

2 Social organization

Introduction

This chapter discusses the social organization of the Orang Laut. It examines how processes of change have altered their social constitution, impacting upon their identity and relationships with others. The discussion traces how Malay rulers of earlier times organized them into status groups, the subsequent and eventual collapse of this framework of social stratification and the present social constitution of the community. In addressing their current social organization from the macro to micro level, the discussion starts from lineage formations and moves in descending order to village community life, the formation and dissolving of households, marriage relations and the nuclear family. Underlying all of this are the interpersonal relations involved within and across their groups that give coherence to their living arrangements and the way they structure their lives.

Suku organization

From historical literature, it appears that the Orang Laut emerged from a common social, political and economic ambit with Malay coastal and island groups. They were incorporated into the Kingdom of Johor in the sixteenth and seventeenth centuries when their ties with the ruler were formalized. They had a very special relationship with the sultan. Their relationship was one of absolute allegiance because of their belief in the ruler's divine and supernatural qualities (Andaya 1975: 37–54). In recognition of their loyalty, they were granted possession of 'the seas and what floated on them by hereditary feudal right from the Sultan of Johor' (Trocki 1979: 56). Moreover, they were granted the special political status of *rakyat* ('subject') of the sultan, and so they became known as Orang Rakyat ('subjects'), Rakyat Johor ('subjects of Johor') or Rakyat Laut ('sea subjects') (Sopher 1977: 90). They were so close to the sultan that the nature of their relationship far exceeded that of ruler and subjects, and was more like one between master and personal retainers. This exceptional relationship and their involvement in the affairs of the Malay court led to the Orang Laut's deep immersion in the complex political-social order of the Malay World. This had a great impact on their identity and way of life. In view of their special political status, they were exempted from paying tax but were expected to assume the role of *orang kerahan* ('nobility's vassals') and perform *kerahan* (traditional or corvée services) to the Malay overlords (Sopher 1977: 90).

Through their relationship with the sultan, the Orang Laut were divided into numerous *suku-suku*. The term '*suku*' means 'group' of people, and these *suku-suku* or groups were feudally organized territorial groups that were obliged to perform various kinds of services for the Malay lords. Such feudal groups were most characteristic of the Riau-Lingga Archipelago and the adjacent coastal islands. In Pulau Tujuh, the group organization was less clear. Nevertheless, each group was obliged to perform feudal services for the local ruler, who was himself a vassal of Bintan or Lingga (Sopher 1977: 269).

Usually each group had a leader bearing the title of *batin* ('headman'), but this was where their similarity ended. Each differed in size and economic and political significance. Those that attached themselves to the centres of power rose in political position. Often, they were also the bigger and better organized ones that were associated with the larger islands strategically located along the main sea trading routes (Andaya 1975: 44). Higher rank and status were accorded to them in the group organization, but they were simultaneously subjected to stricter control by the central power. In contrast, those that lived further away from the centres of power were lower in rank and status, but they retained greater autonomy. They were often beyond the control of the Malay overlords and less bound to feudal obligations. Some of them were not even considered subjects and so they were relatively free to lead much more independent lives. Status and power were determined by the size of the group and their ability to strengthen the political and economic position of the sultan.

As each group was assigned a different task, they were also accorded differing rank and status within an almost caste-like status hierarchy. To put it another way, the type of service that could be demanded differed from group to group and depended on the status of the group within the established hierarchy. The combined roles of all the groups amalgamated into a wide range of services for the court. Most important of all, they were meant to provide for the ruler's personal comfort and welfare. Whenever a group responded to the call of the sultan, they would be reciprocated with food and rations (Andaya 1974: 7).

Attempts have been made to compile a list of all the groups and the services that they provided. However, it has not yet been possible to compile a conclusive inventory. In collating the records so far, variants in similarities and differences in the description of group duties appear. To illustrate the perplexities involved, an example can be drawn by comparing the records of Newbold (1839), Netscher (1854) and Sopher (1977). Newbold (1839) purports to give an account of the Malayan coastal people when in fact he only describes those in the Riau-Lingga Archipelago and three other places on the east coast of Sumatra. He states that 'the Rakyat Laut, or subjects of the sea, or Orang Akkye' detest being called subjects and are extremely proud and scornful of the Malays. This description and his use of the term 'Orang Akkye' in relation to the Orang Laut differed from many other conventional writings on them and this appellation has not been used anywhere else. Although, in later years, Skeat and Blagden (1906) simply assumed it to be Akit without offering any explanation.

In contrast, Netscher (1854: 132–133), a Dutch Resident of Riau in the nineteenth century, carefully lists the following groups in the archipelago: Suku Riau, Mantang, Kelong, Tambus, Sebung, Galang, Gelam, Bulu, Ngeju, Terong, Sugi, Moro, Buru, Karemon, Ongar, Kundur, Kateman, Leboh, Manda, Tegal, Anaki Serka, Bentayan and Pelanduk. Furthermore, he divides the Orang Laut into two categories: the subjects who form 'part of the population that wandered about, living not in fixed villages but on the periphery mostly on boats' and the more sedentary 'native Malayu tribes that are divided into *suku* as in Sumatra' (Netscher 1854: 127–128; 132–133). Later on he notes that that the Suku Buru and Tambus whom he had placed in the second category, live on the sea and 'in their lifestyle have some similarity with the Bajau of Celebes' (ibid.: 140). Sopher (1977: 89–114), as a third source of comparison, documents the following groups as constituting the main ones in the Riau-Lingga islands: Suku Tambus, Mantang, Barok, Barut, Buru, Galang, Sekanak and Orang Posik (or Persik).

By simply comparing the descriptions contained in these three works, it is clear that putting together a complete picture of all the groups is, to say the least, not an easy task. Several reasons account for this difficulty. In the course of time, functions of some groups changed or were modified. Sometimes they were even given new responsibilities. The Malay rulers required an array of services. Intermittently, new or modified tasks were assigned to the existing groups or new ones were created out of necessity to satisfy the most recent demands. Sometimes the Malay overlords would move a group to a new location to provide some needed service or to supply the manpower needed to exploit the area's resources (Sopher 1977: 102). This process went on for centuries. Today's ethnographers are confronted with an inestimable constellation of groups and are constantly awed by the enormity of their task in identifying all of them. Difficulties have been intensified by the nomadic lifestyle of these expert seafarers as well as their being scattered all over the archipelago.

The group organization was enmeshed in a caste-like rank-cum-status hierarchy. The three most high-ranking groups were

the Suku Orang Dalam ['People of the Royal Court'], believed to be late comers, and the leading nobles in the court circles of Johore; the Suku Bintan, members of the old ruling class of the islands, established here in the time of Majapahit's supremacy around 1400; and the Suku Mepar, believed by tradition to be descendants of a Trengganu royal family, who were set in authority over the Lingga Islands and the sea folk by the Sultan of Malacca.

(Sopher 1977: 101)

The unique relationship that these three groups shared with the sultan elevated them above all the other Orang Laut. The Suku Bintan were awarded the highest titles of *Laksamana* ('Admiral') and *Bendahara* ('Grand Vizier'), as they and their followers had served the Maharajas of Srivijaya-Palembang with absolute devotion. In times of war, they would always lead all the other groups into battle. Bintan was also the nucleus of Orang Laut might that empowered the Srivijaya dynasty with maritime strength and prosperity (Andaya 1975: 46).

The next three most prestigious groups were the Suku Galang, Sekanak and Gelam. Of these three, the Suku Galang were ranked the highest because of their intimate contact with the aristocratic Suku Orang Dalam when the latter were placed on Galang Island by the Malay sultans to govern the sea people. Even though the Suku Galang were considered to have interacted extensively with the Suku Orang Dalam, there was still a great hierarchical distance between them and the top-most three groups. The gap between them was emphasized and reflected in their bride-price for marriage. So superior were the three highest-ranking groups that women from these groups commanded ten times more in their bride-price than those belonging to the Suku Galang. Women from the Suku Bintan and Mepar fetched bride prices of 400 and 350 reals respectively in comparison to those from the Suku Galang, who could only expect 44 reals (Schot 1882: 164).

At the bottom of the hierarchy were the Suku Selat, Trong, Sugi, Tambus and Nan. So low in status were they that their women could only hope for 30 reals in bride-price. According to custom, bride-prices were halved for women marrying for the second time. The only exception was that for Galang women, whose price remained unchanged. Undoubtedly, this clearly indicated the difference between the groups and just how eminent the Galang were as subjects of the sultan (Schot 1882: 164).

The Suku Galang controlled the sea lanes at the behest of the Sultan of Johor. They did so with such fierceness that many early nineteenth-century Western writings referred to them as pirates (see, for example, Logan 1850 and Thomson 1851). Women of this group were tasked with gathering and preparing sea cucumbers for the highly lucrative international trade with the Chinese. The Suku Galang constituted a big group that covered a vast expanse of the archipelago which stretched across the islands of Stoko, Rempang, the two Galangs and Karas to Temojong and the cluster of islands in the Bolong Strait, Gelam and Rokan area. By their sheer size, they invariably dominated the other groups (Schot 1882: 164).

A glimpse of how the group organization generally functioned can be caught by descriptions provided by Schot (1883: 472–473 and 1884: 575–576) and Andaya (1975: 44–49). Apart from the prestigious groups described so far, the Suku Bulang, Mepar, Sekanak, Gelam and Bugi made up the rest of the sultan's armed forces. From them, talented and loyal Orang Laut were appointed to the office of *panglima* (title for a military officer, usually the leader of the military force) and *hulubalang* (military officer, warrior). The Suku Posik and Sugi also served as the ruler's fighters (Sopher 1977: 96–97). All of these groups were particularly loyal to the sultan and they obediently undertook expeditions for the ruler (Sopher 1977: 96–101).

In times of war and peace, the Suku Ladi, Gelam, Sekanak, Sugi, Klong, Trong, Moro and Tambus served as rowers for members of the royal household. The important mission of transporting envoys and conveying royal missives to foreign rulers and dignitaries were, however, entrusted only to the Orang Suku Mepar. In addition to serving as rowers, the Gelam provided boats and the Ladi were charged with three other assignments: they performed guard duty for the ruler, supplied wood for houses and warehouses and provided carpentry services. Collecting ebony and eaglewood were, however, the task of the Suku Temiang, Moro, Pekaka and Sugi. Boat building was assigned to the Suku Galang. Several other groups served the royal kitchen. The Suku Nan provided fresh water and firewood, The Suku Kasu (Selat), Trong, Sugi and Moro collected edible seaweeds for royal consumption. The Suku Butu were the sago producers.

At the bottom of the hierarchy of group organization were the Mantang and Tambus. The former were placed above the Tambus because, although they were given the lowly task of weaving mats, they were also assigned the more respectable roles of serving as the sultan's ironsmiths to make weapons. Furthermore, they were given the privilege of rowing the Raja and his deputies for a maximum consecutive period of three months in the year when the nobles embarked on voyages (Sopher 1977: 94). The Suku Tambus were the lowest in rank and hence charged with the lowliest job of looking after the ruler's hunting dogs. They had previously served as the sultan's rowers but were demoted in the late nineteenth century to the point that they were no longer allowed to carry out this task and were confined to taking care of the dogs. Although it has been recorded that the Tambus were originally located in the southern part of Galang Island where there was a Kampong Tambus right up to 1884, there are also archival records which explicitly state that they always lived in boats and had no fixed habitation. They were described as petty pirates who were despised by all the other Orang Laut (Logan 1847: 246; Sopher 1977: 92–94).

Apart from providing for the material needs of the Malay court, several groups were also designated the role of providing entertainment for royal festivities. The practice of singing, dancing and providing music for the Malay sultans is in fact a practice that continues to this day in Malaysia (Mariam Mohommad Ali 2002: 280). Their music can be heard on recordings published by Philip Yampolsky (1996). Up to the present, some Orang Suku Seletar still remember that they gathered forest and mangrove products in addition to accompanying the Sultan of Johor on hunting expeditions (Mariam Mohammad Ali 2002: 280).

The group organization may have appeared to be complex and uncentralized, yet it was precisely this web of constituents that formed the power base of the sultan's kingdom. This structure of organization formed the principal framework by which several regional polities, including the Kingdom of Johor, marshalled island and coastal people under their rule and assembled them into a tributary and defensive network. In other words, all the groups formed the necessary building blocks for the sustenance of sedentary Malay communities (Andaya 1975: 44).

Rapid disintegration of the group organization, which began in 1699 after the regicide of the last male heir of the Malacca dynasty, intensified in the nineteenth cenutry when indigenous rule was displaced by the British and Dutch colonial powers. The trauma and confusion of these events culminated in 'confusion within the ranks of the Orang Laut' (Andaya 1975: 322–323). After the collapse of the traditional power structure, many groups disappeared or changed their roles in order to survive. The few remaining ones, like the Suku Galang, transferred their allegiance to local Malay chieftains, who engaged the Suku Galang in piracy. By the nineteenth century, they were regarded with great disdain as barbaric pirates by the colonial masters in the region, who did all they could to suppress them; many were hastily sedentarized (Sandbukt 1984: 7). The other remaining groups adapted in different ways. When Singapore was founded by the British, the Suku Kallang, for example, who had inhabited the Kallang River, successfully adapted themselves as boat rowers, transporting passengers between anchored ships and the port. They had otherwise been known to make wrappers for palm-leaf cigarettes, provide passenger boat services and gather wood. The Suku Selat, who had plied the waters of the Strait of Batam, the southern coast of Singapore and along the Tebrau Strait on the southern coast of Johor from as early as the sixteenth century and had previously provided edible seaweeds to the Malay courts, adapted themselves to trading fish and fruit with passing ships and lying in wait to pillage vessels that ran aground in the sandbanks and shallows.

Current social organization and identity

Most Orang Laut live in the Riau-Lingga area today. They do not much speak of themselves as Suku Tambus, Suku Galang, etc. When asked, they may even give conflicting answers. Although this is understandable in view of the traditional connotations of servitude that such terms of identification are associated with, it has made it impossible to draft a comprehensive list of the different *suku* that dot the archipelago.

The difficulties encountered are due to several problems. First, almost everyone has a different listing of *suku-suku*. Second, it is not uncommon for them to adopt situational identities and change the way they identify themselves in accordance with different situations. Third, rivals often contradict each other on their identification claims. Fourth, they have been so influenced by the view of the former political centres and system of classification that it influences the way in which they view each other and pressures them to align themselves with the more prestigious *suku*.

Ethnonyms for an Orang Laut tend to be fluid and so situational that they can switch and manipulate to possess a series of identities. They can, for instance, trace their descent from either their mother's or father's *suku*, switch *suku* affiliation through intermarriage or express certain cultural traits (e.g. speaking Bahasa Galang) in order to identify with a more prestigious *suku*. At this point, it would be relevant to consider other ethnographies on the Malay World for a wider view of a highly complex area of human interaction. Parallel processes can be seen at work, for example, with the Semai of Peninsular Malaysia (Dentan 1975), the Orang Lom of Bangka, West Indonesia (Smedal 1989) and the mosaic of indigenous peoples of Borneo (King 1993), who can don and doff being Malay through either abiding by Muslim observances or reverting to earlier non-Islamic practices. Anthropological literature on the Malay World has also indicated that ethnonyms may undergo periodic or episodic transformations, especially when a people are faced with pressures, such as gaining status recognition in their interaction with dominant reference groups. Their latent ethnonyms may reappear with changed circumstances or when new allegiances are formed. Occasionally an amalgamation develops so that the different groups may elect to adopt a new group ethnonym while also retaining their former identities (Hitchcock 1996: 13).

Lineages

Today, the hierarchical *suku* system no longer exists. What remains is essentially what lies beneath this framework of *suku* social stratification. Still discernable is the pattern of several contiguous but more autonomous groups of Orang Laut, a simple division along the lines of kinship into lineages, each with its recognized territorial range. Rather than identifying their *suku* identity, the Orang Laut would more readily identify themselves in toponymic references that signify their place of origin and principal area of settlement. They would for instance, identify themselves as an 'Orang Pulau Nanga' (People/Person of Nanga Island), 'Orang Teluk Nipah' (People/Person of Teluk Nipah) or 'People/Person of such-and-such place name'. These groups or lineages are formed by a process of perpetuation whereby the members of the group are related to each other by common descent. Different lineages own and occupy different islands and moorage areas throughout the archipelago. The actual relationship between members in a lineage can be demonstrated and is not simply assumed. They are a descent group formed on the basis of descent from a common ancestor or ancestress. As a group they have a name and a territory and perform some rituals collectively. Map 2.1 gives an overall view of the spread of the Orang Laut in the Riau-Lingga Archipelago.

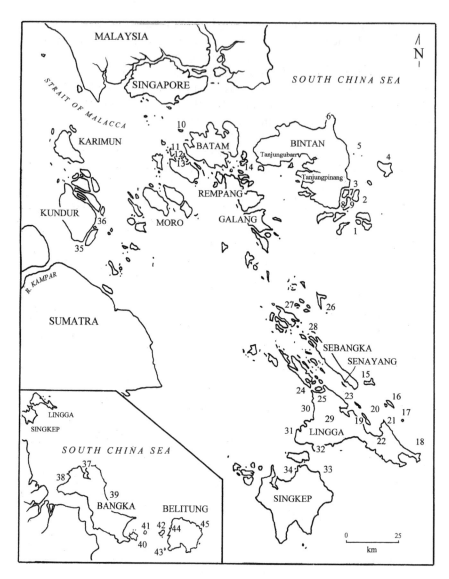

Map 2.1 The spread of Orang Laut in the Riau-Lingga Archipelago.

Key: (1) Orang Pulau Toi, (2) Orang Tanjung Sekuang, (3) Orang Pulau Buton, (4) Orang Mapur, (5) Orang Berakit, (6) Orang Panglung, (7) Orang Tanung Senkuang, (8) Orang Air Kelobi, (9) Orang Pulau Malim, (10) Orang Dapur Arang, (11) Orang Pulau Bertam, (12) Orang Pulau Padi, (13) Orang Pulau Boyan, (14) Orang Nginang, (15) Orang Kentar, (16) Orang Kojong, (17) Orang Pulau Buluh, (18) Orang Mensemut, (19) Orang Sungai Liang, (20) Orang Pulau Hantu, (21) Orang Air Kelat, (22) Orang Pongok, (23) Orang Kungki, (24) Orang Linau, (25) Orang Air Batu, (26) Orang Mamut, (27) Orang Pulau Medang, (28) Orang Limas, (29) Orang Pancur, (30) Orang Tembuk, (31) Orang Lelumu, (32) Orang Mentuda, (33) Orang Penuba, (34) Orang Sungai Buluh, (35) Orang Tanung Batu, (36) Orang Sebele, (37) Orang Mantung, (38) Orang Teluk Kampa, (39) Orang Baturusa, (40) Orang Pulau Lepar, (41) Orang Pulau Liat, (42) Orang Pulau Mendanau, (43) Orang Pulau Seliu, (44) Orang Tanung Pandan, (45) Orang Teluk Pring.

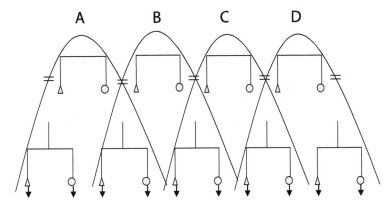

Figure 2.1 Cognatic lineages.

Source: Fox 1966: 47

The Orang Laut have cognatic lineages, which can perhaps be best described as ignoring sex in tracing kinship links. They trace their descent from a common ancestor or ancestress through links of both sexes. The basic principle is that an Orang Laut can be a member of at least two groups: paternal and maternal. This arrangement of the overlapping of groups can be visualized as above in figure 2.1.

In such a system, the mode of group recruitment is cognatic. It equally enables and permits men and women to reproduce the group. It is also a situation which is impartial to whether the females are daughters and sisters or daughters-in-law and sisters-in-law or a mixture, or if males are sons and brothers or sons-in-laws and brothers-in-law or a combinition. It could appear as follows (see figure 2.2).

Daughters can opt to stay and bring in husbands, and sons can likewise

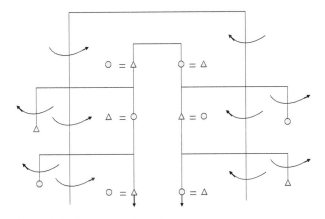

Figure 2.2 Group recruitment.

Source: Fox 1966: 83

decide to stay and bring in wives. The crux of the matter is that there shall be enough but not too many males and females in the unit. The flexibility of this system makes for mobility. It therefore makes it possible for a couple to adjust their marital residence to where they feel would be their best chance to eke out a living or find close relatives with whom they would live and work. Chapter 4 further examines how this bears upon Orang Laut ownership of networks of inter-related territories.

These lineages grow and proliferate. After one generation, they usually increase to comprise siblings and cousins. By their very nature, cognatic lineages inevitably overlap in membership, and an Orang Laut man or woman will be a member of several similar groups simultaneously. From an individual's perspective, their group membership in a cognatic descent system can be demonstrated as follows (see figure 2.3).

For example, ego (the individual) belongs to as many cognatic lineages as he has lineal ancestors. If all four of his grandparents originated from different groups, then taking his cognatic lineage to the grandparental level gives him four and tracing it to the great-grandparental level would give him eight and so on. The Orang Laut preserve their genealogies through an oral tradition. A few are able to recite impressive genealogies that go back many generations, but most are only have precise memories back to their grandparents and somewhat dimmer narrations back to their great grandparents. If an Orang Laut can recall group affiliations of four grandparents, he has four chances of membership and if he can recall the group affiliations of his eight great grandparents, he has eight opportunities of membership and so on. However, chances are slim that he will remember the precise affiliations of his eight great grandparents.

Theoretically, it is plausible within such a system for everyone to be a

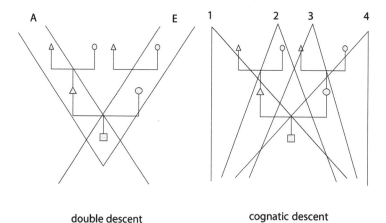

double descent   cognatic descent

Figure 2.3 Cognatic descent.

Source: Fox 1966: 148

member of all the groups in the society. Yet, if this were so, why are there so many Orang Laut groups? This is because, realistically, it is no easy task for any one individual to find a common ancestor in genealogical links to cover the entire system. Furthermore, it is 'rare for cognatic descent groups to be exogamous and indeed impractical' (Fox 1966: 149), and the Orang Laut are not an exception. Endogamous marriages between first and second cousins are preferred, and this avoids much overlapping of groups.

It is simply impossible for an Orang Laut to be permanently resident in two groups at once. There is no rule that all lineage members must reside together at the same place and at the same time. If any one of their lineages were to impose such a demand, it would deprive the other lineages of their members. There are also practical reasons as to why stringent rules are not imposed on members. If a group becomes too large for its territory, the cognatic system embodies contingent and flexible arrangements to ease pressures. Since all of the members belong to other groups as well, they can take up their territorial rights elsewhere so that there can be a more even distribution of the population across their vast maritime estate. Cognatic descent groups and a system which allows for flexibility appear to be common elsewhere in small island communities as well, perhaps due to population pressures on a scarce land area (Fox 1966: 153).

At any one time, a group of members forms the core residents on a particular territory. Other dispersed members can also claim rights to a territory without actually being present. Members possess the right to and may choose to live on a territory whenever they so desire. Moreover, there is no fixed rule to compel members to come at any particular time or even to come and stay at all. Under most circumstances, these lineages form village (*kampong*) communities.

The village community

The village community represents an important focal point of Orang Laut socio-economic life. Members of the community comprise boat-dwelling families and those who have either chosen or have been forced to lead a sedentary or semi-sedentary lifestyle. Families may disappear for fishing voyages that can last a few hours in nearby waters or a few months in more distant waters, but they regularly return to what they regard as *kampong kita* ('our village').

Every community has a leader who today holds the title of *batin, ketua kampong* ('village headman') or *kepala* ('head'). The title used depends on the geographic area. In Pulau Tujuh, for example, *batin* appears to be the chosen term, while in the areas where I did my fieldwork, *kepala* is the preferred title. State authorities assert some control over this matter by issuing a letter of authority through the Social Office or District Office to acknowledge the leader. Several changes in the role of the leader have occurred through time. In the past, the headman of the Orang Seletar was said to have been the chief midwife for his community (Sather 1999: 20). This does not appear to be the case for present day leaders. Instead, they represent and act as spokesman for their community members to outsiders and government authorities, inform or alert local authorities about visitors or unwelcomed intruders in the area and report births and deaths. In addition, they may witness agreements reached by members of the community, mediate internal village disputes or disputes between community members and outsiders and interpret the principles of local customary laws, especially when disputes arise and apologies or compensations are demanded. These additional tasks are performed only if the quarrelling parties invite the leaders to do so. The leader's house is often the central meeting place for matters to be discussed, deliberated and resolved.

Even with a letter of authority, the leader cannot lead by coercion. To be an effective leader, one must have the respect and support of fellow villagers. Otherwise, people will simply regard the letter of authority as nothing but a piece of paper of no import and pay no heed to the leader. Forceful personalities in the village can complicate and challenge the leader's authority. Therefore, it is crucial that the members' support is secured and techniques of persuasion are used to win their favour in volatile situations and stabilize village politics. In a show of co-operative and collaborative spirit, a leader usually invites or allows unwieldy characters to speak at public occasions in order to avoid direct challenges to or criticism of his leadership. Many of these forceful persons possess a good verbal facility and have proven to be even better spokespersons to government officials for the community than the headman. No rule prevents women from being influential, and some women are influential.

The dominant living arrangement of a village community is a compact cluster of closely grouped and raised wooden plank houses built on piles driven either into the seabed or the coast. Houses constructed over the sea are preferred over those built on land. 'Land' means that houses are built along the coast, but never further inland unless the inhabitants have been forcibly relocated by government authorities. Land-based houses always face seaward. In comparison, the entrance door for houses standing over the sea may face various directions. It could face land or sideways or the house could have more than one door (e.g. one door facing land and the other facing sea). A ladder descending into the sea will often be attached to the front door. The Orang Laut secure their boats to the piles below their houses. When they want to embark on a fishing trip, they simply descend the ladder and step straight into their boats. Likewise, upon their return, they secure their boats to the piles below their houses and use the ladder to ascend into their houses. For houses on land, they descend by a ladder from the front door to get to the ground.

Walkways are usually built to connect houses built over the sea to the strand to enable the family and visitors to make their way back and forth between the house and the strand without getting their feet wet. Sometimes, from the main walkway, networks of tributary walkways are constructed to link neighbouring houses. Sizes of walkways vary, as do their stability.

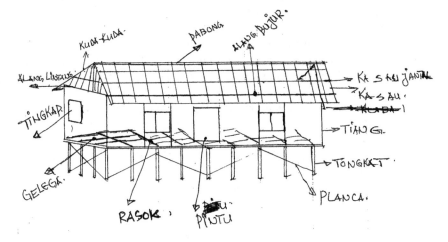

Figure 2.4 Drawing of a house by an Orang Laut from Tiang Wang Kang.

Attached to some houses are uncovered platforms made of wooden planks. Like the houses, they are raised on piles. This open area is an important social space. People come together in this space for various activities, including playing, socializing and working. Children gather to play and adults sit around informally to exchange news on a wide range of subjects, share jokes and enjoy the sea breeze in the open area. It also serves as a workplace for cleaning and drying fish, mending and drying nets, constructing fishing equipment and weaving mats. In this space, behaviour is highly informal and people may act without the constraint that is normal within the house. This space allows people to come together and feel less lonely.

Houses come in different sizes. Some measure approximately five metres long, two and a half metres in width and two and a half metres in height, while bigger ones may be as much as nine metres in length, four and a half metres in width and three metres in height. The interior of most houses consists of one unpartitioned room which is devoid of furniture. Nevertheless, in this general open space, there are designated sleeping spots for each family member. When it is time to sleep, they simply roll out their mats. When they awake, these mats are rolled up and stored in a corner or against a wall. This clears the general open space for other activities. Just as bedding material is stacked in a corner, against walls or in boxes when not in use, so are other personal belongings. Erecting partitions inside a house is an occasional practice which is most common in polygamous families.

The walls of houses are constructed either of wooden planks or *kajang* (movable mat of bamboo laths) mats and roofs are made of either zinc sheets or *atap*. *Atap* are made of dried leaves of a stemless plant, *Eugeissona*, that are woven together by rattan. Zinc roofs last longer but cost more. Zinc sheets are not so easily available in the islands. They have to be imported from towns or bigger islands that may be a distance away, inevitably adding to the cost of construction. The inside of such houses reach extreme temperatures in the midday heat and loud rattling sounds of rain falling on the zinc roof must be tolerated, especially during the monsoon periods. Roofs made from palm leaves cost much less, sometimes even nothing at all. This is because the leaves can be collected from surrounding forest areas, so it only takes is time and energy to weave them together.

There are also houses which are built for temporary or transitional purposes only. Known as *pondok* ('huts'), these temporary shelters are usually constructed by the more nomadic boat-dwelling families who are less inclined to remain in any one place for an extended period of time. On average, a temporary shelter measures approximately 5 meters in length, 2 meters in width and 1.2 meters in height. Predominantly boat-dwelling families may elect to live in these constructions during the monsoon season. Needless to say, living on the open sea in the midst of strong winds, heavy rains and choppy waters can be difficult and dangerous. Sometimes, women may also prefer to deliver their babies in these temporary huts.

Temporary huts are simple shelters that can be erected very quickly. Construction materials for the shelters, such as wooden poles and rattan for securing the poles, are gathered from nearby forests. Mats made of bamboo laths that have served as the roofs of boats alternate as the roofs and walls of houses. Alternatively, the sails of boats may be used as roofs too. When there is no longer the need for these houses, they are quickly and easily dismantled and the material is once again recycled to houseboats.

Deserted houses are feared. When a house is unoccupied for too long, it is said to invite trouble. *Hantu* ('evil spirits') will move in and create problems for the entire village. Therefore, if a house is left vacant for too long, anyone will have the right to move in to prevent evil spirits from entering it. Otherwise, other members of the community will demolish it. No problem or disputes will ensue if the original owner returns. They will simply erect a new house. As everyone in the village is related to each other in one way or another, a special bond may even arise between the previous and present inhabitants of a house. It is believed that a special bond is established between people who have slept and eaten in the same place.

Households

Villages vary in size. An average village comprises about seven to ten kin-related households. Houses are impermanent and their numbers change frequently as marriages, divorces and remarriages are common. When changes in relationships occur, new houses are built or existing ones are dismantled or abandoned. Sickness, deaths, fights, departures for another place, arrivals of kin who would like to set up home for a longer period of time and births leading to the expansion or contraction of the family size are just a few of the many reasons why new houses are erected or old ones disassembled.

The structural continuity of the village thus rests on the development of the families. Below is an example of a village layout and household composition.

As members of the village are kin, they share a deep emotional bonding. They are in contact with one another and live together. Therefore, in times of need, they would be the closest at hand to seek assistance from. This is not to say that all members get along harmoniously with each other, nor that everyone regards the other with great warmth. What is an unspoken rule in the social organization and relations of the people, though, is a network steeped in obligations to help and share. Everyone observes this unspoken rule to show group solidarity. Households and individuals exchange regularly a variety of things, including food, fishing equipment, houses, labour, skills and clothes. They describe these as acts of *bantu* ('help'), *kasi* ('giving'), *tukar* ('exchanging') and *bagian* ('sharing'). These acts are seldom described as *jual*

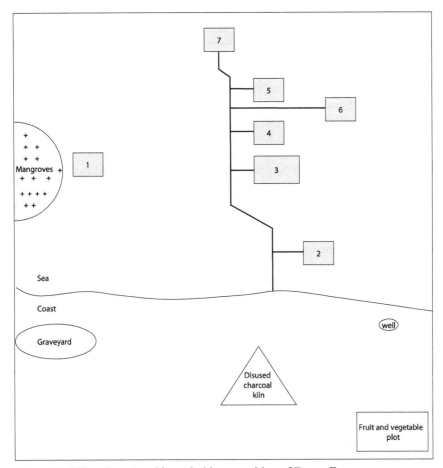

Figure 2.5 Village layout and household composition of Dapur Enam.

Note: The figures in Figure 2.5 correspond to the house numbers in Table 2.1

Table 2.1 Household composition of Dapur Enam

House number	Name of head of house group	Family members	Generation structure/depth	Type of residence
1	Bego	1 son	2	Semi-nomadic
2	Bulat	Wife and daughter	2	Sedentary
3	Awang Ketah	Wife, 2 grand-daughters and 1 grandson	3	Sedentary
4	Awang Ketah's kitchen			
5	Sman	Wife, 4 daughters and 2 sons	2	Sedentary
6	Keling	Wife, 2 daughters and 1 son	2	Semi-nomadic
7	Pehyut	Wife and 2 daughters	2	Semi-nomadic

('sell') even if money is involved as *harga* (cost and price) is never explicitly negotiated. The act of 'giving' is often recalled as 'helping' at a later date. Although these exchanges are a common daily occurrence at individual and household levels, they transpire also at the village community level to demonstrate collective group action.

This network of exchange is particularly important when discussing the social organization of the Orang Laut. The network functions at different levels along a social continuum to unite individuals, households, the village and related kinspeople. These acts are representations of ideal behaviour that imply egalitarian status and generosity between fellow community members. Those who stay out of the system will face immense social pressures. They will be gossiped about, criticized as 'selfish', ostracized and labelled *orang lain* ('outsider') and eventually edged out of the network. As an expression of goodwill and for the purpose of maintaining peaceful social relations – to abate tensions, animosity and envy – networks of helping and sharing are established, maintained and continually renewed. In this manner, social relationships are strengthened because individuals and households in the village community will assist each other in times of need, such as accidents, natural disasters, illness, deaths, periods of scarcity or simply not being able to catch any fish on any given day. The need could also be to win the support of others to promote one's viewpoint or status in matters concerning village politics or to have reliable allies when serious conflicts arise.

Exchanges are preceded by house visits, which is a public way of declaring the social harmony between individuals. It follows that, when social relations break, all communication ties, visits and other forms of exchange cease at the same time. Exchanges of food, particularly seafood such as fish, crustaceans, molluscs and sea cucumbers, express group solidarity. Maritime products constitute the staple food for the Orang Laut, so the ability to gather and catch them is central to their identity. The combination of a person's skill and generosity in sharing earns even more respect from the others. After all,

> [s]taple foodstuffs cannot always be handled just like anything else. Socially, they are not like anything else. Food is life-giving, urgent, ordinarily symbolic of hearth and home, if not mother. By comparison with other stuff, food is more readily, or more necessarily shared ... [Thus] [f]ood dealings are a delicate barometer, a ritual statement as it were, of social relations, and food is thus employed instrumentally as a starting, a sustaining, or a destroying mechanism of sociability.
>
> (Sahlins 1974: 215)

Effectively, these exchanges between individuals, households and the community solidify the group and make it more stable. An elaboration of this socio-economic arrangement continues in Chapter 5.

Marriage and the nuclear family

The village community expands with the setting up of new households. This usually happens when a couple marries and decides to set up a social and economic unit of their own within the village community. Very soon after, when the couple bears children, a nuclear family household is established. Each nuclear family consists of a conjugal couple and their dependent children. Sometimes one or more additional persons may reside with them (e.g. an aged and widowed parent or sick relative who needs special care). Mothers and daughters tend to have close and intense relations. In old age, widows prefer to live with their married daughters than their sons.

A nuclear family often forms one fishing crew, but as soon as the children in the family are able to fish, they form separate crews. It is not uncommon to see a row of boats sailing together, all belonging to the same family. The nuclear family household may choose to reside in a house or in a boat. Some choose to live in a combination of both.

Normally, individuals want and are expected to marry. Marriage is the ideal and norm for all adults. Widowhood is generally followed by remarriage, even in old age. It is inconceivable that anyone should be a bachelor or spinster, and it is considered pitiful if one remains single for too long. Much assistance is given through verbal encouragement, introductions to likely partners and casting of spells to help late starters. Marriage, divorce and remarriage are very common. However, the first marriage is an important rite of passage. To frame this, Arnold van Gennep's (1960) and Victor Turner's (1969) analysis of rites of passage marks three stages: separation from the old status; a transitional state of liminality; and incorporation into the new status. A young and unmarried Orang Laut girl is called *gadis* (maiden, virgin), while a young and unmarried boy is referred to as *bujang* ('bachelor'). Marriage transforms one's social status from youth to adulthood and maturity. Nevertheless, the change of status is not instant. There is a stage of transition as the couple is not fully incorporated into adulthood until the birth of their first child.

The first marriage is usually considered when girls and boys reach puberty. Young people are generally free to choose their own marriage partners. Courtship is accepted, but it is hidden from public view. It is not as though no one is aware of it. Parents may know, but they remain silent if they do not object to the union. Yet, couples are expected to exercise great discretion in their courtship. They arrange to meet secretly, usually in the dark or in a secluded area. Sexual intercourse may even take place. If the girl becomes pregnant, there is a rush to marry. The Orang Laut are not particularly judgmental about sexual exploits as they consider sexual gratification normal, enjoyable and necessary. Nevertheless, all expressions of sexual intimacy, even between a married couple, should be hidden from public gaze.

When a couple wishes to marry, the boy usually requests his father or other related elders, such as his paternal grandfather, his father's brother or his father's or grandfather's close friends, to pay a friendly visit to the girl's parents to make marriage arrangements. Marriage can also be hastened by *lari* (when a couple 'runs off' or 'flees' together), which especially happens when a couple faces objections to their union. Parents may object because they doubt the character of the bride- or groom-to-be or the ability of the boy to take care of their daughter's needs or simply because they think that their son or daughter is worthy of a better match. When couples elope, they usually slip off together in a boat and stay away for a period of time. It could be for two weeks or a few months, but not forever. Before they elope, the couple often lets a close and trusted relative know where they can be found. This is so that as soon as their parents' anger is calmed and they are missed and forgiven, a message can be sent to them so that they can return and be accepted as a married couple. A married person, husband or wife, can also elope with their lover to escape an unhappy marriage. There is little choice for everyone but to accept the marriage of the couple once they elope. As soon as the couple returns, all is usually forgiven and the couple simply functions as a normal couple. Over time, a warm relationship may even develop between the parents and the son-in-law or daughter-in-law whom they once objected to so strongly or between former spouses.

Marriage between kin is preferred. Endogamous unions or alliances between cognatic kin are the most ideal and common form of union. Marriage between cousins, except the children of brothers, is particularly encouraged because such unions maintain and strengthen relations between those who are regarded as close or near kin. Furthermore, *ipar* ('in-laws') become siblings-in-law and so marriages between kin serve to strengthen and reconfirm networks of sharing and helping. Marriage between *orang lain* (non-cognates or outsiders) is possible, but not encouraged. Although marriage between kin is favoured, it must not transgress the rules of incest as explained by an Orang Laut, Bulat, from Dapur Enam:

If they are children of two brothers, they cannot marry. If they are the children of two sisters, then they may marry. The children of a sister may marry the children of her brother. The taboo is on the male line. To cite an example, the children of Joya [my sister] can marry Sman's [my brother's] children. However, Sman's children cannot marry my children. The children of two brothers cannot marry because *sama sama kuat* (literally, similarly powerful or strong [blood-line]). Children may be breast-fed by any of their parents' siblings. However, if Sman's children are breast-fed by Joya, then his children cannot marry her children later on. They may be *saudara* (brother, sister, cousin of the same generation) and friends, but they cannot marry.

A father cannot marry his daughters. Hence, children of the same father, including full- and half-brothers/sisters, and children of brothers or immediate patrilateral parallel cousins cannot marry. It is also prohibited for all those who have been nursed by the same woman to marry each other. Moreover, a woman may not marry a child whom she has nursed.

Polygamy or plural marriage is acceptable in the form of polygyny, that is, a man being married to more than one woman at the same time. Yet, the majority of men are married monogamously. This is because having two or more wives simultaneously can be very expensive. Although women are important contributors to the economy and are able to contribute substantially to their husband's wealth, the starting point of a polygamous marriage is the man's ability to support more than one wife. A polygamous marriage denotes a man's wealth and status. When a polygamous family is set up, it is not uncommon for the entire family to reside as a single domestic household and share household tasks to express household unity. In particular, there is never more than one hearth in the house. Cooking is either done together or the wives take turns cooking for the entire family. This commensality is the prime focus of what it means to be of one household.

In a polygamous marriage, the husband is expected to share his material possessions and favours among his wives according to the principles of equality. Although older wives often have special prestige and the right to be consulted before the younger wives, all wives have equal rights in matters of sex, economics and personal possessions. In everyday living, a household with multiple wives can be difficult.

Although complaint of jealousy may often be heard from those in polygynous marriages, those who complain also speak of the considerable economic and political advantages that such arrangements offer. Because polygynous families tend to have many more members who can fish or bring in other sources of income, an adequate food supply is better assured. Often, there are surpluses that can be sold to obtain extra income. Furthermore, their size means that they have more voices, if not a louder voice, in village community politics. Although wives may complain about each other, they are also aware that they have greater freedom to move around as they can share and help each other with child care and other household responsibilities. The older children of each of the wives can also be called upon to do household chores.

As mentioned, polygynous marriages are not practised by the majority. Rather, divorce before a new monogamous union is more common. Divorce happens frequently. It is accepted and not frowned upon. Men or women can initiate separations and divorces. Reasons for the breakdown in a marriage are normally attributed to the lack of support of providing for the needs of the family or, simply, incompatibility. An indication of an impending divorce is often that the husband or wife returns to his or her natal house. Sometimes, the spouse who has been left behind may attempt reconciliation by visiting the estranged partner to seek forgiveness or discuss ways of making the marriage work. Otherwise, divorce is usually uncontested and the couple simply reaches an agreement on how they should divide their property. Younger children usually remain with the mother, while older ones are given the choice of living with either parent. Fathers are usually expected to make some provision for the support of the younger children who have followed their mothers. However, if they fall back on their word, there are few sanctions except for pressures of gossip, which will hopefully shame them into fulfilling their obligations. There are instances when both parents remarry and their children feel uncomfortable in either of the new household arrangements. Under such circumstances, the children may look to their relatives for help, and a willing relative may take over the task of caring for them.

Remarriages are common after divorce or the death of a spouse. The divorced or surviving spouse generally arranges their own remarriage without much advice from others, who generally regard such unions as a decision made between mature and consenting adults. It is the customary practice that remarriages after the death of a spouse should not take place until a minimum period of one hundred days has passed. Nonetheless, there have been those who have been impatient and persisted in marrying within the hundred-day period. This act brings on wrath, if not fear, for the others. This is illustrated in the remarriage of Lampong, an Orang Laut from Pulau Nanga.

Lampong's wife, Siti, died due to complications in childbirth. The baby died too. The couple had an older son who was a toddler at the time of his mother's death. During this difficult period for the family, the entire community came together to 'help and share' with Lampong. Soon after Siti's death, Lampong sailed to Tiang Wang Kang island to seek a *dukun* ('shaman') who he believed would heal his son, who was crying endlessly for his late mother. However, shortly after his arrival on the island, he met a cousin named Simmon and relayed news back to the Pulau Nanga community that he wished to remarry immediately so that his son would have a mother again. He requested financial assistance too. The news was received with much shock and anger as everyone feared that a remarriage so soon after Siti's death would certainly anger her spirit and the community would face serious repercussions. Bolong, the village head and Lampong's eldest brother, instructed Lampong to return at once to discuss the matter. This infuriated Lampong. Only Lampong's brother, Boat, and his wife, Sarah, sailed to Tiang Wang Kang to offer help. However, while doing so, Boat fell ill. In the meantime, their sister, Suri, recounted all the unusual happenings that had occurred since Lampong's wrong-doing. Suri also spoke of a dream that she had had to everyone in the village:

> Sarah and Boat . . . went to help Lampong to remarry and now they are both ill . . . Boat now knows that he is in the wrong because he acted as the middle-person in arranging Lampong's remarriage. I have already told Lampong that he should wait until a hundred days have passed after Siti's death if he wants to remarry, because this ensures Siti would be far away. He should go to her grave and ask for permission to remarry. If she refuses permission, then he should not remarry. If he asks her, [she] will [appear in his] dream . . . Siti's spirit woke up from the grave and started throwing rocks about . . . The rocks kept hitting the houses. Attan was hit by a coconut that fell from the tree. No one saw how the coconut fell or what caused it . . . A *burung hantu* (bird spirit) . . . spoke in my dream and said that Siti was angry . . . It speaks the truth . . . The *roh* ('soul') of Siti is in the bird.

The community believed that Siti had spoken to them via Suri's dream. They had to act collectively to censure Lampong's behaviour or else the entire village would have to face Siti's wrath. Everyone, including Boat and Sarah, was obliged to withdraw any act of 'helping and sharing' with Lampong as a signal that this early remarriage was an act that defied the social order of their village community.

Formation of new groupings

When households increase in size for reasons of marriage and remarriage and as the children grow up and marry, the unit begins to break up. Some members, such as newlywed sons and daughters, move out to establish new households. These new units are established either in the same territory or elsewhere, such as in their spouse's territory. Nevertheless, those who move away to establish their new households still retain their rights to their original territory and can return at any time to stay or seek help from their lineage relatives. Sometimes the breakaway group may assume another name or become known by another term of reference, but this will not alter the fact that it is nevertheless regarded as a part of the original. Therefore, there may be numerous groups scattered all over that recognize each other, thus forming a network of inter-related territories.

3 History and culture change

The making of a marginal culture

Introduction

Although today the Orang Laut are viewed as a marginalized people living on the outer fringes of society, regarded with disdain and looked upon by the wider Malay and Indonesian society as a backward and unprogressive people, they were not always given such lowly status, as discussed in Chapter 2. Their current marginal status is one that has been culturally and politically constructed through shifts in political rule and their attending social ideologies. This chapter traces the story of historically changing commitments and actions of power holders in macro-polities and the stakes that they held in forging the rise and decline of the seafarers. It covers the time when the mariners were the favoured royal subjects and key players of the Srivijayan empire through to the nineteenth century when they were despised by the European colonialists as pirates to be suppressed. Through a historical survey of the role that the Orang Laut played in the Malay World, the political intrigues at the Malay court, the reasons for their initial loyalty and later wavering allegiance with the Malay rulers, the penetration of migrants with new skills, the introduction of new technologies and the superimposition of European powers, this chapter looks at the processes through which their marginality has been shaped.

The concept of marginality involves 'an ongoing relationship with power' (Tsing 1993: 90). It has at least five implications. First, an understanding of it posits 'a single analytical frame' (Li 1999: 1). That is to say, it calls into focus an assumed single and integrated system whereby margins constitute a part of this encompassing whole (Burling 1965). Second, it involves a social process of construction, maintenance, deconstruction or reconstruction. It is not one that has arisen or lessened naturally (Shields 1991). Third, it is a relational concept that connotes a range of dimensions pertaining to centres, intermediates and peripheries. Fourth, it concerns relationships set within frameworks of rank and order or hierarchies that formulate perspectives on similarities and differences, equalities and inequalities, as well as domination and subjugation. Finally, it deploys a discourse of power. It holds in place a set of discussions about 'distinctive and unequal subject positions within common fields of power and knowledge' (Tsing 1993: 90).

The construction of marginality and the constitution of a marginal culture serve particular agendas, including hegemonic projects (Li 1999: 1). They are constructed around ascribed or 'imagined' (Shields 1991: 47) differences. The imagined differences, once enacted, form prejudices for decision makers who design and interpret policies that bear upon centre–periphery relationships. The social construction of Orang Laut marginality as a hegemonic project has therefore involved a process through which they have become subject to simplified, stereotyped and contrastive descriptions that have rated them according to criteria defined at the centre. An inquiry into the context from which these perceptions have been formulated and the purposes that they have served suggests a medley of questions. What political agenda is accomplished by designating certain populations as 'marginal'? To what extent is 'minority production' a 'technology of rule' (Stoler 2001: 167)? What or who is being threatened or contested and why?

To define particular areas, peoples and practices as marginal or backward and in need of 'development' is not simply a description of a social world but a discourse of power. This chapter thus discusses the 'representational strategies and the social categories' that power holders of the macro-polities at particular political moments relied on to 'represent' those whom they wanted to penalize under certain circumstances (Tsing 1999: 161). This is no wonder at all because the European colonizers brought new concepts of borders that entailed fixed boundaries and demarcations when they arrived in the Malay World. Their notion of territorial power defined by centres and peripheries as well as clearly defined borders was in contradistinction to indigenous notions of power (Anderson 1972; Tambiah 1976: 112). Coupled with this were perceptions of 'external territoriality' and 'internal territoriality' whereby the colonial masters imposed their sovereignty on the people within their newly defined borders (Bryant 1998: 35–38). The Orang Laut consistently contested the rule of the colonialists and, in fact, have never relented on their claim of sovereignty over the Malay World. Even so, colonial representational strategies and the social categorization of the Orang Laut have committed them to a path of marginalization.

Historical records (cf. Schot 1882) reveal that the Orang Laut were once highly appreciated in the royal Malay courts. One aspect conspicuous in historical and contemporary studies of them is 'the ease with which the various *suku* . . . formed new settlements on other islands, or moved the general area of their operations over a period of years from one island to another' (Sopher 1977: 101). Their ability to move about and open up new areas constituted the most crucial factor for generating wealth in the pre-colonial Malay World. They created numerous micro-polities resulting in a broad network of constituents that eventually formed the power base of the Malay ruler. They were undoubtedly the building blocks for the sustenance of sedentary Malay communities. Their continual migration was not always due to their habit of mobility. Sometimes, under the direction of Malay overlords, they moved to new locations to provide some needed service; they were the manpower required to exploit the resources of those areas (Sopher 1977: 102).

To appreciate the significance of the Orang Laut's mobility, it is necessary to grasp an understanding of the Malay conception of a state. The territorial domain of a Malay polity was not determined by a fixed centre or boundaries. It was constantly in flux and it could be considerably expanded and strengthened by the mobility of its people. The centre was defined by the personage of the ruler. The centre moved according to where the ruler established residence. Hence, a centre never remained in any one location for long because the ruler would often look in all directions for security (Wolters 1999: 27–28). The sultan symbolized the unity of the state and his very existence ensured the continuation of his sovereignty. Likewise, the periphery of the polity was not demarcated by fixed boundaries. It was defined by tracing a community leader's relational propinquity to the royal family by blood, milk or adoption (Day 2002: 76). Therefore, the Malay rulers recognized that it was strategically important to accommodate the uncentralized Orang Laut micro-polities within the Malay state. They did this by employing the *mandala* system of loose governance. The close relationship between the Orang Laut and the Malays rulers was one of mutual benefit. Undeniably, the Orang Laut served the Malay rulers with profound loyalty. At the same time, they had much to gain from this alignment with the rulers and from their way of governance.

Seventh to thirteenth centuries

Key members of the Malay state

Chinese and Arab sources reveal the close bonds between the Orang Laut and the Johor royal house dating as far back as the rule of the Srivijaya-Palembang Empire from the seventh to the eleventh century. This special relationship was re-established in the late fourteenth century when Prince Parameswara founded Malacca (Wolters 1975: 9–12; Sopher 1977: 339–344). At least three assets constituted the power base of the Srivijayan ruler: the attractive port facilities of Srivijaya; produce, trade goods and labour that flowed from the interior communities to the city; and the Orang Laut. However, the Srivijayan Empire would not have existed without the all-important key role played by Orang Laut. Their loyalty to the ruler determined the rise and fall of the Empire. Years of sailing and living aboard small boats along the Sumatra coast gave them an intimate knowledge of the east and west coasts of the Malay Peninsula. They achieved the indisputable reputation of expert seafarers who could move swiftly in the intricate channels between the islands or narrow rivulets. They guarded the shipping lanes for the Srivijayan ruler and steered traders towards the Srivijayan port 'whether they wished to or not'

(Cribb 2000: 76). Because of them, the Srivijayan rulers were able to keep 'a broad suzerainty over potential rivals along the coast' while maintaining centralized power (ibid.: 76). They also played the crucial role of forming links to other sea peoples in the South China Sea (Wolters 1975: 13; Andaya 1997: 492).

According to a reconstruction of the early history of the Malay kingdoms by Wolters (1975: 77–78), there was a Srivijayan prince who was known particularly for his war exploits. Archival sources clearly reveal that it was the Orang Laut who played a major role in the successes of this prince. Upon succeeding his father as ruler of Palembang, the prince rejected the title of Sangaji that the Majapahit overlords imposed upon the Palembang rulers. Instead, he assumed the new title of Parameswara, proclaimed his kingdom's independence from Majapahit and appealed for the re-establishment of the glorious Srivijaya-Palembang past. His confidence in so doing rested on his power in the 'neighbouring lands' that covered much of the Orang Laut domain.

The loyalty of the Orang Laut to Parameswara was crucial for the rise of the Srivijayan Empire. Much of our knowledge of what follows is based on scattered European accounts (i.e. Pires 1944, II; De Barros in Pires 1944, II), contemporary reconstructions and analysis (i.e. Wolters 1975; Andaya 1975; Kwa 2005; Sopher 1977) as well as metaphorical tales (i.e. *Sejarah Melayu*). From these sources, it appears that when Parameswara was consecrated in the *abhi ṣeka* rites to mandate his *daulat* ('spiritual prowess'), he won for himself and his successors the undivided loyalty of the Orang Laut. Tomé Pires, the Portuguese apothecary who wrote the *Suma Oriental* by studying local documents and collating stories recited or read in Malacca at the time of the Portuguese occupation of that city, referred to the Orang Laut as *Celates*. Written between 1512 and 1515, the work was a study intended to deepen Portuguese understanding of their Malay enemies and assess trade possibilities in the region. Pires (1944, II: 250–264) described the *Celates* as 'obedient to Malacca' and 'thieving corsairs who go to the sea in small *praus* robbing when they can'. De Barros, a fellow Portuguese, concurred with Tomé Pires, in recapitulating the *Celates* as 'persons who live on the sea, whose occupation is robbing and fishing with the ruler's favour' (Pires 1944, II: 316).

The importance of the Orang Laut was affirmed in history again when Parameswara's ambitions in Palembang were abruptly halted with the arrival of a Majapahit punitive expedition which aimed to eject him from the region. To safeguard his people, Parameswara placed 1,000 men and their wives on junks and lancaras, while he remained on land with some 6,000 men to fight the Javanese. When the battle was lost, according to the *Suma Oriental* (Pires 1944, II: 233), only the Orang Laut could be trusted and relied upon to carry out the honourable task of ensuring the ruler's safety as he fled Palembang.

A select group of Orang Laut followed the prince in his flight. In times to come, they formed his core support group and were specially charged with the task of establishing a new centre to ensure the continuity of the empire (Barnard 2007: 36–37). The prince sailed first to Bintan, according to one version of the flight found in the *Sejarah Melayu* (Malay Annals), a semi-historical account written with the royal commission of 'setting out the descent of the Malay Rajahs and their ceremonial custom' based on stories that were originally transmitted orally and eventually compiled in the sixteenth century (Brown 1970: x, 2). The prince's voyage to Bintan was to seek refuge on an island with a large concentration of Orang Laut. The island, one of the largest in archipelagic Riau, had one of the most powerful naval forces because of the Orang Laut. It was also there where links with the wider Orang Laut world could be set up to safeguard the Srivijayan Empire. The ruler of Bintan had also previously recognized Parameswara's consecration and vision for Palembang and had provided most of the ships and manpower for the Srivijaya-Palembang Empire (Andaya 1975: 45). Upon Parameswara's arrival at the island, the Queen of Bintan, Wan Sri Benian, dispatched a fleet of 400 ships to greet him. Very soon after, the prince married the queen's daughter, a strategic move to assume the loyalty of all the Orang Laut. Subsequently, he was installed by Wan Sri Benian as her successor (Brown 1970: 18–10). In spite of all this, the prince had no desire to locate his kingdom on Bintan; instead, he wanted to establish a new capital elsewhere.

The Orang Laut, who had pledged unquestioning allegiance to the Palembang prince, found a new centre. It was only with their help that Parameswara, now also known as Sri Tri Buana, founded a new settlement on Temasek, which he renamed Singapura (Wolters 1975: 77–78; Kwa 2005: 126). The role of the Orang Laut in ensuring the survival of the empire was clear. As a token of gratitude and to strengthen his bond with them, Parameswara elevated their social rank by appointing them his royal retainers, thereby reinforcing their loyalty to the ruling house. Like Palembang, Singapura was previously under a Sangaji and a part of the Majapahit's *mandala*. Barely eight days after setting foot on Singapura, Parameswara assassinated the Sangaji. For the following half decade, he stayed in Singapura and the ruler of Bintan faithfully remained by his side from 1391 to 1392 and 1396 to 1397 (Andaya 1975: 46). Parameswara was expelled from the island when the father-in-law of the murdered Sangaji commanded a powerful expedition from Patani to avenge the death of his son-in-law. Once again, the Orang Laut had to find yet another new centre when Parameswara was forced to flee.

From Singapura, the Orang Laut led Parameswara to Muar. There, they cleared the jungle for food cultivation, fished and 'sometimes robbed and plundered the sampans that came to the Muar River to take in fresh water' (Pires 1944, II: 232). Trade remained the main concern for the prince, which he could easily pursue while situated in Muar, which was one of the choicest of Orang Laut sites and an important waterway linking the Muar River to the Pahang River via the Penarikan passage. Important products were transported from inland areas via the Muar River for international trade. Bertam and Malacca were other sites discovered by the Orang Laut following Parameswara's departure from Singapura. They advised their lord that these were potential locations that could serve as the centre for perpetuating the

Srivijaya-Palembang Empire. To convince their master of these choice spots, they presented him with a basket of fruits and a tree growing near the village at the foot of Malacca Hill.

There have been at least two interpretations as to what this meant. One view is that the Orang Laut wanted to show their master that these were fertile areas for cultivation and animal husbandry and were therefore able to support sizeable populations. Another view is that it was a metaphorical message to indicate that these were excellent trading locations that would reap great abundance (Timothy Barnard, verbal communication). Whichever interpretation is adopted, Parameswara's interest was aroused and he decided to examine the sites. He promised that if the sites proved suitable, he would move. Nevertheless, a quarter of his followers would remain in Muar to continue their work there. It is recorded by Pires that the Orang Laut answered:

> We too belong to the ancient lordship of Palembang; we have always gone with thee; if the land seem good to thee, it is right that thou shouldst give us alms for our good intentions, and that our work should not be without reward.
>
> (1944, II: 234–236)

The prince agreed, and the Orang Laut declared openly before all that 'if the land seemed good to him and that if he wanted to go there, he should do so and call himself king, and thus he could give them honor and assistance' (ibid.).

The sites recommended by the Orang Laut pleased Parameswara so much that he moved to Malacca to establish it as the centre for perpetuating the Srivijaya-Palembang traditions (Andaya 1975: 46). Although the prince had the support of the overall Orang Laut population, there would always be a small and select group who constituted his core supporters and accompanied him everywhere (Timothy Barnard 2007: 38). Thus, the 18 Orang Laut who had formed this core group and suggested the sites of Bertam and Malacca were rewarded handsomely; they and their sons and wives were made hereditary nobles.

This unique relationship between the Orang Laut and Srivijaya-Malay rulers remained impenetrable for centuries up to the colonial period. Two of the most important offices of the Malacca kingdom, namely the Laksamana and the Bendahhara, were always held by Orang Laut until 1509, just two years before the Portuguese took control of Malacca. The highest non-royal office in the land was that of Bendahara. It was initially occupied by an important Orang Laut who was such an important leader that he rose for no one but the ruler (Brown 1970: xi, 62, 160). The office of Laksamana also bore much power and influence in the kingdoms of Malacca and Johor as it was in charge of the ruler's fleet of ships. None other than the Orang Laut ever manned the fleets of both Malacca and Johor. Hang Tuah, the most famous Laksamana in Malay history, is thought to have been a member of the *Sakai*, a reference in the *Hikayat Hang Tuah* ('The Story of Hang Tuah') to the Orang Laut.

According to Pires (1944, II: 236), Paramaswara arranged for his son Iskandar Syah to marry 'the principal daughter of the mandarin lords who had formerly been *Celates*'. This was a reward for the girl's father, who was an Orang Laut leader. Subsequently, the father himself was also appointed Bendahara. The *Sejarah Melayu* provides further details on what happened later. When this Orang Laut father died, he was succeeded as Bendahara by his son, whose daughter then married Raja Kecil Besar, the son of Iskandar Syah (Brown 1970: 41–42). Henceforth, the tradition of the rulers of Malacca marrying the daughters of the Bendahara was established. These were favoured arrangements to ensure the absolute loyalty of the Orang Laut, which was crucial at a very precarious time in the history of the Malay kingdom. Constant threats from Majapahit and the more immediate danger of Ayudhya, which had become the leading entrepot in the region and regarded Malacca as a challenge to its status, necessitated the Orang Laut's services to protect and defend the Malay kingdom. After Iskandah Syah succeeded his father, he moved with his Orang Laut father-in-law and 300 followers to establish residence in Malacca to participate in international trade. He resided on Malacca Hill and the Orang Laut settled on the slopes of the hill 'to act as his guard' (Pires 1944, II: 233). The Orang Laut assisted in establishing the port city of Malacca. Within three years the population around the port increased to 2,000, thus transforming Malacca into the most important entrepot in the region.

Elsewhere in the Malay World, the Orang Laut continued to rule the seas. Bintan, in the thirteenth century, was described as possessing 'ships armed for war' as well as the Orang Laut, presumably from the Suku Bintan, who were called 'black pirates' who manned these ships and were told to 'rob the people but do not take them captive' (Sopher 1977: 342).

Sixteenth to seventeenth century

Honours received

The fact that the Malaccan royal house claimed their sovereignty through their lineage to the illustrious Srivijayan-Palembang Empire (Wolters 1975: 77–78) ensured the rulers of the allegiance the Orang Laut had so long ago pledged to the Palembang prince. Thus, the descendants of both sides pre-served the unique bonds that Parameswara had cultivated with his faithful Orang Laut followers, whom he had ennobled. For their continued loyalty, the Malay rulers regarded them as his sea subjects (Sopher 1977: 90).

Services provided

Throughout the sixteenth century and the during first third of the seven-teenth century, the Malay polity faced 'constant harassment and attacks from the Portuguese and the Achnese' (Andaya 1975: 48). However, the Orang

Laut were always able to support and restore the ruler to power throughout 'the changes of the kingdom' (Andaya 1975: 48). When Malacca fell to the Portuguese in 1511 and the capital was seized, the Orang Laut once again rose to the task of ensuring the continued survival of the Malay polity. Their command of the seas enabled them to fetch the deposed ruler and transfer him to a new capital (Wolters 1975: 12). With their protection and help, Sultan Mahmud Syah initially fled upriver to the royal residence in Bertam, next to Muar, then upriver via the Penarikan route to Pahang, then to Bintan (Pires 1944, II: 281) and finally to Kampar (Brown 1970: 186). Except for the flight to Pahang, it was a retracing in the opposite direction of the very route taken by Parameswara in founding the Malacca kingdom. This route was taken for the very same reason as that which had prompted Paramesswara to do so before. These were areas with heavy concentrations of Orang Laut populations who gave unquestioned loyalty and support to their Malay lord. Bintan remained the heart of the Orang Laut lands, and it was easier for the sultan to call upon them for help than from anywhere else on the Peninsula itself. On Bintan, when the Portuguese descended upon the Sultan and he sought refuge in the jungle, an official instructed his son to 'go and collect all the people living on the coast, and we will then go and fetch the Ruler', and so the son hailed 'the coast tribesmen who thereupon assembled'. These 'coast tribesmen' were Orang Laut who then led Sultan Mahmud to Kampar (Brown 1970: 186). They had to be summoned and assembled as they were normally not based in the capital city, but were scattered all over the place. They were often deliberately scattered in order to transport traders to the new capital to help establish Bintan as the new location of the entrepot. With the aid of the Orang Laut, Sultan Mahmud Syah re-established himself at Kampar.

Guardians of the sultan

In spite of the conquest of Malacca by the Portuguese in 1511 and attacks from the Achenese that led to great upheavals in Malay sovereignty, the Malacca kingship and power structure survived into the late seventeenth century (Andaya 1975: 15). Throughout the seventeenth century, the Orang Laut steadfastly remained the guardians of the sultan. This was clearly demonstrated by an episode in 1691. At that time, there arose a political struggle for power between the Bendahara Tun Habid Abdul Majid and the Laksamana named Paduka Raja. Paduka Raja and his family had managed to monopolize all the key positions of power in the kingdom, hence leaving the Bendahara family to a minor subsidiary role. At that time, the Sultan of Johor was only a child and placed under the regency of the Laksamana. Governing the kingdom on behalf of the young sultan, the Laksamana procured great benefits for himself and his family. As long as the Laksamana retained official custodianship of the child, who physically remained with him, no one dared to oppose the Laksamana and his position was secure. One day, the Bendahara faction abducted the child. Upon discovering what had happened, the Laksamana and his family understood the gravity of the situation and wasted no time in gathering their accumulated riches and the kingdom's regalia to escape by boat. In the meantime, the Bendahara ordered the *nobat* ('royal drums') and *nafiri* ('royal reed pipes') to be played as a signal to the Orang Laut crew of the Laksamana's boat that the ruler of the kingdom was now under the care of the Bendahara. Immediately, the Orang Laut abandoned the Laksamana and his family and switched sides to protect their young lord. No amount of gold or treasure could induce the Orang Laut to remain loyal to the Laksamana once the ruler was not physically with him and under his custodianship (Andaya 1975: 158). Their abandonment of the Laksamana sealed the fate of him and his family. The last details recorded of this episode depict the desperate attempts of the Laksamana to ward off attacks by the Orang Laut by firing coins from his cannon when his ammunition had been exhausted (Andaya 1975: 157–160).

Maritime subjects par excellence

The Orang Laut also fulfilled a function in the Kingdom of Johor in the seventeenth century similar to that which they had performed in Srivijaya-Palembang from the seventh to the fourteenth centuries. Regional and international trade in the Malay kingdom flourished, especially between 1641 and 1699 (Andaya 1975: 15), primarily for two reasons which can be directly attributed to the Orang Laut. First, their talents in attracting trade. The Malay rulers recognized the rivers, straits and islands off eastern Sumatra as Orang Laut territories (Wolters 1967; 1979; Andaya 1997: 487). Second, it was their skill and intimate knowledge of these areas that provided the much needed mother-of-pearl from the shell of the sea turtle (Hawksbill turtle/ *Eretmochelys imbricata*), edible seaweeds and *tripang* ('sea cucumber') for trading, particularly with the Chinese. As demand for these products intensified, they improved their skills in searching for them in places unknown to others throughout the Riau-Lingga Archipelago. Through their raiding activities, they also supplied slaves for trade. As their forefathers had guarded the sea lanes to bring in trade for the Malay ruler at the port of Srivijaya, they once again protected the Malay polity's commerce by patrolling and guarding the sea lanes for the sultan. Depending on the allegiance of the trading ships, they would either protect or 'demonstrate the ability . . . to wreak a fearful punishment' (Andaya 1975: 51) on ships passing through their home waters. Ships calling at ports under Malacca-Johor control, especially the port in Riau, were protected and led safely to dock. For these friends, they opened sea lanes, guided them through treacherous rocks and reefs, guarded them from attacks, showed them sheltered anchorage points in bad weather and led them to fresh water supplies (Wolters 1967; 1979; Andaya 1997: 487). Pires (1944, II: 233) recorded that in comparison, ships that traded with the Portuguese and Dutch in Malacca were considered traitors and viciously attacked, often with lethal blowguns and poison darts. The Portuguese incurred extensive damage when they attacked Malacca in 1511 because the Orang Laut were a major part of the defense of the city and their Malay lord.

Motivations for these Orang Laut attacks have been grossly misunderstood as criminal piracy in early colonial European literature. Mohamed Sharom bin Mohamed Taha observes that

> [t]he negative portrayal of the Orang Laut by early European literature might also be attributed to the fact that the Orang Laut fought against the Portuguese before the sacking of Melaka in 1511 and continued to be hostile to merchant ships trading with Melaka even after its capitulation.
>
> (2003: 17)

Contrary to colonial renditions, the maritime assaults were duties loyally and fearlessly carried out by the 'Maharaja's maritime subjects *par excellence*' (Wolters 1975: 10) under the direction of the Malay overlords. These activities fulfilled the wishes of the rulers for the specific purposes of creating and maintaining the Malay kingdom. The raids ensured a constant flow of trading activities for the success of the Malay entrepot. Indeed, these maritime forays were a crucial part of the Malay system of governance. The rulers had much power and wealth to gain by marshalling this nomadic and semi-autonomous sector of population onto their side (Pires 1944, II: 233–238). Therefore, their raiding activities at the behest of the Malay overlords were not only considered legitimate, but also greatly honoured. By Malay royal command, these loyal subjects were rewarded a percentage of the booty, and brave leaders were bestowed titles and incorporated into the official hierarchy of the Malay polity. This partnership resulted in the culmination of Johor's power in the seventeenth century. Ports under their control thrived from the constant increase in trade. Port facilities were constantly improved with the best equipment available, and the Orang Laut offered protection from possible Jambi and Dutch attacks. All this, which hinged on the pivotal role played by the Orang Laut, facilitated trade for the Johor kingdom.

Another constant threat faced by the Malay rulers was the power and influence of the *Orang Kaya* ('Malay nobles'). These were Malay nobles who formed their own centres of power (Andaya 1975: 39–44) and were supported by their own troops. However, according to Dutch calculations in 1714, the sultan was supported by an overwhelming number of 6,500 soldiers made up principally of Orang Laut (Andaya 1975: 50).[1] The influence of the Orang Laut in the Malay kingdom was considerable, as they provided the much needed fighting manpower to uphold the Kingdom of Johor (Andaya 1975: 50).

This chronological account of Orang Laut contributions towards establishing the Malay sultanates emphasizes their loyalty to the Malay rulers. They were indeed loyal, but their allegiance was also motivated by more mundane concerns (Chuleeporn Virunha 2002). As a people dependant upon the sea for their livelihood, they were aware that the sea yielded little during the monsoon seasons and, for survival, they had to engage in other activities, such as trade, to maintain steady sources of food supply. In return for their services and allegiance, the Malay lords conferred upon them patronage and protection which legitimized their trading forays and raids. Moving goods around in trade and carrying out legitimate raids proved to be beneficial for lord and vassal alike. By acknowledging the Orang Laut's seafaring style as vital for generating prosperous trading activities, the Malay lords saw their sphere of influence widen. In opening up more and more trading ports in the area, the Orang Laut gained the space and opportunity to develop new sources of food supply. Thus, this was a mutually beneficial exchange system; the rulers expanded their influence while the seafarers acquired necessities and luxury goods beyond their own ecological niches.

Close alliance with the sultans also meant that the Orang Laut were protected from the many surrounding threats, namely competing centres of powers that would enslave them, attacks by other fighter groups that threatened to subjugate them and even from their own internal group rivalries. Ambitious Orang Laut also perceived a close relationship with the Malay court as strategically opening opportunities for them to rise in their maritime careers. Rulers were willing to bestow honour, prestigious titles, high offices and wealth to those who served with the utmost loyalty. Power relations between the Orang Laut and the Malay state were open for negotiation. Those who rose in the ranks catapulted in political power and authority. In many ways, this enabled the seafarers to display their might and standing in the Malay World.

The decline of the Orang Laut

The role of the Orang Laut in Malay history began to decline with the death of Sultan Mahmud Syah in 1699 and the subsequent disintegration of the Malacca-Johor dynasty. Without an heir apparent, the impeccable lineage and sovereignty of the Malacca-Johor dynasty was discontinued abruptly. Sultan Mahmud Syah was assassinated by the Orang Kaya; Bendahara was the principal protagonist. Serious repercussions reverberated throughout the next two centuries for the Johor Kingdom and its loyal subjects. His death crushed the people's long-held belief in the sacredness of the ruler and affected their unquestioning loyalty. The Orang Laut were no exception.

The Orang Laut reacted strongly against the assassination and refused to acknowledge the new Bendahara dynasty. In their oral traditions, the Orang Kaya and Bendahara had asked them to collaborate in their plot. Almost all Orang Laut pledged loyalty to the sultan and would not participate in this heinous crime, described contemptuously in the Malay Annals as *derhaka* (treason toward a Malay ruler). Only one group, the Bintan Penaung led by Laksamana Bintan, collaborated. Henceforth, they were cursed for generations because of their complicity in the assassination (Mariam Mohd. Ali 1984: 89–91). Following the death of Mahmud, the usurper Bendahara installed himself as Sultan Abdul Jalil Riayat Syah. The Orang Laut, who condemned the regicide, refused to accept the new sultan. All centralized control over the Orang Laut was thus lost, and that special relationship between the sultan and Orang Laut dissipated (Andaya 1975: 189; 259).

In breaking ties with the new Sultan, each group of the Orang Laut went its own way and followed its own leaders, the headman and the Raja Negara Selat (Andaya 1975:189; 259). This led to an increase in undirected raiding, a predicament for the Dutch that was interpreted by them as a great increase in 'piracy' (Andaya 1975: 185). Raids by Orang Laut were no longer centrally co-ordinated. Instead, the head of a group would approach some individual of means, such as an Orang Kaya, and offer him their services. If the individual accepted the offer, he would then fund the expedition on the promise of receiving two-thirds of the booty (Andaya 1975: 51–52).

The desertion of the Orang Laut meant that the Bendahara dynasty had to seek a new pillar of support. Some years into the rule of the Bendahara, the Siamese launched an offensive, forcing the Bendahara to move his capital to Riau. The Orang Laut was frustrated and doubted the legitimacy of the Bendahara as ruler. Along with various dependencies, such as Selangor, Kelang, Siak and Indragiri – which altogether constituted approximately a quarter of the Malay population – the Orang Laut rebelled against the new Bendahara dynasty (ibid.: 191). Among the rebels there was an adventurer from Minangkabau, Raja Kecil, who claimed to be the son of the assassinated Sultan Mahmud Syah and, therefore, the rightful heir to the Malacca-Johor dynasty. This proclamation gained him massive support from the Malays, Minangkabau and Orang Laut (Barnard 2003: 55). As recorded in the *Tuhfat-al-Nafis* and *Hikayat Siak*, the Orang Laut stood united once again under the leadership of Raja Kecil (Andaya 1975: 267). Facing this situation, the Bendahara ruler sought help from the Bugis, who were highly skilled warriors and seamen from southwest Sulawesi (Trocki 1979: 9). In the end, the Bendahara was victorious. Some Orang Laut groups returned with Raja Kecil to Siak, pledging allegiance to him, while others stayed in Riau under their own leaders. Raja Kecil did not continue to command the allegiance of all Orang Laut, in spite of his claim to be the rightful heir of Sultan Mahmud Syah, because he had killed some Orang Laut leaders. Some groups even defected to Raja Sulaiman of the Bendahara dynasty (Andaya 1975: 294).

In the eighteenth century, the position of the Orang Laut started waning as the Bugis penetrated deeper into the Malay courts. The Bugis and other Sulawesi groups had in fact been coming to Malay areas since the last quarter of the seventeenth century. Initially, the Malay rulers perceived the commercial and military skills of the newcomers as great assets that could be used to realize their own ambitions for greater economic and political power. Thus, in the Strait of Malacca, the Bugis formed semi-autonomous communities governed by their own leaders. They paid nominal allegiance to the Malay overlords, but as their population increased along with their economic and military strength, they asserted their interests by involving themselves in the politics of the Malay states. Through intermarriages with Malay aristocrats, they established direct rights and links with the Malay courts. It was this ascent of Bugis power, both politically and economically, that led to the decline of the Orang Laut in the Malay states. The navigational and fighting skills of the Bugis made the Orang Laut redundant in the eyes of the local lords, and very quickly Bugis mercenaries were hired from the island of Bawean, north of Java, to replace the Orang Laut (Benjamin 2002: 45). By the mid-eighteenth century, the Orang Laut were no longer a united force and were still reeling from the unprecedented assassination of the sultan.

With the rise of the Bugis, the Orang Laut were forced to turn to maritime raiding under the leadership of various leaders to regain influence in the Malay World (Trocki 1979: 26). In the absence of a strong centre of authority that was respected by everyone and united the entire kingdom, fragmentation of the state happened quickly. The sultanate also weakened considerably because of the added stress of the emerging Dutch presence. Consequently, the different localities of the kingdom broke up into dissenting factions. Each locality wanted not only to control its own area but also to vie for political power at the perceived centres of Siak and Riau. To attain political supremacy and prestige, they conducted maritime raids against one another (Barnard 2001: 342). Furthermore, this was also a way with which the Malay leaders upheld the *adat* of the Malay idea of masculinity, courage and honour (Barnard 2001: 338). The absence of a powerful and all-uniting centre meant that each local leader could gather some men and conduct raids against his rivals. All this resulted in a period of protracted warfare between the Malay, Bugis and Minangkabau leaders. Each leader attempted to destroy the trade lifelines of their rival fiefdoms in order to reign supreme (Andaya 1975: 321). The Malay lords contracted the services of the fragmented groups. The Orang Laut could no longer be a united force as they were divided by conflicting claims to the throne and were uncertain of where their loyalty laid. The political importance of the Orang Laut dipped to its lowest ebb in the latter half of the eighteenth century as the Bugis and Illanun were now also available and could be engaged for maritime raids and other services too (Barnard 2007). The Orang Laut were no longer regarded as indispensable to the Malay lords.

Nineteenth century

By the nineteenth century, the Orang Laut were divided along complex lines of allegiances. Different groups acknowledged different Malay chieftains as their overlords. Examples of this fragmentation and the resulting mosaic of allegiances follow. Some 10,000 Orang Laut who comprised the Suku Gelam, Seletar and Kallang alligned themselves with Temenggong Abdul Rahman of Singapore, a prominent sea-lord of the time (Trocki 1979: iv). They fished for him and served as his boatmen (Sopher 1977: 105). When the British colonist Stamford Raffles stepped ashore on Singapore, Wa Hakim, an Orang Laut who witnessed the landing, reported that the Temenggong requested Batin Sapi, an Orang Laut leader, to fetch Tengku Long from Bulang for the occasion (Cameron 1882: 285–286). In contrast, the Suku Moro, Sugi and Gallang still aligned themselves with the Sultan of Lingga. Other groups pledged their loyalty to other rulers in the region. Some of these Orang Laut even joined the Illanuns and roamed the waters of the Reteh, Saba, Jambi and Indragiri rivers on the east coast of Sumatra (Warren 1981: 156–157, 159–161; Logan 1849a: 586–587).

The local chiefs and their respective entourages battled to extend their realm of influence. They traversed the Strait of Malacca and carried out raids to assert their political power. Although the Orang Laut faithfully carryied out their loyal duties to their respective local lords on such raids, the British misunderstood these raids as criminal acts of piracy. British perceptions were not to be cast aside lightly. As their expansion into the Malay Archipelago grew rapidly in dominance, their negative perceptions of the Orang Laut had serious repercussions. Identified as one of the main pirate groups and described as a class of thieving Malays by British colonial authorities, the Orang Laut came to be viewed as common criminals. These grossly misinformed and negative perceptions were derived from the earlier writings of other Europeans as well, particularly those of De Barros and Tomé Pires, who labelled these seafarers pirates and plunderers (Mohammed Shahrom Bin Mohamed Taha 2003: 26). In John Crawfurd's widely read *A Descriptive Dictionary of the Indian Archipelago and Adjacent Countries*, for example, there is a description of the Orang Laut:

> Next to these come the Malay pirates, once the most formidable of all, but now comparatively few and feeble ... and the principal parties concerned, the orang laut, or 'men of the sea,' the same race that is denounced by De Barros, as a people whose habitual occupation was 'fishing and robbing'.
>
> (1856: 354)

As the British penetrated the region, they were keen to learn more about its history and peoples. Nevertheless, in their efforts to deepen their knowledge, they consciously exercised a bias towards European works over other writings, in particular and rather ironically, Malay writings. According to Maier (1988), there were two groups of British colonists: an earlier group comprising Scottish merchant scientists who served in the first half of the nineteenth century and a later group of scholar-administrators who came at the turn of the twentieth century. Inspired by ideas of Scottish enlightenment, the first group held that all human beings were equal and possessed similar capacities if – and only if – they were given opportunities for improvement and civilization (Maier 1988: 37). This group regarded the Malays as 'intrinsically different ... They needed guidance because they were inferior. The gap between East and West should be confirmed rather than bridged' (Maier 1988: 119). The second group of scholar-administrators rejected any possibility of the

Malays attaining the benefits of civilization. Hence, the Malays could not be taken seriously at all (Maier 1988: 51–57). Darwinian theories influenced many of them to believe that there was a natural hierarchy of races in the Peninsula, with the British at the top and the Malays at the bottom, and thus some degree of inequality and segregation had to be maintained (Lian 2001: 864). By the turn of the twentieth century, British social and cultural superiority was firmly established.

Therefore, while such attitudes prevailed, in the view of the colonists 'the Malays themselves, like all people in the same state of society, have no history' (Crawfurd 1856: 250) and '[t]hey [native historical events] teach few lessons, because they belong to a system altogether distinct from our civilization. All that we learn from them [the natives] is the nature of barbarism' (St. John 1853: 43). From this perspective, only European works were worth consulting and translating for publication in their journals to propagate knowledge about the area and its people. Chinese and Arab works did not receive more than minimal attention (Mohammed Sharom Bin Mohammed Taha 2003: 26). It therefore came to be the case that all Malay conceptions of respect and honour for the role played by the loyal Orang Laut in their long history were rejected and dismissed as barbaric. Instead, antagonistic European impressions of the seafarers as maritime plunderers were embraced. This is shown below in sample quotes from colonial publications:

> The Lanuns were afterwards joined by the Rayats or sea-gypsies who made a large portion of the population of the Johore kingdom, and who lived entirely in their boats. These piratical tribes formed at last settlements on various points of the island, from which they incessantly harassed the natives of the island, carrying desolation in all quarters, and intercepting the supplies which the Sultan of Palembang sometimes sent to his unfortunate subjects in Banka.
>
> (Logan 1849a: 587)

> [They] took to the sea, and reinforced probably by more than a mere sprinkling of mere Malay adventurers, developed into the famous piratical race which under the generic name of Orang Laut became for a space the terror of all who sailed these Eastern seas.
>
> (Skeat and Ridley 1900: 248)

Explanations for maritime raids by the Orang Laut were framed in terms of criminal acts of piracy, an evil addiction and hereditary practices of an uncivilized people who observed barbaric native customs and traditions because of their inability to differentiate between right from wrong (Tagliacozza 2000: 71). In a letter by Sir Stamford Raffles to Lord Minto concerning the state of the Indian Archipelago, he wrote that piracy was embedded in Malay habits (Logan 1849a: 583). This view is echoed in Logan's (1849b: 635) comment that the Orang Laut 'are attached to [piracy] as to an industry which they have inherited from their ancestors so that it is impossible to convince them that it is criminal to addict themselves to it'.

In the eyes of the Malays, there is a clear distinction between *merompak* ('robbing') and legitimate raiding to uphold Malay customary laws for their lords (Barnard 2001: 338). The *Tuhfat-al-Nafis* clearly separates the actions of the Orang Laut under the leadership of Malay overlords from those of a robber (Trocki 1979: 57). In consciously rejecting the emic perspective to appreciate Malay culture, the British were unable to discern this. Instead, as indicated in the writings of the colonists below, all that they could see was that the Orang Laut were feudal vassals of various nobles who were providing them with arms, provisions and information to conduct piratical expeditions in order to procure surplus income to support their chieftaincies (Trocki 1979: 40–48).

> The receivers of the stolen goods, or at least the sharers in the booty, have been many of the native princes, who, far from thinking piracy any discredit, have looked on its gains as a fair and regular branch of their incomes.
>
> (Crawfurd 1856: 354)

> These chiefs did not go upon the cruise themselves but fitted out the expeditions, furnishing the necessary provisions, stores and arms, and received repayment after a successful cruise with a profit of 100 per cent.
>
> (Logan 1849a: 585)

As the above excerpts reveal, the Malay lords were likewise implicated in this piratical system. In fact, it was soon to be pronounced that 'the vice of piracy was stamped on the whole [Malay] race' (Thomson 1851: 143). The colonialists were aware that what they deemed as piracy was 'celebrated' (Logan 1850: 46) and admired (Thomson 1851: 143) by the Malays. What they failed to understand was the context within which maritime raiding occurred.

The tension between the emic and etic perspectives concerning maritime raiding was captured in the following newspaper article:

> The pirates of the eastern archipelago are communities recognising no superior jurisdiction; barbarous states not emerged from that grade of civilization, in which predatory warfare is considered honourable, the profession which the European pirate known to be lawless and immoral, is legal, and highly esteemed in the communities to which they belong.
>
> (*Straits Times*, 7 February 1850 cited in Hooi 1957: 45)

Opinions differed over the threat posed by the maritime raiders. By some, the boats of the raiders were considered just too small and simple to surpass the might of the bigger and more sophisticated European ships (Thomson

1851: 141). Some thought otherwise as they recalled the fates of the British schooner *Wellington* in 1812 and the Dutch cruisers *Iris* and *Doris* in 1823, which were overpowered by the raiders (St. John 1849: 259). Nevertheless, as British residents in the colonial settlements were bombarded with reports about maritime raids, their anxieties intensified (Tagliacozzo 2000: 70–71).

Finally, the British and the Dutch decided to cooperate to end these maritime attacks. Different strategies were adopted throughout the nineteenth century. These included making negotiations and alliances with local chiefs, using gunboats and steamships to attack the pirates, improving the sophistication of European naval technology, increasing resources such as weaponry and naval technology to curb piracy, judging the Orang Laut as criminals under the European system of law, signing treaties and delineating international boundaries. On the one hand, for instance, Sir Stamford Raffles was persuaded that a system of alliance between the local Malay lords and the East India Company would control piracy and restore the rights of native sovereignty (Tarling 1962: 69–70). Therefore, the British conscripted Temenggong Daeng Ibrahim, who controlled the Suku Galang, a group particularly known for their exploits in maritime raiding, to control his seafaring followers and surrender the slaves they had captured. On the other hand, the Dutch government and Sultan Riau-Lingga signed several treaties concerning jurisdiction over the islands that dotted the area. The British and Dutch signed a treaty in 1824 delineating international boundaries for more effective control over their territories. This treaty inevitably affected the control of the sultan and the local Malay lords over their people, including their Orang Laut followers.

Previously, there had been absolutely no need for the Malay rulers to know or determine which of the islands in the region were under their authority because the Orang Laut simply took care of such matters. Now, as the British and Dutch intensified their imperial command, the powers of the Malay rulers diminished and the Orang Laut enforcement necessary to their power base was also substantially crippled. Consequently, local Malay lords had no choice but to re-establish their political power around sedentary agricultural and mining communities, making the Orang Laut and their once all-important role redundant in the colonized Malay World (Trocki 1979: 85). In the ominous words of Trocki (1979: 85), 'Britannia ruled the waves, and no Malay ruler would ever again build a state based on the Orang Laut'.

The anti-piracy campaign was synonymous with an offensive against the Orang Laut by the Europeans. It was conducted with the propagation of an exaggeratedly uncivilized image of the Orang Laut, a people who had to be controlled and schooled to be civilized. The following excerpts from sample British publications present a quick overview of how the sea peoples were viewed through British lenses:

> The Orang Laut of the present day are not known to be addicted to Cannibalism, tho' it is extremely probable they were in former times, as

they yet retain all the characteristics of the most savage life. They rove about from one Island to another . . . They subsist wholly by Fishing and are very expert at striking Fish with the Spear. They live principally in small Canoes. Sometimes when the weather is boisterous or their little barks require repair, they erect temporary huts on the seashore. They are almost all covered with ringworms and scorbutic eruptions and have altogether a most squalid, wretched look. They are sometimes, when chance throws them in the way and they have become a little civilized, employed by the Malays to pull an Oar, at which, from their continual practice, they are very expert. Their Religion is, as Symes says of the Andamaners, 'the genuine homage of nature,' offering up a hasty petition to the Sun and the Moon.

(Anderson 1965 [1824]: xxxv–xxxvii)

Of a Creator they have not the slightest comprehension, a fact so difficult to believe, when we find the most degraded of the human race in other quarters of the globe, have an intuitive idea of this unerring and primary truth imprinted on their minds, that I took the greatest care to find a slight image of the deity within the chaos of their thoughts, even however degraded such might be, but was disappointed. They neither know the God nor Devil of the Christian or Mahomedan, though they confessed they had been told of such, nor any of the demigods of Hindoo mythology, many of whom were recounted to them.

(Thomson 1847: 343–344)

They [the Orang Laut] are less civilised than the *Orang Darat* [land-dwelling Malays], probably because most of them have not yet embraced Islamism. They are mostly fishermen and formerly pirates.

(De Bruyn Kops 1855: 108)

Henceforth the stereotyped picture of the Orang Laut as bad pirates, savage cannibals, less civilized than the Malays, among the most uncivilized pagans in the world and sea nomads who do not possess any territory was promoted. In the words of the British, the Orang Laut were a 'national disgrace' (St John 1849: 260). It was therefore the nation's 'sacred obligation to ameliorate the moral condition of the native' (Logan 1849b: 466). With both this thrust to civilize the pirates and their increasing spheres of political control penetrating into the peripheral realms such as the Riau-Lingga Archipelago, where the Orang Laut roamed, the British, like the Dutch, pressured the seafarers to abandon their uncivilized nomadic life at sea and assimilate with the larger sedentary and more progressive Malay community (Tagliacozzo 2000: 75).

Indeed, the anti-piracy campaigns ensured the safety of international traders plying through the region. Trade flourished and the region prospered (Campo 2003: 199). This only seemed to confirm the negative perceptions of the Orang Laut perpetuated by the colonists. As the colonial regime expanded, the Orang Laut were alienated and the traditional group organisation collapsed. They became thoroughly marginalized and viewed with disdain as a coarse people living in the periphery of Malay society.

The legacy of the European colonists in cultivating a negative perception of the Orang Laut has had a lasting impact. To this very day, it has affected their acceptance in other communities. Terms of condescension referring to them or representing them stubbornly persist. Ironically, the harshest ostracism the Orang Laut face today is from the very people whose history and world was mapped out by them: the Malays and the Malay kingdom. In spite of this dismal situation and even after the British and Dutch formally divided the kingdom of Johor in 1824 into two different worlds, the Orang Laut continued to view the Malay World as an integrated whole. It is a world they were so instrumental in constructing. They firmly hold to the view that the Malay World which they founded is not a bounded territory, but an expandable network of genealogically related places that evolved politically in the course of their maritime pursuits. No imperial power has been able to obscure the historical reality of the Malay World that existed for centuries.

4   The inalienable gift of territory

There are endless controversies over whether or not the Orang Laut own and may claim rights over areas that have historically formed their sailing routes, fishing grounds and moorage points. The first major impediment to understanding and accepting their notion of ownership lies in the supposition that they do not have any kind of fixed attachment to places. This assumption is based on the all too often negative characterization which stresses what they lack rather than what they have, namely an enduring relationship with their environment and an interest in increasing its productivity. This controversy suggests at least two ways in which the terms 'ownership' and 'rights' are understood: first, in terms of a permanent settlement that maintains exclusive privileges within an area with clearly defined boundaries; second, in terms of the movement across spaces to engage in and maintain social relationships with sites, navigational routes and surfaces for the practical business of gathering or harvesting maritime resources and rotating harvests to ensure the sustainability of resources. The latter means that the social relationships of ownership and rights are firmly anchored in the landscape via movement.

The next obstruction in acceding to their claims of ownership and rights lies in the presumption that title deeds in the form of paper documentation are necessary to legitimize one's assertions. This of course overlooks the long history of customary rights based upon oral traditions that are passed down through the generations. These oral traditions not only offer a subtle and richer discourse of the past that leads into the present, but they also show how spaces are mapped from an indigenous perspective. They show how '[p]eople do not simply experience space as abstract,' reflecting 'peoples' lived social relationships and the histories of their interactions' (Vandergeest and Peluso 1995: 389) with their coastal and maritime spaces.

The principal aim of this chapter is to introduce Orang Laut concepts of ownership and rights to maritime spaces. They revere their life and living spaces. To them, these spaces are inalienable gifts received from their ancestors. Their occupancy is a thing that cannot be separated, removed or alienated from them. It is indestructible and cannot be given away, exchanged or sold.

An inalienable gift is a highly treasured heirloom. A variety of things ranging from ancestral estates, human relics, precious stones and ritual para-phernalia to seemingly mundane objects like fishing spears, cooking utensils and combs can be accorded the status of inalienable gifts. Yet, they are dis-tinguishable from their mundane counterparts. They are artefacts and objects with extraordinary qualities; they have inherent supernatural powers. These items are spoken of as containing *ilmu*. The word '*ilmu*' is an Arabic-derived term which means 'knowledge' or 'science'. Wilkinson (1959: 421) lists the different kinds of knowledge that the word *ilmu* refers to: 'learning [based on divine revelation], science, magic, any branch of knowledge or magic'.

Inalienable gifts are devices for establishing and reproducing social rela-tionships between the givers and their recipients. They are sacred and removed from circulation, becoming unambiguously singular in kind. Tekong, an Orang Laut from Pulau Nanga, explains the meaning of inalienable gifts:

> They can be gongs, *keris* (wavy double-bladed daggers), *tanah* ('terri-tory') and whatever else. They can be given to us even before we are born. These are things from people of the past. These are very good things to own and they are also our customs, origin, and descent (*adat, asal, dan keturunan*). These things contain *ilmu*. Therefore, if anyone attempts to steal these things from us . . . [they] can make trouble for whoever tries to disturb it . . . There must not be quarrels over the ownership of the thing. No one can buy these things. They are things that look after us.

In an in-depth study on the meaning of inalienable gifts, which are at times also known as inalienable wealth, Annette Weiner (1985: 210) writes,

> Inalienable possessions are imbued with affective qualities that are expressions of the value an object has when it is kept by its owners and inherited within the same family or descent group. Age adds value, as does the ability to keep the object against all the exigencies that might force a person or a group to release it to others. The primary value of inalienability, however, is expressed through the power these objects have to define who one is in an historical sense. The object acts as a vehicle for bringing past time into the present, so that the histories of ancestors, titles, or mythological events become an intimate part of a person's pres-ent identity. To lose this claim to the past is to lose part of who one is in the present. In its inalienability, the object must be seen as more than an economic resource and more than an affirmation of social relations.

For the Orang Laut, the inalienable gift of ancestral estates is highly charged with a passionate, emotional and spiritual bonding between their ancestors (the givers) and themselves (the receivers). It bears the personhood of their ancestors beyond any spatio-temporal bounds and is invested for an indefinite time. Therefore, their ancestors who are the givers of these gifts continue to this present time to have a hold over them.

Through a complex network of inter-related territories that are collectively owned by different kin groups, the Orang Laut assume responsibility for protecting and taking care of their gift of ancestral estates. It is their responsibility to *piara* ('adopt') the gift. As with adopting a person, they take care of, maintain, raise and protect the gift that their ancestors have given them to the best of their abilities. Rituals are periodically carried out to empower the continuing life between their ancestors and themselves. Therefore, every claim of ownership is a 'linear projection of past into future, rather than as a sequence of isolable events each frozen in the instant of the present' (Ingold 1986: 138). All claims constitute 'part of a continuous process, expressing an intention or promise for the future through the fulfilment of past obligation ... A contemporary patch of land is thus seen as having a time dimension represented by ancestral estates' (ibid.: 138).

Ownership and rights

Before proceeding any further in discussing Orang Laut ownership of and rights over territory, we must first clarify their understanding of terms such as *tempat* ('space'), *punya* (possess, own and rights to) and *tanah* ('territory').

The most crucial point to keep in mind is that space, for the Orang Laut, is a continuous expanse defined by movement, perception and behaviour or activity. A space, or a number of spaces together, can constitute a territory in which people are dispersed. Territories need not be demarcated by borders or boundaries. That is, there is complete openness along the so-called 'borders', if any are to be spoken about, with no restriction on the flow of individuals crossing them. Visitors are free to access the resources in a given territory so long as they assume a subordinate position to, and recognize the privileges of, their hosts. To accomplish this, the visitor only needs to seek the permission of the host before entering a territory.

Ownership is an act of appropriation by which persons exert claims or rights over resources dispersed in space. The claim of rights specifies a system of relations whereby people as owners, subjects or productive agents direct their purposes. Otherwise said, the claim of ownership and rights regulates relations of production and delineates access to and control over the resources of an area. However, this does not mean inaccessibility or a closed-door policy to others. The mandate of ownership here is to enforce 'an ideological separation between "givers" and "receivers" in the case of food-sharing, and "hosts" and "visitors" in the case of territorial admission' (Ingold 1986: 134). This separation is essential 'for the expression of generosity, and for the satisfaction and prestige that accrues to those who – from time to time – are in a position to play host to their neighbours' (ibid.: 134). As such, ownership and claims of rights engage nature in a system of social relations.

The Orang Laut claim two different kinds of ownership and rights. First, when they speak of owning and having rights over a place (e.g. over a certain island), they mean to claim affiliation to this place which serves as their home base and for which they are the custodians. Second, they also claim rights over the sea routes by which they travel to other islands and along which they engage in fishing, their mode of production. Therefore, ownership and rights for them cover a network of places which include their home base as well as dispersed fishing sites, navigational routes and moorage points.

Oral traditions, title deeds and collective rights

Ancestral estates are never given specifically to any one individual Orang Laut. They are gifts that are given to a whole group. As such, the gift is inherited collectively and the recipients assume collective rights over the territory. No one disputes such claims. Different groups acknowledge each other's respective territories. Likewise, other local non-Orang Laut communities respect such areas as belonging to the mariners.

Territorial boundaries are generally open to movement across them. Nevertheless, all movements, including arrivals and departures, must be made publicly known and not concealed. For new arrivals or for those who wish to take their leave, there are correct procedures to observe. Of greatest priority upon arrival is a visit to the headman of the territory. The following information must be reported: the number and names of persons who have arrived, the anticipated duration of the stay, place of residence, moorage points and proposed fishing areas or other intended ways of procuring food and earning a livelihood. Claims of genealogical and kinship ties, no matter how distantly related, serve as the basis for a friendly welcome, hospitality and protection, if the need should arise. Within the correct order of procedure, it is also expected that those who are about to leave inform the headman of their intentions. Taking one's leave in the correct manner is an expression of gratitude for the host community's generosity, and such a gesture ensures that future visits are possible and would be warmly welcomed. Proper respect shown to the host community at the time of one's departure also shows the gratitude and honour due to the host.

Unannounced and secretive arrivals and departures upset and alarm people. Such movements fuel gossip, suspicion, anxiety and hostility. The aspirations of those who attempt to conceal their movements are bound to be suspect; they harbour malicious intentions and are up to some mischief. Referred to in hostile terms as 'outsiders', they are seen as dangerous intruders. Anxieties about them abound not because they are thought to be poaching the territory's resources, but because they are very likely practitioners of evil *ilmu* and are out to attack, harm and poison members of the resident group. This suspicion, in turn, motivates unfriendly behaviour towards such interlopers. As a precautionary measure, retribution in the form of casting *ilmu* attacks on these intruders to ward off their anticipated attacks must be meted out before the resident group suffers great harm. Quite regardless of any possible evil intentions, unannounced movements of interlopers pose a potential threat, perhaps quite accidentally, to disrupt the daily lives and work habits of others. Such disruptions, for example, could include drying up fresh water holes and disturbing fishing activities at critical moments.

On my first visit to the Orang Laut on Pulau Nanga, the Malays in the neighbouring island of Sembur instructed me to inform Bolong, the headman of the Orang Laut there, of my presence before I commenced my fieldwork. Similarly, while making known my presence to Bolong, the latter reminded me that he was only the headman for that particular territory. He stressed that should I row over to Teluk Nipah to seek Meen's permission. Meen was the headman of the opposite territory. Such advice to make my presence known and show deference to the respective headmen of different terri-tories was always given to me by the local inhabitants throughout my entire journey through the archipelago. It did not matter if I was travelling between rival territories. Although rivals have little to recommend of each other and much to criticize of the other's evilness, they would all remind me to respect the headman of their rival's territory if I persisted in travelling there.

The ownership and rights over a territory are contingent on the exclusive story that each group possesses. The crucial aspect is the claim that they were the first to recognize the potential of the area as a moorage ground or as a place to be cleared of jungle shrub for settlement. This is how an Orang Laut brother and sister, Boat and Suri, explained their family's claim over Pulau Nanga.

Boat:
My father's ancestry can be traced back to Pulau Tujuh. It is near Singkep ... My father rowed from Pulau Tujuh to set up a settlement here. He used to live at sea. Although he cleared up this island, which was formerly all jungle, he would also from time to time live at sea. We do that too. My father went to Dapur Enam to get married. He went everywhere in his boat and that was how he met our mother in Dapur Enam. We were all born in Pulau Nanga. My grandfather's name was Bubong. My father's name was Apong. Bubong's wife was Jolok. Jolok was also from Pulau Tujuh ... We are all siblings here. There are no other families who are mixed with us here ... We do not want to have other families with us either ... It may cause difficulties. It is much easier if we are all one family. If we have difficulties, we can help one another.

Suri:
My father, Apong, used to live on the sea. Then after much interaction with the Malays, he took up a religion. He then cleared the jungle – Pulau Nanga was formerly all jungle – and built his house here ... Sometimes, my father would live on land. Sometimes, he would live on the sea. Our father was the first to live on Pulau Nanga. Therefore, our ancestry is in

Pulau Nanga. This is our territory. No one can buy or take Pulau Nanga away from us.

Other local communities in the archipelago support this brother and sister's claim of territorial ownership. They support and confirm the fact that Apong, an Orang Laut, was indeed the first settler on Pulau Nanga, and thus the owner of the island. In fact, the Chinese in the area recount that even before Apong had set up house on the island, his boat-dwelling family had long moored their houseboats in the surrounding waters of the island.

In recognition of Apong's descendants' territorial claim to Pulau Nanga, Meen, the headman of the Orang Laut in Teluk Nipah, explained how his family had in turn come to claim ownership of Teluk Nipah. In his account, Meen also explained why his family chose not to go to the neighbouring Pulau Nanga in spite of its being inhabited by Orang Laut:

> My mother's name was Nenah and my father's name was Gebak. Their ancestry is Daik. We were facing difficulties there, so we rowed over here to fish. We were boat-dwelling. We set up a small cultivation plot and built a house on Teluk Nipah. There were already people on Pulau Nanga. Apong had settled there, but our origin is different. Although there were Orang Laut on Pulau Nanga, we did not want to settle here. We are different. We were the first Orang Laut to settle on Teluk Nipah. This is our territory.

Meen's claim of his family's territorial ownership of Teluk Nipah parallels that of how Apong's descendants have come to state their claim over Pulau Nanga. Both of them possess stories that relate how their ancestors were the first people to realize the potential of the territory as a living space, and that it was their ancestors who gave life to these territories by transforming them into habitable places.

Next is another story that is exclusive to the Orang Laut of Pulau Dapur Enam, transmitted orally through the generations to verify their collective claims of territorial ownership. Here again, their story is acknowledged and accepted as their title deed by the other inhabitants of the area. This is the story as told by Awang Ketah, the founder of Pulau Dapur Enam, and his daughter, Joya.

> Awang Ketah:
> I was the first to settle in Dapur Enam and I have been here for many years. Dapur Enam was formerly all jungle and I cleared it to build my house. Later, a Chinese boss named Chin Keong arrived. I helped him build his house and construct his charcoal kiln business. A number of Malays then came to work for him. The Chinese boss was very close to me and provided me with most of my needs. He was very grateful to me and when he died he left everything to me. The Malay workers left after

the demise of their boss. However, my family stayed as this territory is ours.

Joya:
Our ancestry is from Dapur Enam. My grandfather Umur lived in a boat. He was the first to move into Dapur Enam. He would return to live in his boat whenever he needed to fish for food. My father, Awang Ketah, was the first to settle in Dapur Enam. When the weather is bad and the sea is choppy, one can feel very dizzy in a boat. Hence, my father decided to move to land . . . All the houses in Dapur Enam belong to my father's children. My father feels that it will be a pity if he should ever have to leave our territory . . . We have many family graves in Dapur Enam. Although my father has moved to land . . . From time to time, he returns to live in his boat.

Awang Ketah's claim of territorial ownership was readily endorsed by Kelit, a Malay man in the neighbouring island of Abang, and Su Lang, the daughter of Chin Keong the late boss of Awang Ketah.

Kelit:
Awang Ketah was indeed the first to open up Dapur Enam and to settle on it . . . In the early days, he was still single and did not have any children, so he used to row and sail all over the archipelago and would only return to Dapur Enam once in a while. Now, he has a house and stays in Dapur Enam for longer periods because his children and grand-children live there too. The Malays moved out of Dapur Enam after the death of the Chinese boss, but Awang Ketah's family remained on their territory.

Su Lang:
My father had set up a charcoal industry on the island, but he was only able to do so with Awang Ketah's permission and help. Although my father was the boss, Dapur Enam belonged to Awang Ketah. Awang Ketah was the first to clear up Dapur Enam and establish a settlement there. Awang Ketah and his family were very good to my family. They not only worked for us, they also protected us [from pirates] . . . When my father died, my brother did not want to continue with the charcoal indus-try in Dapur Enam. Our family gave all our charcoal kilns and fruit trees on the island to Awang Ketah. After my father's death, all the workers moved away. Only Awang Ketah and his family remained as they owned Dapur Enam.

In recognition of Awang Ketah's descendants' claims over Dapur Enam, other non-related Orang Laut will neither moor nor set up house on the island. Today, visitors to the island are expected to inform Awang Ketah's second son, Sman, who has since been appointed head of the territory, of their arrival.

These stories, as exemplified by the narratives of Boat and Suri of Pulau Nanga as well as Awang Ketah and Joya of Dapur Enam, are the Orang Laut's title deeds to these territories. They are tantamount to an assertion and a certification of their group's collective rights to a particular place or places. These stories must also be understood as revelatory myths whereby local spirits revealed to particular Orang Laut groups the areas' potential. The indigenous knowledge possessed by these groups about their particular areas is therefore not just technical know-how, but sanctified understanding. The right of the Orang Laut groups to live and work in their territories rests on their unique understanding of a particular place through their own experience or as revealed by the local spirits of that place. Although they are the owners of these territories, they allow anyone and everyone free access to their sea, coast and land. There is the unwritten understanding that, as they hold the territorial rights to these areas, anyone who wishes to enter their territory should first seek their permission.

Custodianship and responsibility

One of the most important outcomes of these title deeds is that they modify and organize maritime spaces into areas for custodianship. That is, these rights charge the owners with responsibilities for the area and its resources that their ancestors have given them. They must *jaga* (guard, preserve and look after) their territories, as well as maintain and reproduce the resources in them. These territories cannot be left idle as idleness would anger their ancestors and harm would befall the surviving descendants and receivers of the gift. Since these gifts of territories were first supernaturally revealed to the founders and consequently given a social life through the creative actions of their ancestors, these areas are imbued with a power that germinates with the people who possess it. This power is charged through periodic ritual renewal – as gifts demand repayment – to ensure its continued social life. These rituals must therefore be performed to manifest the performers' responsibility for custodianship. This transformation of objects into subjects and vice versa resonates with the central Maussian theory of gift exchange: Things transform into 'parts of persons' and persons 'behave in some measure as if they were things' (Mauss 1954: 11). This 'conception of the consubstantiality or interconvertibility of places and people, according to which each appears immanent in the other' also clearly indicates that there is a 'spiritual quality' in staking such claims of territorial ownership (Ingold 1986: 139).

The *upacara menebar beras* (ritual of scattering uncooked rice) is one of the most important rituals regularly performed throughout the year by the Orang Laut to look after their territory. No fixed dates in the calendar year determine when it should be carried out, and the timing is not dictated by the number of times that it should be performed each year. Such decisions largely rest on the member of the community who is appointed to look after the territory. The person in question is always a respected shaman known to possess the particular power and spells, and hence the presumed effectiveness for looking after territories. Certain events or circumstances will prompt the shaman to carry out the ritual. Such events or circumstances include the arrival of unannounced visitors, communal unease about the intentions of a visitor, the unexplainable illness of a member, conflicts with rival communities and the appearance in dreams of spirits or ancestors asking to be fed. Apart from these happenings, even when all things appear to be calm, the shaman will at times perform the ritual simply to honour his group's responsibility as custodians of the territory.

In the ritual of scattering uncooked rice, the shaman casts a spell over some uncooked rice. The rice is first scattered around the territory and then inside the territory. As discussed earlier, a territory is not demarcated by any visible boundaries or enclosed with a clearly marked entry point. Thus, the boundaries where the shaman scatters the uncooked rice around are abstract boundaries symbolizing 'perimeter defence' and 'social boundary defence' (Cashdan 1983).

As one's territory intrudes into the realms of air, water, earth and jungle, the territory must simultaneously show signs of reverence to the spirits that live in all these realms and be protected from the outside spirits. The goodwill of all these spirits must be enlisted to maintain the life of their territory, which is inextricably linked to their own well-being.

Lacet, a member of the Pulau Nanga Orang Laut community, explained how his older brother, Ceco, was responsible for protecting their village and territory:

> None of us knows everything. Hence, each of us is responsible for different things. My elder brother Ceco protects the village ... from other people who might try to harm us and disturb our territory. From time to time, Ceco will *jampi* ('cast spells') and give uncooked rice to our territory.

The explanation offered for Ceco's act of scattering grains of uncooked rice is twofold. First, the rice is scattered in the open space that separates the front and back of Pulau Nanga from the other island communities. This ritual was thus performed to both protect the Pulau Nanga Orang Laut from outsiders who might enter and cause harm and induce the intruders to leave the territory. Second, the rice is scattered to appease the spirits who govern the territory to ensure the well-being and continued productivity of the area.

Studies on the symbolism of rice in the Malay World maintain that rice is perceived as a gift from God for the survival and well-being of human beings (Ismani 1985: 122). It symbolizes the 'condition of primordial plenty' (Jacobsson 2005: 30), indestructibility, procreation or regeneration as well as the 'atonement for transgressions against local *adat* prescriptions' (ibid.: 31).

A rice ritual is thus a 'temporary re-enactment of the blissful state of primordial plenty on earth' (ibid.: 30). It also wards off misfortunes (Ismani 1985: 123) and promotes the reproduction and 'prosperity of human beings' (Jacobsson 2005: 37).

According to Lacet, the Malays in particular constantly harbour evil intentions to harm the Orang Laut. It is therefore necessary to protect their territory. Lacet explained that they were saved from being harmed by a Malay woman who had entered their territory through the guardianship of their ancestral estate and Ceco's efforts in honouring the ancestral estate:

> You must wonder why we did not want to have anything to do with the Malay woman who came to sell us her cakes. She had a basket full of cakes but none of us touched a single one . . . Ceco had been warned in his dreams. We would have been harmed if we had touched or eaten her cakes. It did not take the woman too long to be aware that we knew of her evil intentions, and she became afraid of us. That is why she retreated so quickly . . . The Malays like to harm us . . . We have to be cautious . . . We told the Malays that our territory and ancestors could strike back and they knew that we were not lying.

Thus, as Lacet explicated, they have to be watchful because of the constant threat of possible attacks from the Malays. They must observe appropriate rituals and cast spells to ensure their protection and survival.

Spells and custodianship of things

The casting of spells constitutes an integral part of the Orang Laut's responsibility for custodianship. This is also reflected in the manner in which they assume custodianship of other things. They possess different categories of things, which are valued differently: non-adopted things, adopted things and inalienable (adopted) things (Chou 2003: 74–84).

Non-adopted things include anything owned or used by the Orang Laut. These things possess very little or no supernatural power. Few inhibitions are felt about selling or bartering these things. Within this category are: things that bear their craftsmanship, such as their fishing spears, hand-carved ornaments and hand-woven baskets; things derived from their practical activity, such as maritime products they have caught or collected, plants they have cultivated or food and drinks they have prepared: and things that have been brushed over and been in contact or close association with their bodies, such as their clothes and footwear.

Adopted things are those that the Orang Laut consider important to them.[1] When a non-adopted thing, such as fishing equipment, a house they live in or a small cultivation plot, begins to assume some importance for them, it will be transformed into an adopted thing. This is accomplished by casting a spell over this thing and carrying out rituals of adoption. This adoption endows the thing with supernatural powers which then increase its usefulness or effectiveness. It is believed that in order to reap plentifully from one's possession, one has to adopt the thing well and take good care of it. More importantly, when a thing is adopted, a bond is established between the person and the thing. This is because an Orang Laut's decision to adopt a thing entails a decision to merge it with his or her own identity, which is the soul or inner essence of the person. To adopt something in an irresponsible manner endangers one's own well-being. If the owner is negligent in caring for an adopted thing, he or she may be harmed, attacked or polluted by the invasion of a spirit that would, by way of the adopted thing, either consume or cause danger to the well-being of the owner.

An adopted thing is therefore also a metonymic sign (Leach 1989: 12) for the owner. That is, a contiguous relationship exists between two things, A and B, because they belong to the same cultural context. A part of A stands for the whole of B (metonymy) (Leach 1989: 14). If the sign is destroyed, the owner will also be injured. Therefore, one of the most feared dangers is permitting adopted things to fall prey to the evil intentions of outsiders. Outsiders could sabotage a person through these adopted things which bear directly on the owner's well-being.

However, this is not to say that adopted things cannot be alienated. It is common for them to be exchanged, given away or even sold. The bonds between the owner and the adopted thing and the meaning and power of an adopted thing can be transformed via the way in which it is circulated. A greater social value in an adopted thing and a deeper bond with the owner are created if the thing is circulated as a gift or a bartered object, rather than sold for money. This is because reciprocity is involved in these two forms of exchange. That is, a thing given or bartered away is often replaced by another thing or some kind of service.

As adopted things are metonymic signs that embody the identity of the owner and can establish a spiritual bond between the transactors, it follows that they can only be given or bartered away between members of the same group. When the thing enters the sphere of inter-community circulation, selling for money becomes the preferred form of exchange. The logic of money is that it is bound up with impersonal markets. It is something that can enter into definite and quantitative relations independent of the persons engaged in the transaction. When money is used for exchange purposes, contracts of exchange can be fulfilled immediately with no further obligations between the transactors. As money reconstructs the meaning of things and establishes new identities, it is a kind of safety valve to distance things from their owner, thereby reducing the dangers involved.

The highest order of valued things is inalienable (adopted) things. Examples include ancestral estates and other heirlooms such as musical instruments and stones. These things are deemed to be imbued with inherent powers and an inalienable value. The word 'adopted' in this category of things is bracketed to highlight the point that even though the owners of such things usually make the effort to adopt them, their efforts do not endow, but rather enhance, the inherent supernatural powers of the things. The longer these heirlooms are kept within a family or its descent groups, the more their value increases. Age adds value and so does the ability to keep these things against all urgent needs and demands that might force a person or group to release them to others. Keeping them is therefore a creation of value (Weiner 1985:210). Orang Laut ancestral estates are thus, to put it succinctly, systemic to their being and thus their most precious possession.

Network of inter-related territories

The Orang Laut organize their space around a network of inter-related and collectively owned territories based on kinship ties. They have three types of fishing grounds in relation to their anchorage sites or land-based settlements. First, families tend to fish most intensively in areas closest to their home anchorage as these anchorage sites give a partial sense of territorial definition. Second, families may fish in areas that are several days' travel away from their village. Third, families may fish in distant areas that are several months' travel away from their village.

The sites of production chosen, especially those of the last category, are usually kin-related areas that they consider *tempat kita juga* ('our place too'). Map 4.1 is a cartographical approximation of this web of inter-related, kin-infused, cultural–economic territories. It not only shows the inter-related and collectively owned territories based on kinship ties, but also the maritime resources that can be found in these areas. All places marked with the same number form a particular network. These ethno-geographical maps are held by individual Orang Laut as mental images which are shared and transmitted orally and through experiences with others. The map is plotted on the basis of field research spanning almost a decade and a half, yet it depicts only a fraction of a more expansive picture of how social relationships are firmly anchored in the landscape. New anchorage sites and centres of settlements are constantly being discovered, established and linked to the existing network of communities.

Orang Laut mobility is not random. It is a widely held and mistaken view that they are a people who move around on a whim. The guiding principle of their navigational routes involves inter-related territories which are based on kinship ties and related resource sites. This network translates into spheres of sustenance. As discussed in Chapter 2, they select among kin-related territories based on their cognatic lineages to seek a maritime harvest. Likewise, they identify themselves with particular moorage communities. Within an anchorage, related families moor together in a cluster, often tying their boats one behind the other and securing them to one or possibly several common moorage posts. Network number 3 on Map 4.1 shows the inter-relationship of the Tiang Wang Kang Orang Laut with those in Pulau Setengah, Air Lingka, and Pulau Nanga. It is common practice for those from Tiang Wang

Kang to sail all the way to Pulau Nanga off Galang when they want a good harvest of sea cucumbers and jellyfish. While en route, they have the choice of stopping at Pulau Setengeh and Air Lingka. Correspondingly, during the season of *comek* (a variety of cuttle-fish) fishing and for work in the mangrove and charcoal industry (to increase their income), those from Pulau Nanga undertake a journey in the opposite direction, to Tiang Wang Kang. Network number 2 on Map 4.1 traces the mobility pattern of the Orang Laut in Teluk Nipah to Bertam, Pulau Cakang and Berakit. A journey in the opposite direction similarly occurs for those in Berakit going to Pulau Nipah.

These networks link the Orang Laut in space and time. The time aspect is that a routine sequence of resource use based upon the change of seasons brings the Orang Laut of a certain group together. On the side of pragmatics, these movements from settlement to the sites of production mean that populations of Orang Laut are never so concentrated in any one place as to exhaust the area's resources. This arrangement functions as a mechanism for population dispersal for the purpose of evenly spreading out resource consumption over a wide area. The association of specific Orang Laut groups with specific parts of the archipelago parses resources in such a way as to give each group the best yield for the season. It keeps in check over-exploitation in terms of over-harvesting and over-utilization of resources. The seasonality of consumption ensures that different sites have fallow periods when the resources can be renewed. Attention devoted to the conservation of resources ensures sustainable consumption.

The relationship between the Orang Laut and specific areas is more than just toponymic identification. While moving between these inter-related territories, they identify these sites as 'our place too'. In identifying certain territories as such, they assume responsibility as custodians of the territory. The Orang Laut of Dapur Enam, as indicated in Map 4.1, amply illustrate this point. Their main resource zone is located in the Dapur Enam area. They are recognized as the custodians of this zone, but they do not necessarily dwell on the island of Dapur Enam all the time. Some live on the island permanently, while many others live in boats more than on land. Moreover, an individual may be variously more land-dwelling or more boat-dwelling at different times in their life, depending on factors such as personal preference, state of health and season. The number of Orang Laut on Dapur Enam, therefore, constantly changes. Whether they are land- or boat-dwelling, the island is their home base to which they have the right to return if and when they wish.

While most voyages carried out in this network of inter-related territories are undertaken for purposes of fishing, the seafarers do not just spend all their time fishing and travelling to and from fishing sites. Within a given network, a family may stop over at a place for a night, several days or an extended period of time. Reasons for the stopovers vary and include visiting kin, exchanging news, selling their produce to the middleman of that area who may buy their catch at more attractive prices and finding extra paid work. Trips are also explicitly made to visit relatives within the network. Such

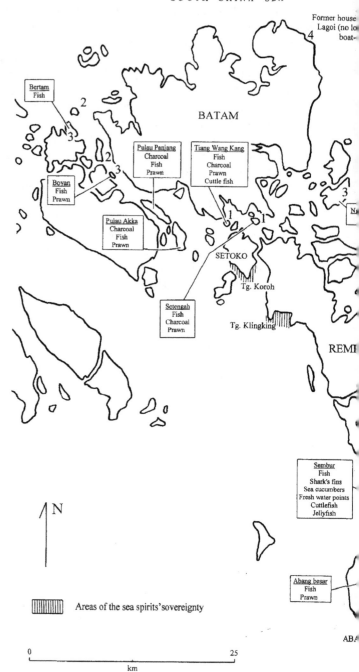

Former house
Lagoi (no lo
4   boat-

Bertam
Fish

BATAM

2

3

Pulau Panjang
Charcoal
Fish
Prawn

Tiang Wang Kang
Fish
Charcoal
Prawn
Cuttle fish

2

3

Boyan
Fish
Prawn

3

N

Pulau Akka
Charcoal
Fish
Prawn

SETOKO

Tg. Koroh

Setengah
Fish
Charcoal
Prawn

Tg. Klingking

REMI

↑ N

Sembur
Fish
Shark's fins
Sea cucumbers
Fresh water points
Cuttlefish
Jellyfish

Areas of the sea spirits'sovereignty

Abang besar
Fish
Prawn

0   25
km

AB

Map 4.1 The network of Orang Laut inter-related territories and their identification
of resources.

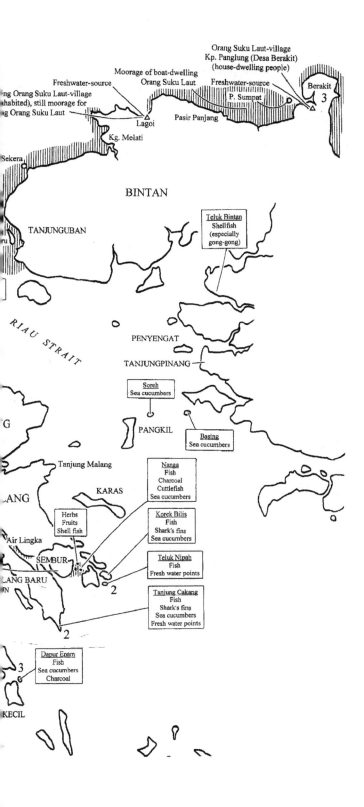

Orang Suku Laut-village
Kp. Panglung (Desa Berakit)
(house-dwelling people)

Moorage of boat-dwelling
Orang Suku Laut

Freshwater-source

ng Orang Suku Laut-village
habited), still moorage for
g Orang Suku Laut

Freshwater-source

P. Sumpat

Berakit

3

Lagoi

Pasir Panjang

Kg. Melati

Sekera

BINTAN

TANJUNGUBAN

Teluk Bintan
Shellfish
(especially
gong-gong)

RIAU STRAIT

PENYENGAT

TANJUNGPINANG

Soreh
Sea cucumbers

PANGKIL

Basing
Sea cucumbers

Tanjung Malang

Nanga
Fish
Charcoal
Cuttlefish
Sea cucumbers

KARAS

Herbs
Fruits
Shell fish

Korek Bilis
Fish
Shark's fins
Sea cucumbers

Air Lingka

Teluk Nipah
Fish
Fresh water points

SEMBUR

ANG BARU

2

Tanjung Cakang
Fish
Shark's fins
Sea cucumbers
Fresh water points

2

Dapur Enam
Fish
Sea cucumbers
Charcoal

3

KECIL

trips may be made to attend a sick relative, await a birth or participate in a marriage celebration. It is also common practice, for example, for the Orang Laut of Pulau Nanga to travel to Tiang Wang Kang each year over Christmas. Christian missionaries visit Tiang Wang Kang during the Christmas season, hold church service on the island and offer the Orang Laut a sumptuous feast with lots of presents for the children.

Sometimes a family journeying through a network may have no intention of stopping for long at a particular point. However, should a family arrive at a destination to find that a relative has unexpectedly passed away, they are obliged to stay a few days. As a mark of family solidarity, they are expected to make a contribution towards the expenses for the burial as well as other pre- and post-burial ceremonies. On these stops, be they for happy or solemn occasions, the family may decide to stay in their own boats or set up temporary settlements next to their relatives' homes. Some will also stay with their relatives. They are never refused a welcome, and food and drinks are shared between the families. This hospitality is reciprocated when these relatives visit the family at the other end.

Territorial ownership by spirits

The Orang Laut also fish in other pre-established locations that they do not identify as 'our place'. This can be understood within their overall concept of territorial ownership. They consider these other territories to be borrowed areas that are owned and governed by *hantu laut* ('sea spirits') and *hantu darat* ('land spirits'). To enter and use these spaces, it is necessary to present prestations to the spirits. Due respect must be paid to the spirits and they must not be neglected. Anyone who wrongs them or trespasses upon their sacred territories faces imminent and ferocious attacks. Their *semangat* ('souls') will be consumed and the essence of their being will be gorged out. Spirits do not devour the physical part, such as the flesh, of a person. Instead, they consume the essential part that constitutes the life or death of a person.

Two Orang Laut children from Pulau Nanga, Halus and Bagong, rowed me in a boat to educate me in recognizing and respecting the various forms of territorial ownership by spirits. Halus pointed out some coral reefs and said, 'These [corals] here are the house of fishes. These are not dangerous areas'. Further on, Bagong pointed in the direction of jutting rocks and cautioned me,

> Over there is a sea spirit's house. It is best to avoid that area so as not to disturb the spirit. There are spirits in all the capes. Sometimes when we come home, we fall ill immediately with a fever because the sea spirit has hit our head.

Certain features can be associated with territories identified as places or houses of fish, sea and land spirits. The houses of fish are usually coral reefs and small rocky or pebbly areas. These are important breeding grounds for various varieties of fish. The Orang Laut use spears for fishing and therefore prefer shallow-water fish. Areas that they identify as the habitat of fish can be approached without fear of much danger.

In contrast, the houses of sea spirits are linked with accident-prone areas such as bays and capes. Sacred territories are also danger zones where fishing activities are avoided. These areas must be approached with great respect. For instance, while travelling with a group of Orang Laut from Tiang Wang Kang to Pulau Nanga, I was instructed to observe *sembah di teluk tanjung* (worship and pay homage in bays and capes) and remain silent to show respect to the spirits as we negotiated a cape. Conversation and noise were taboo. This observance of silence and great attentiveness also served to avoid mishaps until we had cleared the zone. On other journeys I undertook with other Orang Laut, we not only remained silent as we negotiated these sacred territories, but we also had to observe the rule of not looking backwards. The belief was that if we looked backwards, we would not be able to move forwards to complete our journey successfully. Looking backwards would anger the spirits who would cause us to backslide to our starting point.

Where land is concerned, the Orang Laut tend to concentrate on using coastal areas. The land spirits are seen as concentrated further inland, particularly in forested areas. Therefore, if possible, these areas are avoided. Should one intentionally or unintentionally fail to avoid these areas, the spirits would be angered. The offender would be severely punished, even to the point of death.

Ross, an Orang Laut woman in Pulau Nanga, suddenly fell gravely ill. Her condition deteriorated with each passing day. She suffered from occasional fits and shivers and her entire body was turning yellow. She could neither eat nor drink and was rapidly loosing weight. She suffered from such pain in her abdomen that she could not even stand up. For days, she simply lay on the ground rolling around and moaning in excruciating pain. Her pain was so intense that no one could even lift her to seek medical treatment. There were many speculations as to what she could be suffering from, and a variety of spells were cast over her by her siblings in the hope that she would be cured. However, her condition only worsened. One evening her sister-in-law, Suri, had a dream and related it to Ross and her siblings:

> The spirit – *nenek hantu* ('grandmother spirit'), *hantu darat* ('land spirit') – appeared in my dreams. The spirit asked for a white chicken. The spirit told me this in my sleep last night. The spirit said that Ross had relieved her bowels at the place of the spirit.

Ross confirmed that she had indeed relieved herself at such a place. This was not the spot where she would usually do so. However, on that particular day, she had been in a hurry and simply relieved herself at what she thought was the first convenient spot. She had no intentions of polluting the sacred territory, but from Suri's dream, she realized that she had profaned and polluted the territory belonging to the land spirit. To appease the spirit, a white chicken was immediately offered to the spirit to plead for Ross's recovery. Chickens have an important ceremonial function (Dietrich 1998: 239) as their 'souls ... are in permanent contact with other souls in the environment' (Schefold 1970: 82). Their souls create 'lasting relations' (Dietrich 1998: 240) and contacts with the other souls. Thus, one is able to receive enlightenment (Schefold 1970: 83) or attain insights about what is going to happen in the future through the chicken (ibid. 1970: 82). Thus, in the case of Ross, it was hoped that the ceremonial conjuring of a chicken's soul would bring about the prevention and cure of a disease. Indeed, Ross was finally cured.

This observation leads us to an important realization concerning the overall Orang Laut concept of the spirit of things and territorial ownership. Ownership of a territory is expressed and conceived of as a spiritual bond. A territory is a source of wealth and an authority unto itself. That is, a territory is not passive. It is imbued with magical, religious and spiritual forces. It is strongly linked to persons or ancestors, a group and the spirit world. While these territories remain accessible to everyone else, what imposes absolute obligations on those who enter these spaces is the fact that spirits who own these territories are also embedded in these places. To show any form of disrespect means the trespassers will suffer the pain of losing their right to enter these places. In this system of ideas, one clearly and logically realizes that one must give back to another person, and not destroy, what is really part and parcel of his own nature and substance. This is about law and morality. Not according proper respect to the rightful owners is tantamount to declaring war; it amounts to a direct rejection of the bond of alliance and commonality.

Inalienable gift

The gift of territory is an inalienable gift for the Orang Laut. During the second phase of my fieldwork in 1994, when development projects were threatening to encroach upon the Orang Laut territories in the Galang area, they expressed their anxieties over losing their ancestral estates. Boat, a member of the Pulau Nanga community, was among the first to discuss this matter with me:

> This place is going to be developed [by the government] ... This is our territory and not [that] of others. Hence, we want to know what they intend to do with our territory. If the government wants to buy this territory from us for Rp. 20,000 or Rp. 50,000, we will not accept it ... This is because our father, mother, and siblings are all buried here. We do not want to disturb them. It will only make us ashamed to move ... Just think, if your father has died and you have buried him ... Who is going

to look after his grave if we move? His spirit will disturb us. We may not be able to see [our ancestors], but they are able to see us . . . We just want to live and die on this island, which is ours. This place has been earmarked for development . . . They are building a highway to link Batam to Senyentong and it will pass through Galang. Perhaps if this development happens, they will close our territory and we will no longer be able to work as fisher-people. We will have to work in the motorbike service or drive cars or taxis. If we do not know, we will have to learn. If we can go on fishing, then of course we will be fisher-people. However, we do not want to move. Our father gave us this territory.

Keeping an inalienable gift and preventing its potential loss is essential for the Orang Laut to retain their social identity. An inalienable gift – in this case, the gift of ancestral estates – serves as the physical and spiritual point of origin for individuals and groups alike. Moreover, holding on to such a gift perpetuates its identity into the future. The loss of their ancestral estates, through development programmes, theft or warfare, will ultimately diminish their ancestral identity and rights in addition to their social and political influence in the present. Such loss reveals a weakness in their group's identity and casts anxieties over their ability to survive in the present, let alone to carry on into the future. The loss of such a gift not only marks the destruction of their past, but it also collapses the very foundation of their future. This is because

the immaterial characteristics that inalienable objects absorb, such as one's mythical or sacred origin, one's antecedents through past generations, one's hierarchical position vis à vis others, give to these objects a force that carries beyond the social or political exigencies of the moment. An individual becomes more than what he or she is because the self is enlarged and enhanced by the power of the past.

(Weiner 1985: 212)

When a thing and a person are linked together, the thing acquires the life force of the person, thereby becoming something more than the thing in its material sense. The person whose life force is embedded in the material thing has now transcended his or her mortal life, whereas the thing, having been so charged, constitutes the means of immortality for the person. Therefore, the Orang Laut make every effort to retain some part of these possessions which distinguish who they are in relation to the past. As the social identity of a person encompasses the sum total of his past, present and future, inalienable gifts such as ancestral estates carry the qualities of sacredness and constitute the social self in relation to the past and the future. This is 'the capital stock of substance belonging to a family' (Garnet 1975: 89).

Inalienable wealth also 'takes on important priorities in societies where ranking occurs' (Weiner 1985: 210). This cannot be more true than in the hierarchical world of being Malay in the Malay World of Riau. As Weiner explains,

[p]ersons and groups need to demonstrate continually who they are in relation to others, and their identities must be attached to those ancestral connections that figure significantly in their statuses, ranks, or titles. To be able to keep certain [things] that document these connections attests to one's ability to keep oneself or one's group intact. Alternatively put, to give up these objects is to lose one's claim to the past as a working part of one's identity in the present.

(1985: 210)

For the Orang Laut, to lose their territory is a material admission of their own mortality and a sign of the weakening of the group's identity and power.

The history of the past is concentrated in a thing which, in its material substance, defies destruction. The dynamic surrounding all this is that the thing adds a new force to each generation. An ancestral estate is about its own genealogy and it cannot be replaced by another spatial configuration. Its age, name and associated history make it unique and therefore not wholly replaceable.

The fusion of individuals or groups with things means that the latter are no longer just material or physical objects. Now, they appropriate the power of personification and attest to a person's or group's immortality. This fusion also collapses 'distance in space-time' (Salmond 1984: 120). Hence, ancestral estates are agents of transmission, extending the presence of a person into situations where the material object stands for the person. Different Orang Laut groups have their own stories and genealogies to tell about their ancestral estates. However, all of their respective ancestral estates compress their histories and historical rights into a document for, and about, the present. Ancestral estates procure the identity, disposition and spiritual essence of their owners, perpetuating the virtual existence of their owners. When the bones of ancestors or ancestor founders are buried in the territory, ultimate inalienability is confirmed. From the viewpoint of the Orang Laut, each ancestral estate or territory is a fixed point in their network of histories, relationships and economic production sites. Territories are things that bear references to individuals within families and are a group possession held in a trust by the group. These territories belonged to particular ancestors and were passed down along particular descent lines that possessed exclusive stories to ascertain their rights. These territories embody history and manifest life simultaneously. They reverberate the past into the present.

The Malay World

The Malay World as a whole thus emerges as we piece together the oral histories of the Orang Laut. These narrative accounts are significant not only in providing clues to the undocumented past, but also in constituting a linear projection of the past into the present and future as a continuous process of defining the region. Thus, the Orang Laut offer us another perspective or understanding of the region. It is not just a contemporary patch of land. The

Orang Laut's oral histories show us how places are strung together via genealogy and a network of kinship-infused cultural–economic units and how a region is formed with a time dimension that is represented by ancestral estates. The region is based on life- and living spaces. As a whole, the region is about the social identities of a people who own and belong to the place. It is an inalienable heirloom that a people has inherited from their ancestors, and as such, a thing that cannot be separated from what constitutes the very essence of the people's being and identity.

The Orang Laut know full well that the region they regard as theirs has been subjected to ownership claims not only by the Muslim Malays and the current government of the Republic of Indonesia, but also by various other external foreign agencies, both past and present. They are aware that various border constructions symbolizing and signifying various eras of political domination by the Malay sultans, British, Japanese, Portuguese, Dutch and the Republic of Indonesia have been imposed on their maritime world. They recall these changes with great clarity and relate their life experiences under these regimes in all sorts of ways. Some, like Buntot, a woman from the island of Teluk Nipah, recount experiences of hardship and fear during the Japanese occupation in comparison to the relative comfort of being left alone by the Dutch colonizers:

> I can still recall what life was like during the Japanese Occupation. They entered Pulau Korek . . . very close-by. I cannot remember the exact year, but they also attacked Singapore. Life was extremely difficult during the Japanese Occupation . . . Things were all so expensive . . . The Japanese were always arresting people. For us females, we could not leave the house at all . . . The Japanese took away all the young people . . . We wanted them to leave . . . [During] the era of the Dutch and the Republic of Indonesia, nothing much happened. It was better during the era of the Dutch . . . The Portuguese did not come here at all. They just left us alone and food was very cheap . . . People did not suffer under the era of the Raja.

Others, such as Imah, Suri and Yang from Nanga Island, speak of how the changing impositions of borders signalled various changing realities and everyday practices in the region. These women perceived the changes by way of different currencies that had been used in the archipelago.

> Imah:
> We have changed currencies five times already in my lifetime. The first time, we used *uang dollar* (dollar money). As a matter of fact, initially, we did not use any money. We just bartered goods.

> Suri:
> Formerly, people did not use any money. They had enough to eat, so they

simply bartered goods. In my lifetime, I have experienced four changes of currencies.

Imah:
First, it was the dollar money.

Suri:
Then *cap burung* ([currency with the] seal of a bird).

Imah:
Dollar money . . . the currency with the bird seal . . . rupiah.

Yang:
Pakai layah ([currency] with [the seal of a] sail, now it is the bird.

Imah:
There was also *duit sen* (money [in the form of] cents or coins).

The Orang Laut understand that money is a symbol (Hart 1986) of the overarching political authority or state which issues it. In these terms, they explain that the introduction of a new type of currency often indicates a change in the central political body. More often than not, as the women emphasized in their conversation, a change in the seal on the currency meant the replacement of one central political authority by another.

Borders for the Orang Laut are by-products of particular historical circumstances (Chou 2006b). While politicians and geographers debate over the definitions of the spatial dimension of borders and boundaries and their roles in nation and inter-state relations, the Orang Laut perceive borders from a different perspective. For them, borders and boundaries are temporary markers that alter over time, connoting the rise and fall of different political realities. Yet throughout the rise and fall of different borders and political realities in the region, they do not view these realities as superseding their perspective on what constitutes their maritime world.

The Orang Laut see the region, instead, as a borderless social space whose breadth and width is defined by the gift of ancestral estates, their network of inter-related territories based upon social relations and the extent of their mobility. They call this social space the Malay World. However, this is not the Malay World defined either by the Muslim Malays or by politicians; it is a social space unified by their history and genealogy. For them, this region is an area of unbroken historical tradition that overrides all other temporal borders and boundaries. The ultimate sovereignty and custodianship of the region lies with them. It is a place that has been discovered, formed and expanded by them. The cohesion of the entire region has long been based on their network of genealogical and kinship ties which continues to prevail in spite of the interference of modern political borders. It is an arena of and for shared activity, experiences and meanings highly charged with mental and emotional associations. The region is, for the Orang Laut, 'my homeland', 'my place' and 'my territory' (Chou 2005). The inter-connections of such places are thus far so little understood that there is constant danger of an outsider unwittingly causing temporary or permanent damage to vital social processes, thus destroying the value, order and coherence specific to the place.

5 The fishing economy

Introduction

It is the common belief that the Orang Laut are passively dependent on their natural environment for their resources, not producing but simply extracting whatever they can from an unmodified environment in order to survive. It is this belief that has constituted a major reason behind the Indonesian government's denial of the Orang Laut's territorial rights. The focus of this chapter is on the fishing economy of the mariners. It demonstrates the meaning of production for them, and in particular that fishing is a planned action that is willfully performed in fulfillment of intended purposes. Not only do they organize their fishing activities and form working partnerships to obtain the best yield, but it is not uncommon for them to plan for amounts of catch to be harvested from the sea prior to their fishing trips. More than ever, they have an increasing desire to plan for surpluses so that they will be able to engage in external trade to increase their income. Furthermore, this chapter presents the wealth of their fishing assets: their ownership of means of production, range of material and intellectual technology to obtain their catch. Finally, it shows that economic decisions maximize their livelihood and are deeply embedded in social relations.

There is no distinction between religion and economy for the Orang Laut. Spiritual matters and economic production are inextricably linked, and there is no way of comprehending the two as separate concerns. It could be said that from their point of view 'both moral and physical movement, the religious journey and the economic quest for food, are part of the same process: namely living' (Ingold 1986: 153). Fishing is an activity full of religious meaning, which always involves some kind of exchange between human beings and the spirit world. As creatures from the fishing world are believed to possess magical or supernatural powers, the economic act of securing them for subsistence or deriving an income from their sale must be coupled with some means to satisfy or appease the powers or beings of a supernatural order at the same time. Therefore, the mechanical aspect of fishing is only one side of the equation. Indeed, success or failure cannot be interpreted without magico-religious considerations. Care must be shown for living creatures that are caught.

Diet and income

The sea in the Riau Archipelago shelters a rich and varied marine life that has supported the mariners' communities for as long as they can remember. Their diet and income derive from a range of fish species, crustaceans, molluscs and other maritime creatures. Families typically consume part of their catch and sell the rest to outsiders or distribute it within their community as food gifts. Most transactions involving fish with outsiders are conducted in cash or through bartering. Sometimes, surplus fish are sun-dried either to be sold or stored for consumption at a later date, especially in the monsoon season when going out to sea may prove treacherous.

Fishing forms a central part of Orang Laut identity. From birth, a child learns through participant observation, observing and imitating the skills of fishing and living at sea as exemplified by other members of the family or group. Children grow up to be skilful spearers of fish and other maritime products.

Adults and children distinguished for me by name 138 varieties of fish which can be found in the waters of the surrounding region. Of these, I was told that around 40 varieties form part of their regular diet. Not all of the varieties identified as edible can be sold. This is simply because the potential of these species as food resources has not yet been fully realized by others, thus no market demand exists for them as yet.

In addition to fish, a catch can also include a smaller number of marine mammals and reptiles, including sea cows, turtles and sea snakes. Some of these are eaten, if only rarely, while others, such as sea snakes, are poisonous and belong to the taboo food category.

The Orang Laut also collect by hand, scratchers, traps and snares what they identified for me as 17 varieties of crustaceans, including several varieties of crab, shrimp, prawns and lobsters. (See Appendix for fishing gear used by the Orang Laut.) Another 46 varieties of molluscs and other edible sea creatures such as octopus, cuttlefish, squid and jellyfish were identified as forming part of their catch. Squid and octopus are caught by using spears, nets and wooden prawn-lures, and several varieties of giant rays are hunted with harpoons. Besides fish, crustaceans and molluscs, the Orang Laut diet also includes sea urchins, sea cucumbers, sea grapes and seaweeds.

The number and variety of marine resources that are harvested is ever increasing. New resources are continually being identified by these marine experts. During my first stretch of fieldwork in 1991, they showed very little interest in jellyfish. However, four years later, they informed me that jellyfish had become a lucrative trade item for them because a Chinese merchant from Singapore arrived and and informed them that he would buy jellyfish. The Chinese value jellyfish for its peculiar tender, crunchy and elastic texture. They are particularly eager to buy this marine product which is sold in dried disc form measuring as wide as 35 centimetres in diameter.

Apart from fish, crustaceans, molluscs and other sea creatures, the Orang Laut also occasionally supplement their diet with hunted animals such as mouse deer and wild pig, as well as fruits and berries gathered from surrounding islands. Some of them, in particular those who spend more time in their coastal settlements, have even attempted to grow tapioca.

The Orang Laut are aware of the market value of different maritime products. Some yields from the sea, such as sea cucumbers, are valued as trade commodities with high economic worth. They are therefore collected and dried for market sale, often smoked over a small fire, but rarely eaten by themselves. *Gamat* and *nabi* are two types of sea cucumbers that are commonly harvested from the sea. *Nabi* is a higher grade of sea cucumber and fetches a higher price than *gamat*. The Orang Laut often wait until they have collected a sizeable number of sea cucumbers before they sell them in bulk. They know that they can reap higher profits if they sell them directly at the Pasir Panjang Barter Trading Station in Singapore rather than sell them to the Chinese middlemen in Riau. However, they are also aware of the strict policing of the Indonesia-Singapore border by Indonesian immigration officers.

> Boat, an Orang Laut man from Pulau Nanga:
> [Many of us] go into Singapore [to sell our sea cucumbers, crocodiles and turtles at the Pasir Panjang Trading Station]. However, if we meet the marine customs officers, we encounter difficulties ... It is the marine customs officers from Balai [Indonesia] that we fear most. We are not afraid of the customs officers from Singapore.

> Den, an Orang Laut man from Pulau Nanga:
> The officers from Balai keep watch even at night and they will send us back to the customs officers at the jetty in Tanjung Pinang [Indonesia].

> Boat:
> The officers from Balai are always asking for *uang kopi* ('coffee money').

> Den:
> The customs officers from Singapore are not wicked. They will simply ask us if we are bringing in any drugs. If we do not have any drugs on us, they will not give us any trouble, and let us in. They will only arrest us if we have any drugs on us ... [The customs officers from Balai] will ask us if we have any money. If we do not have any money, we will be arrested. They want money. If it is a big boat, they want S$200 to S$300.

Boat:
We are talking about Singaporean currency. If it is a small boat, then they may ask for S$100.

Den also showed me documents that he had to obtain to legitimate his mobility.

Den:
I even have the cards for entry into Singapore here with me. [I had to pay] S$30 for these permits. [Even with these papers, the customs officers at Balai] will give trouble . . . if you do not give them any money, you will be arrested and taken to Balai or Blakan Padang [in Indonesia].

The imbalance of opportunities existing on the two sides of the Indonesian-Singaporean borders has created compelling reasons for the Orang Laut to attempt border crossing. However, because of the dangers that they face in making cross-border sales of their maritime products, they will only consider undertaking such a journey to Singapore if they are able to collect several sacks of sea cucumbers. If their cross-border negotiations should fail, they would immediately be categorized as illegal immigrants and arrested. Hence, what is known as 'coffee money' has become a means by which to move across borders and an important feature of international business. When crossing national boundaries to trade, the seafarers find themselves operating under complex socio-cultural, economic and legal conditions.

To consider coffee money indiscriminately as a bribe could be erroneous as it impedes exploring and understanding it objectively in terms of its cultural context, ensnaring one in ethnocentrism and moral absolutism (Jacoby, Nehemkis and Eels 1977: xi). Coffee money can be understood as 'lubrication payments' (ibid.: 5), 'speed' and 'grease money' (Lui 1985) to both prevent detrimental action or inaction and hasten favourable action. The Orang Laut know they need the co-operation of governmental functionaries to issue the visas and permits necessary for business. Almost all businesses operating here, including that of the Orang Laut trading activities, cannot escape this payment. It is understood that, because of low wages in Indonesia, officials need extra income. Even Singapore, which is recognized for its honesty, is known to be a 'smuggler's den' (Jacoby, Nehemkis and Eels 1977: 18). A former Indonesian trade minister once commented that up to 80 per cent of Indonesian exports were smuggled out of the country by Singapore-based syndicates (Alexander 1973: 221). In an area such as Southeast Asia where contraband trade has prevailed for centuries, the pervasive habit of coffee money offers smugglers and custom officers mutual advantages: It establishes goodwill and is regarded as a token of appreciation from the smuggler to the customs officer who has helped people adjust to the business climate by offering specially rapid service to increase efficiency and curtail uncertainties.

The coffee money expected of the Orang Laut is usually mutually determined. As Boat and Den explain, the amount expected is largely intuitive. For a small boat, the customs officer may ask S$100. This amount increases to as much as S$300 as the size of the boat increases. The idea is that a small boat carries a smaller load of goods for trade and therefore cannot afford a bigger payment. The way for a government official to induce payment is by imposing obstruction or dawdling about his official duties. The Orang Laut are anxious to avoid such delays because they can be very costly as maritime products go stale easily. These payments at the border can also be read as a symbolic expression of the difference in status between the one who requests the favour and the one who grants it.

Apart from sea cucumbers, which are highly valued trade commodities, there are also other types of maritime resources which are valued even more highly by the seafarers because they are believed to possess 'special qualities'. That is, they are either delicacies or possess supernatural powers. These are prized resources which are seldom sold to outsiders. If possible, those who are not in need of an income prefer not to sell them. Turtles are such an item. Their meat and eggs are considered delicacies.

It is difficult for the Orang Laut to understand the concept of protecting endangered species such as turtles. Apart from the flesh, the other parts of the turtle are also highly valued. Turtle shells are believed to possess supernatural qualities, and are carved into bangles, pendants and combs. Concerned for my safe journey through the archipelago, Suri and Boat, a sister and brother from Pulau Nanga, gave me a turtle-shell bangle. Boat explained why I should wear it for protection:

> Evil people can poison you by offering you food and drink spiked with poison. If you wear this bangle and hold a glass of beverage spiked with poison, this bangle will crack immediately to warn you of the impending danger.

Although turtles are prized items for the Orang Laut, they will always release and return to the sea a turtle whose shell is inscribed with a person's name. The inscription on the turtle's shell serves as a sign of someone else's claim on the turtle. It would be *dosa* ('sinful') for anyone else to kill such a turtle, and misfortune would befall them.

As an item for trade, turtles are also in great demand in Singapore. However, this trade in turtles has met with some complications in recent years. Even though pressures from conservationists have forced Singapore into prohibiting the import of turtles, the demand for them in Singapore has not ceased. The trading of turtles can no longer take place openly, but ways and means have been devised to smuggle them into Singapore. Turtles are also in great demand by the Chinese in Riau, who regard turtle meat as a delicacy and an aphrodisiac.

The *dugong* ('sea cow'), which is now close to extinction, is another maritime resource much sought after by the Orang Laut. They consider sea cow meat a delicacy, but above all it is the *mata dugong* which is the most valuable. The term *mata dugong* literally means 'the eyes of the sea cow', but for the Orang Laut it connotes the tear drops of the sea cow. The tear drops are used to concoct powerful love potions. Everyone in the Malay World, including the Chinese, believes these love potions to be the most powerful of all. Ah Nia, the wife of the Chinese boss on Pulau Abang, describes them:

> The Orang Laut are famous for collecting and transforming the tears of the sea cow into potent love potions. They wait for certain special days to do this. Collecting the tears of the sea cow on special days increases the potency of the love potion. These days are the birthdays of the sea cow, and this is determined by looking at the moon. On such days, the Orang Laut will lay out a comb, cotton wool, and a few other things for the sea cow to pretty itself. The tears of the sea cow are absorbed by the cotton wool, from which they are later squeezed out to make the love potions.

Everyone fears being affected by these love potions, yet at the same time, everyone also wishes to own them in order to possess the power to control others. Often even urban Malays in Tanjung Pinang asked me if there was any possibility that I could obtain some love potions for them from the Orang Laut so that they could cast spells on other people. Sea cows, like turtles, are classified by conservationists as an endangered species. However, this has not persuaded the Orang Laut to release any captured sea cows.

Spiritual owners

Be it the common fish species or the highly valued sea cow, the Orang Laut believe that their production levels of these maritime products are based upon the cordial relations they establish with the spirits who govern and own the territories the Orang Laut enter. Territorial ownership, as discussed in Chapter 4, confers on the owner much respect and the right to require others to seek permission before accessing their territory. As elaborated by Ceco, an Orang Laut man of Pulau Nanga:

> Boat asked the spirit to give him fifty to sixty fish. The spirit gave him even more. After that, [Boat gave] one kilo of glutinous rice to the Tua Peh Kong (a Chinese spirit) near Teluk Nipah. Not to the Tua Peh Kong here, but to the Tua Peh Kong there because Boat entered there and not here. Others give an egg on a plate to the spirit. Yellow glutinous rice is given to the land spirit and white glutinous rice is given to the sea spirit. There are also different sorts of spells for different spirits. Let me explain. The sea spirit takes care of the sea . . . It owns everything in the sea. If we cast spells, we will get the fish. If we do not cast spells, the fish will leave.

Maritime creatures have a spiritual owner, master, or guardian. The *hantu laut* ('sea spirit') is regarded as the superior ruler and spiritual owner of the maritime world. Supported by an army of spirits, it controls the maritime world with formidable powers. Known to have a vacillating character, the sea spirit can manifest its power by bringing about either good or evil. It directs its horde of spirits to withhold or present all maritime creatures to human beings. It can cause great peril in the sea, damage fishing gear and destroy all maritime creatures and human beings or offer protection and prosperity to everyone.

The transfer of control of maritime creatures from the supernatural spirit to human beings marks a ritual inversion rather than a trend towards the secular. On a fishing trip, a prestation of maritime creatures is made by the spirit to human beings. In the ritual after the fishing trip, human beings must present an offering of glutinous rice and spells to the spirit. In both cases, before embarking on their fishing trips the Orang Laut call the spirit to send maritime creatures to them; in the ritual after the catch, they call upon the spirit to accept prestations from them. Veneration of sea spirits is observed all over Indonesia. A practice deeply rooted in Southeast Asian history that pre-dates the coming of Islam to the area, it is also part of a wider pattern of belief in chthonic beings in Southeast Asia (Wessing 1997: 98–99). Symbolic parallels can be drawn with the enigmatic Nyai Roro Kidul, the Spirit Queen of the southern coast of Java, and Batara, Guru of the Sea in the Malay Peninsular, who grant assistance if due respect is paid.

Veneration is focussed not on the maritime creatures per se, but on their spirit owners; guardianship over maritime creatures is transferred from the spirit owner to human persons. As Ceco pointed out, Boat could decide and had decided upon a certain amount of catch that he wanted prior to his fishing trip. Upon obtaining the catch, Boat was obliged to make an offering of glutinous rice and spells to the spirit that had allowed the fish to be caught. Boat explained his offering to the spirit as follows:

> It depends on where you want to go – land or sea. You have to request permission from the spirit here. You place glutinous rice, bananas, and rice wine in your boat for the sea spirit. This will appease the spirit. The spirit wants to eat. If you do not give glutinous rice to the spirit, you will fall ill. We give [this prestation] three times [and place it in] the middle of the boat . . . Later, when you go fishing, you recite spells and ask the sea spirit to help give you fish. The sea spirit will help you. The spirit will not harm you. We also give this food [prestation] once every fifteen days [to the spirit]. Three times we give food at the front and at the back of the boat. Two times we give food in the middle of the boat. There is mutual help between the sea spirit and us . . . There are many spirits in the sea. There is Hantu Gin Bisu, Tok Putih and Raja Blaer. There are many more.

The types of food prestations named by the Orang Laut vary, as do the ways in which the prestations are made. Basically, the food types symbolize the various aspects of life. They are either placed on sacred places or cast into the sea. In this case, Boat's offering of bananas of the variety known as *pisang raja* ('king's bananas') denotes the royal status of the sea spirit, rice wine and rice symbolize fertility and the cigarettes are meant to enhance the enjoyment of the meal. Other possible food offerings include, for instance, eggs to symbolize fertility, a live white chicken plus a black roasted chicken to respectively represent life itself and the sacrifice of it, coconut to connote body and soul and incense to indicate a spiritual link to ancestors. In the past, the Srivijaya king paid tribute to the sea spirit by casting gold bricks into the sea. In Riau, a virgin was sacrificed (Jessup 1990:68). Over time, these practices have undergone much simplification.

The emphasis is not on what the prestations comprise or how they are presented, but rather on the cordial relationship of giving and reciprocity between the spirit and humans, an integral and necessary part of the creative cycle of renewal. Humans depend on the spiritual master for food and trade commodities, and the spiritual master depends on the acts of prestation to ensure the reproduction of the maritime world.

After each fishing trip in which an Orang Laut has cast spells to call upon the sea spirit for help in obtaining a catch, it is necessary to *buang hantu laut* ('throw out the sea spirit'). That is, they have to respectfully request the sea spirit to go away peacefully; otherwise, the power of the sea spirit might consume their inner being. To make the request, an Orang Laut must carry out the following ritual. They must visit the spirit's place and offer it food or whatever else it might desire, cast spells to ask the spirit to leave and wash their bodies. Water must be thrown over their heads and it must cascade all over their bodies and onto their feet. If this is not done properly, the spirit will devour them. If this happens, a person will suffer dizziness, fall ill and go insane.

Although the Orang Laut has to ask the sea spirit to leave, they by no means discontinue their respect for the spirit. On the contrary, specific spells such as the example below, venerate the power of the spirit:

> Toss the betel nut and let it roll
> Roll and split in half
> Whatever tide you send, and whatever wind you bring
> Oh! Proud, mighty, and powerful sea spirit.

The sea spirit must be recompensed for all that it has provided; otherwise it will consume human lives.

The Orang Laut have many stories to tell of the sea spirit that governs the maritime world. Because of the vacillating character of the spirit, it has appeared to them either as 'male or female' and 'young' or 'old'. In some sightings of the spirit, it has shown itself to resemble human beings. In other instances, it has revealed itself as being extraordinarily 'pretty', 'white', 'red-eyed', 'fierce', 'angry' and with 'very long hair'. The spirit is described as immortal and reigning over life and death. The seemingly contradictory description of the spirit's appearance actually and precisely depicts its cyclical birth and death and hence, immortality. It is an ever-rejuvenating character. It resembles a young and beautiful maiden when the moon waxes; it grows old and ugly when the moon wanes.

There are times when an Orang Laut community may call upon one of their members who is especially credited with mystical power to visit the spirit guardians of the maritime world to appeal for help. The visit can be done by way of dreams. In their dreams they communicate pleas for a good catch of fish to alleviate them from poverty and a promise to recompense with food offerings if their requests are granted. There are also times when individual members of the community claim that they have been visited in their dreams by an emissary from the spirit world. Based on these encounters, the Orang Laut have described the spirit world as comprising a multitude of spirits who have very different personalities and can manifest themselves in different ways.

The intermediary role of the sea spirit and her troop of spirits are clear. The ambivalent character of the sea spirit connects realms of good and evil, beneficence and destruction and life and death. Good communication with the sea spirit is necessary to ensure harmony. Today, managers of a four-star Samudra Beach hotel at Pelabuhan Ratu in Java are fully convinced by the beliefs of the expert seafarers. They have reserved a room specially to venerate the sea spirit after a sea storm wrecked their opening ceremony. Muslim purists, though, condemn these beliefs and practices because they deem the sea spirit a counter image to Islamic socio-cultural order. All spirits are associated with the devil even though they know that these spirits can do good and render aid to the impoverished.

Intellectual and material technology

The Orang Laut deploy an outstanding wealth of material and intellectual technology to obtain their catch. This technology is ever improving and it bears testimony to Orang Laut ingenuity. Their *sampan* ('boat') is an important example of the material technology that facilitates their nomadic lifestyle and economy. It functions as a vehicle for them to traverse the archipelago, an important site for production and a house in which to live. Some Orang Laut construct their boats themselves, while others choose to buy them.

If the Orang Laut want to make a boat, they make it by digging it out of the trunk of a *sialang* tree (*Koompassia excelsa* or *tualang*). The symbolism and significance of making a boat from the *sialang* tree was explained to me by Tengku Nasyaruddin or Pak Tenas Effendy, as he prefers to be known. Pak Tenas Effendy is a Pekanbaru-based Malay intellectual and cultural expert who has provided intellectual, moral and personal support for the indigenous peoples of Riau. He has been able to make a significant contribution to the documentation and conservation of the indigenous peoples' history which has shaped Malay cultural identity. The following is a paraphrase of his explanation of the symbolism of the Orang Laut boat made from the *sialang* tree, which has since been translated by Geoffrey Benjamin (Tenas Effendy 2002: 364–383).

The *alam secara makro, alam yang luas* ('macrocosm') is the source of human life. The macrocosm also merges to be one with human life. The macrocosm makes itself one with human life through symbols. The *sialang* tree is one such symbol. This imagery is reflected in the mantra *koto bumi selebe dulang, koto langit sekombang payung* ('the earth is as broad as a tray, the sky is as an unfurled umbrella'). This connects the concept of the microcosmic *alam dalam di'I* ('world within ourselves'), or the *semangat* ('soul') which lives in the physical body, with the macrocosm.

The *sialang* tree is also known as the *Pohon Alam* ('World Tree') to the Orang Laut. It is for them a symbol of their *jati diri* (very life and identity). Different parts of the World Tree are controlled by different *akuan* ('spirits'). The uppermost parts come under a spirit called *Bughung Putih Yono Belayeh* ('White Bird of Changing Colours'). The middle section to the ground is controlled by the *Akuan Sidi* ('Spirit of Effective Charms') and the underground spirit is called *Akuan Sa'ti* ('Spirit of Supernatural or Divine Power'). If this is superimposed onto the living human body (see Figure 5.1), the body from head to the shoulders is controlled by the White Bird of Changing

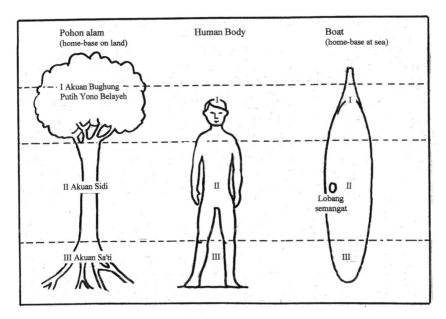

Figure 5.1 The *Pohon Alam*, human body and boat.

Source: Adapted from Tenas Effendy 2002: 374

Colours, from the shoulders to the knees by the Spirit of Effective Charms and from the knees downwards by the Spirit of Supernatural Power.

There are four stages in the making of a boat. In the first stage, a suitable tree is chosen in the forest. A *kemantan* ('shaman') chants a special mantra over the tree to request permission from all spirits before any work can begin. Then the tree is cut down and the trunk is shaped into a long rectangular log.

The second stage sees the shaping of the rectangular log into boat form. A *labang nyao, nyawa* or *lobang semangat* ('soul hole') is made in the mid-section in preparation for the reception of the tree spirit. Then the boat is moved to its *pangkalan asal* ('original starting point'), also known as *persumpahan* ('place of oath-taking').

Here, the third stage begins with the performance of four rituals in order to bring about the transfer of the soul of the *sialang* tree into the boat through the soul hole. The soul, which now resides in the boat, is still that of the *sialang* tree and so remains even after the boat is finally taken to sea. Further work is then done to take the boat to its final shape.

Finally the boat, fully prepared, is moved to its *pangkalan tambatan* ('initial mooring-point') from whence, in the fourth stage, it is launched either into the river or sea.

Through this series of rituals, the soul of the *sialang* tree is transferred to the boat; when the boat is on the sea, the soul is thus moved from land to sea. As the boat maintains a continuing unity with the soul of the *sialang* tree and its *pangkalan asal* and *pangkalan tambatan* ('birth and mooring places'), it is always linked to the land, no matter where it travels. In one of their oaths, the Orang Laut say that *yang laut pulang ke-laut, yang darat balik ke-darat* ('what is of the sea returns to the sea, and what is of the land goes home to the land'). This is why the Orang Laut refuse to be separated from their boats. If they were compelled to live on land, they would not be able to fulfill their vows. No matter how far they may sail in search of food, they retain their links with their original starting-point and mooring-point, and through these, to their original oath. The area through which they travel is therefore not just a wilderness, but a continuum of significant places.

Sometimes the Orang Laut also own other types of boats which are employed for different purposes. Below are sketches of these boats by an Orang Laut from Tiang Wang Kang. He has taken on the name Joshua as it was given to him by a Christian missionary. I have retained the names of the boats as spelt by Joshua.

The boat is both the production site and living space for the Orang Laut. Spearing is done from the bow, which occupies approximately one-fifth of the length of the boat and has a permanent deck higher than that of the mid-section. This platform-like deck is usually kept free of obstacles and meant to give a person wider vision and freedom of movement. The bow is also where food is prepared, with the cook sitting or squatting in the lower mid-section. The mid-section occupies three- fifths of the length of the boat, with a removable plank deck under which fish can be stored. This is where the

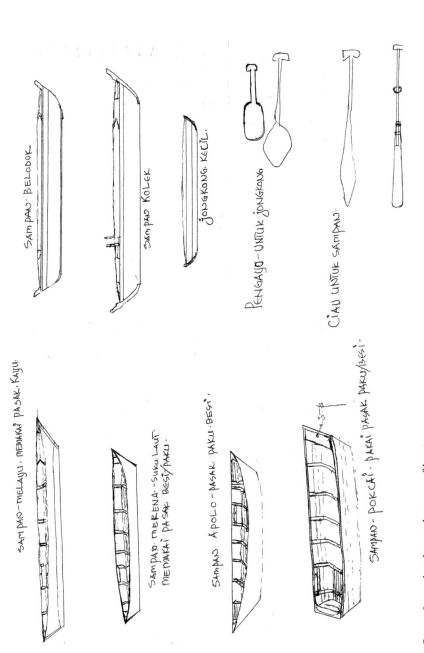

SAMPAN BELODOK

SAMPAN KOLEK

JONGKONG KECIL.

PENGAYO-UNTUK JONGKONG

CIAU UNTUK SAMPAN.

SAMPAN-MELAYU. MEMAKAI PASAK-KAYU.

SAMPAN MERENA-SUKU LAUT MEMAKAI PASAK BESI/PAKU.

SAMPAN APOLO-PASAK PAKU-BESI.

SAMPAN-POKCAI PAKAI PASAK PAKU/BESI.

Figure 5.2 Orang Laut sketches and names of boats.

Key: (1) Sampan-Melayu memakai pasak kayu, (2) Sampan Merena-Suku Laut memakai pasak besi/paku, (3) Sampan Apolo-pasak paku besi, (4) Sampan-Pokcai pakai pasak paku/besi, (5) Sampan Belodok, (6) Sampan Kolek, (7) Jongkong Kecil, (8) Pengayo-untuk Jongkong, (9) Ciau untuk Sampan family members eat, live and sleep. When there are small children on board, this section can be separated from the bow and stern by inserting two planks vertically across the beam. A mat made of strips of palm leaves is pitched over this section as a roof against bad weather. Rowing and steering are done in the stern, which is usually not decked. This is also where people vacate themselves while others keep their eyes forwards.

Apart from their boat, the Orang Laut also possess a wide array of other material and intellectual technology to procure their catch (see Appendix). Spear, harpoon and lance fishing are methods specific to the Orang Laut. The *serampang lima mata* ('five-pronged spear') is one of the most commonly used spears for small- and medium-sized fish. The five-pronged spear consists of four single-barbed spikes set at right angles to each other. The spikes curve towards each other. The fifth spike, which is also the central double-barbed tine, is much shorter. All five spikes meet at a common base which is attached firmly to a long wooden or bamboo shaft called *gagang*. For larger catch, such as the sea cow and turtles, the *tempuling* ('single-pronged harpoon') is used. A single iron-hinged barb is fixed to a long wooden or bamboo shaft. A line is also attached to the head of the harpoon. After the prey has been speared, it is usually not disabled immediately. The shaft is therefore detached from the prong and the retaining line is played out until the prey surrenders. Unlike smaller fish, the larger prey can be hauled in only after a tug-of-war.

Most of this fishing gear is self-constructed. Hence, the Orang Laut are able to own their own means of production. They obtain wood for the shafts of their spears, lances and harpoons from the forest or mangrove swamps. They only need to buy the metal spearheads, prongs and nylon lines. The Orang Laut find these things affordable. They manifest great resourcefulness in constructing their own fishing devices. There are times when they do not have to buy any material at all. They look along the shoreline for washed-up materials, such as tyres or broken umbrellas, with which to construct their fishing gear.

Panching ('hand-line fishing') is another fishing method employed especially during periods of high tides. One or two barbed hooks are attached to a single line. The hooks are concealed in bait. A weight is attached to enable the line to sink into deep water. Relatively large deep-water fish which are in demand by Chinese buyers are caught by this method. The materials needed to construct hand-lines are relatively cheap. Hence, the Orang Laut are once again able to own their own means of production. They also use self-constructed fish traps, stakes and fish stupefacient.

Some Orang Laut have adopted net-fishing. However, most of these nets are expensive and they have to be acquired on credit from Chinese middlemen. They must promise a particular Chinese middleman that they will sell their fish exclusively to him until they have paid off the debt. This usually places them in a long-term relationship of debt with the Chinese middleman. It also prevents them from travelling afar as they are obliged to return to the middleman. Therefore, there have been pressures for those who have adopted this method to move ashore.

Other initiatives have also been taken by the Orang Laut to expand their fishing methods. During my second fieldtrip in 1994, I observed how those in Teluk Nipah had started to construct their own fish farms next to their coastal homes. They were breeding the fry they had caught to reach sizes that would fetch them a good market price. They also explained that live fish were in high demand by seafood restaurants and could thus be sold for higher prices.

Of spiritual matters

As discussed in Chapter 4, the Orang Laut adopt or care for things and cast spells over things that are of importance to them. These things include their fishing equipment, such as their boats and fishing spears. They explain that spells cast over their material technology for fishing endow their fishing gear with supernatural powers. Meen, an Orang Laut from Teluk Nipah, elaborates on why he must adopt his boat:

> I adopt my boat because I would not have enough to eat if I were to catch only Rp. 1,000 worth of fish per day. Therefore, in order to travel further away safely and to catch as much as say, Rp. 40,000 worth of fish, I need to adopt my boat properly. I must feed and spellbind it to attain a good catch. To adopt my boat, or as the Chinese would say, to *kong* ('cast a spell on') it, I have to feed it with glutinous rice and nuts by placing them in front and at the back of the boat.

Besides their material technology, the Orang Laut never fail to emphasize their possession of *ilmu* as a necessary aspect of their technology for fishing. The Malays often speak of the Orang Laut as possessing the most powerful and extensive *ilmu*. By this, the Malays usually mean *ilmu hitam* ('black magic'), since they regard the Orang Laut as an evil and backward people who lack religion. The Malays believe the Orang Laut's *ilmu* enables them to prepare powerful love potions, diagnose and cure various illnesses, safeguard pregnancies, deliver babies, construct houses, hunt and gather jungle resources to supplement their diet, pull in crowds when they entertain with their musical instruments, propitiate or intimidate spirits and control souls. The Orang Laut are also seen as dangerously beholden to spirits that control the maritime world, which is supposedly why they have such powers and know-ledge of this world. The Orang Laut dispute this accusation. They talk of their good *ilmu* to help them navigate through the maritime world, control the winds and storms that can threaten people out at sea and maintain the general well-being of a person. They explain that they have 'not received any school education', but learn by 'retaining everything in [their] memory'. An Orang Laut likened their acquisition of *ilmu* to 'people who attend school'.

Ilmu is central to the Orang Laut. From a young age, through observing and participating in their parents' fishing activities, children acquire the extensive *ilmu* necessary to be adept fishing-people. Their *ilmu* comprises learning how to construct their fishing gear and cast spells to make the tools effective and powerful. Their *ilmu* also includes learning how to: understand and control the winds, currents and tides that govern the sea; be knowledgeable about rich fishing grounds, mangrove swamps and danger zones (in contrast to secure zones); and navigate their way through the archipelago by studying the position of the sun, moon and stars. The need to control winds is especially important for navigational purposes. Winds can also send ripples across the surface of the sea, making it difficult for them to gauge the position of the fish for their spear-fishing.

Through their *ilmu*, the Orang Laut also establish a link with the spirits that control the maritime world. This link is mediated by deciding to adopt spirits and learning how to cast spells to communicate with the spirits and facilitate the release of their maritime creatures. On such spell is:

> Darting obstructions
> Come hither
> Come in from the right, come in from the left!
> May all the fish be sucked into the centre of the water!

Consequently, the Orang Laut enmesh themselves in a network of exchange and obligations to reciprocate with the spirits concerned. The emphasis is on the necessity of intertwining spiritual matters and economic concerns.

Working partnerships

When discussing Orang Laut technology, it is important to stress the importance of their working partnerships. The Orang Laut fishing technique involving the spear, harpoon and lance requires them to work in pairs. Men and women form working partnerships when using these methods. Spear fishing is most productive in clear shallow waters during the night and other periods of calm. The man stands vigilantly at the head of the boat with a spear. The moonbeams and lighted kerosene lamp in front of the boat lure the prey close to the surface of the water. The woman rows the boat into position. Through this partnership, couples catch fish, turtles, sea cows and crocodiles. To find smaller fish, they station their boat a short distance from the shore when the tide is receding. The tides carry medium-sized fish, which the Orang Laut are able to see and spear with remarkable speed and accuracy. The success of Orang Laut *ilmu* and casting of spells rests on, among other factors, whether the fishing-couple *cocok* ('match'). The powers in a couple's *ilmu* and their fishing abilities are further heightened when a correct partnership is formed. This partnership is so significant that a couple's productivity as a fishing-couple reflects their compatibility as husband and wife. Thus, the success of a marriage is measured by the ability of the couple to fish together. Such importance is placed on the significance of a working partnership that if a man and woman go out fishing in the same boat, they are immediately considered married. The division of labour between a couple is complementary rather than a differentiation based on gender (Chou 1995). As all couples are quick to point out, it is necessary that one partner to control the boat and the other to be ready to aim at the catch.

While working among the Orang Laut in Pulau Nanga, I was told that Bolong achieved great success in his fishing activities while in partnership with Temah, his first wife. The extent of this success was so great that they were able to return with crocodiles, which they could sell to the Chinese middlemen for high prices. To the Orang Laut, only those who are most powerful in *ilmu* are able to cast spells to hypnotize crocodiles towards their harpoons, but the success of a person's *ilmu* and spells depends partly on whether the couple are compatible. Bolong and Temah were accorded much prestige and status for their *ilmu* and fishing abilities. In the case of Bolong's second wife, Tri, the fishing partnership was not as successful. As a result, people spoke of her with less respect. Bolong was also considered to have lost some of his *ilmu* and status in the community due to his marriage to Tri.

When Lampong from Pulau Nanga was planning to marry Simmon from Tiang Wang Kang, I asked the former's siblings what they thought of the wife-to-be. They replied that they would have to wait and see if the two could form a good partnership to bring in lots of fish, and that it would only be then that they would be able to ascertain if the couple were compatible. Lampong's siblings reflected that Lampong and his late wife, Siti, were so compatible that the couple often had excess catch to sell and share.

From time to time, Orang Laut women spear for fish. On such occasions, their children are involved as rowers. Husbands acknowledge the spear-fishing ability of their wives. They often mention that they felt assured that their wives were able to feed the family during times when they were too ill to fish. The women are just as speedy with the rejoinder that they are not afraid that they or their children would ever die of hunger in the event of a divorce.

There are other types of fishing methods that do not call for a working partnership. The method of hand-line fishing is one example. Hence, Orang Laut men and women may embark on individual trips when they carry out hand-line fishing. Youngsters engage in hand-line fishing from house platforms and village jetties.

The working partnership of Orang Laut men and women also encompasses the handling of money. Both men and women are actively involved in processing and selling their catch. The surplus catch that is above their own needs is brought to the fish merchant by the couple together. Sometimes the couple will tie their boat to a wooden stake by the shop and bring their catch to the merchant together. Other times, either the husband or wife will keep watch over their boat while the other goes to make the sale. Usually, the partner who feels more tired stays with the boat. However, on other occasions, the responsibility of making a sale falls on the partner who has a warmer rapport with the fish merchant and is thus in a better position to negotiate a better price. Lampong describes this process:

> My late wife Siti used to sell our fish. She could say things that would please Acuk, our Chinese boss . . . Acuk would not only give her good prices for our fish, he would also be willing to help every time she asked for anything. She could obtain many things on credit from his grocery store. But every time I brought the fish to Acuk, I was not so successful in the negotiations. Therefore, whenever possible, Siti rather than I would bring the fish to Acuk.

As fishing is a family-oriented activity and money is generated jointly (male and female together), there is no concept of individual or male money. However, women take charge of and manage all money earned from selling their surplus catch. Different groups of Orang Laut offer different reasons for this. Some maintain that a man who brings money out to sea will not be able to catch anything. Others disagree and simply say that different groups observe different customs. The reason that all agree upon is that women are considered more disciplined in holding on to money and less likely to gamble it away. Women are also said to be better able to decide on the family's material needs. The Orang Laut either barter or sell off their surplus catch for a variety of things including rice, prawn paste, sugar, salt, coffee, tea, cigarettes, alcohol, clothes, gold jewelry and electrical goods. Both men and women have equal rights to engage in this cash-based economy.

Sharing and helping: strategies in production

The Orang Laut system of sharing and helping is yet another strategy entrenched in their mode of socio-economic organization, as discussed in Chapter 2. This system is configured within the networks of social obligations based upon the onus to display a sense of group unity through their network of exchanges. It is the self-interest of calculating individuals that is the motivating factor in fulfilling their social obligations to share and help.

Embedded in the obligations of sharing and helping is the knowledge of an unsigned contract between the giver and the recipient; the latter is obliged to reciprocate either immediately or at some time in the future. The Orang Laut's principle of helping and sharing is an extension of their system of either balanced or generalized reciprocity. The former entails either a direct and immediate exchange of goods or services with a simultaneous return of goods of equal worth or an agreed upon exchange of things or services over a stated period of time. There is the clear expectation of reciprocity in the short term and within a particular time frame. In generalized reciprocity, which is much more commonly practised, gifts are putatively altruistic – there is no apparent expectation of a return gift. The giver appears to enjoy giving, simply desiring to offer assistance or perhaps feeling obliged to do so for some reason. The giver does not usually calculate how or when the recipient will reciprocate. Yet, it is not so much the case that the giver does not expect reciprocation, nor that the receiver does not feel obliged. The expectation of reciprocity in terms of its timing and amount depends on several other considerations. These considerations include the value and amount of things given, the time when the giver might need a return gift, the kind of thing the giver might need and whether the recipient is in a financial, social and/or political position to make a return gift. The delay in fulfilling an obligation is in essence the deliberate calculation of events on the part of the giver. Givers invest in an idea of reciprocity that may bear fruit in the future in a time of need. Production is hence embedded in a set of social relations. To reiterate Sahlins:

> The obligation to reciprocate is diffuse: when necessary to the donor and/ or possible for the recipient. The requital may thus be very soon, or then again, never. There are people – the widowed, the old, the impaired – who in the fullness of time remain incapable of helping themselves or others. Yet the obligations to them of close kin may not falter. A sustained one-way flow is a good pragmatic sign of generalized reciprocity. Failure to reciprocate, or to give just as much as was received, does not cause the original giver of things to stop: the goods move one way, in favor of the have-nots, for a long time.
>
> (1968: 82–83)

The delay in reciprocity is, if anything, a unifying force that establishes a bond of trust between the giver and recipient and aims at waiting without forgetting.

Halus, a young Pulau Nanga Orang Laut boy, was left behind by his father to fend for himself and his two younger brothers. His mother had died some years before. Halus had a boat, which he described as '*saya punya*' ('my own'). The boat was in fact something that his cousin, Jais, had given Halus to *bantu* ('help') him. However, it was an unspoken obligation on Halus' part that he would share part of his fish earnings with Jais. Likewise, it was a tacit understanding that Jais would be entitled to a share of Halus' catch procured with the boat. Also, Halus was obliged to be Jais' fishing partner if the latter's wife was unable to join him. However, should Halus partner with Jais, there was also an unspoken understanding that part of the fish earnings would be shared with the former. In short, this was a cyclical network of sharing and helping to increase the production level of each.

Households and individuals may not always be so willing to share and help, but the obligation to do so weighs heavily upon them. This is because these acts are representations of ideal behaviour that imply egalitarian status and generosity between fellow community members. Those who stay out of the system will face immense social pressures. They will be gossiped about, criticized as selfish, ostracized, labelled 'outsider' and eventually edged out of the network. Networks of helping and sharing are therefore established, maintained and continually renewed. In this manner, social relationships are strengthened so that individuals and households in the village community will assist each other in times of need. Such need could also include winning the support of others to promote one's viewpoint or status in matters concerning village politics or having reliable allies when serious conflicts arise.

Pui's family in the Pulau Nanga Orang Laut community caught a huge turtle. News of her family's catch spread around the community immediately. After Pui had cut a sizeable portion of the turtle for her own family's consumption, the rest was soon given to some of the other members of the community. These were either given at Pui's initiative or on request by those who had come to ask for a portion. While there was still enough turtle to be shared, none of those who came to ask Pui for a portion were refused. Pui was obliged to share and not to turn down anyone.

The distinction between 'share' and 'sharing' is crucial (Bodenhorn 1989: 83). A share is any of the part, portion or quantity into which the ownership of a piece of property or possession is divided. Sharing is a 'social relation' (Ingold 1986: 125) that is a 'wilful regulation' of distribution depending 'upon the capacity of the subject to reflect upon his own existence in the dimension of time' (ibid.: 115). The division of certain catches, such as a huge turtle, into 'shares' is expected among the Orang Laut. The members of the community to whom Pui had to distribute the shares were not pre-determined, although she was under great pressure to share with her extended family. She could decide with whom she would share the meat, but she was obliged to accede to the request of any kin who approached her.

Yet, each individual in this network also tries to gain as much as possible for himself. It was not uncommon for the Orang Laut to make calculated attempts not to share their surplus. These attempts are concealed to prevent accusations of selfishness and avoid being edged out of the community's network of sharing and helping. It is not unusual to hear complaints that there are members of the community who fake illness to obtain shares simply because they are either too lazy to go out to fish or are greedy for more.

The ideal objective of the system of sharing and helping, however, remains to allot small portions of delicacies and other resources for the person and the rest of the community over the same length of time. Everyone profits from this network of social obligations and is involved in economic ties that make the Orang Laut group more stable than it might be otherwise. The co-residence of kin is thus to some extent a function of the sharing of food and help. A person can rely on the other members of the community to share their resources because sharing ultimately benefits the person sharing. The crux of the network of social obligations is that relations among the Orang Laut are both social and material.

6 Modernization and development

The Islamization process

Introduction

The New Order regime of former President Suharto defined the State as the chief architect of all *pembangunan* ('development') programmes. Within the state of Indonesia, 'development' emphasizes the 'need for guidance by those in power and knowledge, in this case the government officials who elaborated the notion in the first place' (Hobart 1993: 7). Based on this definition, the government deems that one vital aspect for the modernization and development of Indonesia lies in the necessity to direct its people in cultural development for nation-building.

At official, regional and local levels, the Orang Laut have been and are viewed and defined as a *kurang halus* ('less-refined') people in comparison to the wider Indonesian society. In government files, they are still categorized as a people forming isolated minority communities who live a backwards life. From the perspective of state planners, they lag behind the wider society not only in the areas of social, economic and political development, but also in cultural development. The official opinion is that they are not able to adapt to modern conditions and, therefore, are not able to participate in the process of nation-building (Gatot Soeherman 1993: ix–x). Therefore, it is necessary to subject them to directed change in order to enable them to become viable members of the wider society. The change must encompass several aspects. First, the cultural, social, economic and political gap between the Orang Laut and the majority must be diminished. Second, as Indonesian society is defined as comprising dominant and subordinate groups, change must be directed from the top or representatives of the dominant group. Third, changes must be designed to serve the needs of Indonesian nation-building.

Strategies for development are formulated within the framework of the country's motto, '*Bhinneka Tungal Ika*' ('Unity in Diversity'), and the national ideology of Pancasila, the Five Moral Principles: (1) belief in one supreme God; (2) democracy through consensus and representation; (3) the unity of the nation; (4) social justice; and (5) just and civilized humanity. These principles are not necessarily complementary to each other. There is recognition of cultural diversity but also, perhaps, the greater principle of national unity. This contradiction between the State's professed commitment to a multi-ethnic state and its overarching policy of unifying and standardizing can be most clearly seen in the religious sphere. In accordance with the national ideology, Section 1 of Article 29 of the Constitution states that religious practice 'shall be based upon belief in One Supreme God'. However, Section 2 allows that 'the State shall guarantee the freedom of the people who profess and exercise their own religion'. The State accepted practitioners of the major world religions of Islam, Catholicism, Protestant Christianity and Buddhism in the beginning (1945). Later, in 1962, the practice of polytheistic Balinese Hinduism was also deemed acceptable. The multiplicity of other indigenous religions, however, has never been given such special consideration. No special legal provisions have been instituted to accommodate the social and cultural differences of these peoples. Instead, the State expressed its intolerance of indigenous religions by enacting an open policy of directed change intended to convert all Indonesians to one of the five recognized religions (Colchester 1986: 96).

In realizing the national ideology, the State finds it necessary to guide the Orang Laut towards cultural development by: (1) getting them to live in permanent settlements; (2) increasing their production capacity; (3) expanding their social network beyond the family; (4) enhancing their rational and mental capacities; (5) uprooting their tribal worldview and way of life; (6) inculcating norms similar to the rest of the country; (7) increasing their consciousness of state and nation; and (8) developing a monotheistic religion (Departemen Sosial 1981 and 1986). Converting the Orang Laut from their ancestor-spirit worship to one of the State's five mainstream religions is perceived by the State as one of the most important measures in raising their living standards and integrating them into a united nation. The State believes that this will form the basis for solving all other problems. The transformation of the Orang Laut, in the interpretation of the State, is necessary because

> It involves national and humanitarian prestige. The fact that there are still isolated and remote peoples, developing at too slow a pace, can affect a nation's prestige and the dignity of man in that country. Therefore the problem must be tackled.
>
> (Achmedi 1972: 20)

The transformation of tribal communities from *masyarakat terasing* ('isolated communities') to an assimilated and acculturated part of the population thus encompasses not only the introduction of economic and social changes, but also cultural manipulation. A few of their cultural traits, such as their dances and music, are viewed by officials as worthy of preservation; officials think that others, like their ancestor-spirit worship, should simply be forgotten (Departemen Sosial 1981 and 1986).

Encouraged by the government's stand in religion, missionary organizations of the five recognized religions intensified their guidance and promotional activities in areas regarded as under-developed have since played an important part in promoting national development and integrationist policies (Colchester 1986: 96; Persoon 1998: 293; Rigg 1997: 121; Stoll 1982). In this chapter, the case study of the Islamization and re-education programme of the Orang Laut on Pulau Nanga will show that this guidance is oftentimes accompanied with benevolent material aid. Whatever good intentions there may have been in dispensing material aid by state authorities, missionaries and representatives of multi-national corporations, it has also led to unexpected and unfortunate consequences. Where previously the Orang Laut had to cultivate their own skills to find the necessary raw material or to rely on their own means to provide for themselves, they are now increasingly expecting others to provide them with necessary supplies of food, clothes and wooden planks. Calculated conformity by the Orang Laut may have been clever in obtaining these things which would have to be earned otherwise, but it has also whittled away their sense of self-reliance and bred a new dependence on others. Fishing has always been a family activity for them. Now women are discouraged from fishing and scorned when they do. They are also discouraged from taking pride in their fishing skills. This has broken up what used to be efficient family production units. As many men and women are unable to attain other jobs due to a lack of relevant skills or simply because they are afraid of being ostracized by others, they fall into dire straits. The result is that throughout the day, some just stay in their newly given houses, doing nothing and not even coming out to interact with others. They only leave their homes at night to do some fishing. There are others who do not even leave their homes at night but have taken to the bottle. Alcoholism and addictive gambling have become rampant problems in these communities of displaced seafaring members. In a domino effect, this has led to a sequence of other problems: mountains of debts are incurred; incidences of wife-beating increase; family relations break down; and many alcoholics and gamblers end up destitute. It is not uncommon to hear members of the younger generation complain about feeling 'dizzy' when they are out at sea. They no longer feel comfortable riding in a boat. Even though these social problems are on the increase, no appropriate programmes have been put in place to assist those who have been displaced by conforming to the greater vision of the Republic of Indonesia.

A common identity

Of great importance in the Indonesian government's *pembangunan nasional* ('national development policy') has been the need to forge a common national and cultural identity for all citizens. As articulated by Turner (1997: 648) in his study of the Suku Petalangan, another indigenous group on mainland Riau, '[r]egarding the issue of continuing cultural identity, the governmental position does not acknowledge the linkages that exist between cultural identity and territorial tenure'. The aim has been to shape this national identity according to what has been defined as national culture in the Constitution of 1945, Article 32. The national culture is to take the form of an ersatz 'mixture of selected traits of those Indonesian cultures regarded as superior such as Javanese, Sundanese, Buginese-Macassarese and Malay, enriched by Western values of humanism' (Lenhart 1997: 595).

When the first Five-Year Plan of Development commenced in 1969, a crucial alteration in the basic formulation of government policy that had further negative impact on tribal peoples, including the Orang Laut, occurred. Tribal peoples were not to be considered as constituting distinct cultures, 'but were rather classified according to their overriding common cultural pattern – that is, by their alleged "primitive" nature' (Colchester 1986: 91). Colchester gained this information from pages 2 and 3 of a 24 October 1975 Department of Social Affairs internal document:

> The Indonesian Government has been and is of the resolve to transform the societal status of said isolated communities, so that these communities will become normal communities, as well-developed as, and on par with, the rest of Indonesian society ... Direct activities of development ... are also initiated in various forms, affecting all aspects of life, such as programmes in human resource development, religion, the material culture, societal life, and dwellings.
>
> (Colchester 1986: 91)

Following that, the State pursued policies of acculturating and assimilating the Orang Laut into a united nation-state.

In the wake of a series of presidential decrees and ministerial directives, the Department of Social Affairs was created and a programme known as *Pembangunan Masyarakat Suku Terasing* (PMST) ('Development of the Isolated Tribal Communities') was initiated. Under the Department of Social Affairs, a special office known as *Pembinaan Kesejahteraan Masyarakat Terasing* ('Development of Welfare of the Isolated Tribal Communities') was set up to take charge of the programme. Projects of directed economic, social and cultural change aimed at the assimilation of the minorities were designed and launched. The adaptation of these communities to the regional majority society was slated as a first step. Adaptation was considered a precondition to helping these communities develop a political maturity that would facilitate their integration into the national society. The ultimate goal was to enable these tribal communities to become an integral part of the economic and social life of the country, and for them to accept new values with which to cope with modernization. The State gave provincial, district and sub-district authorities the power to translate the policies for the directed change into action in ways appropriate to local conditions. In Riau, the measures of directed change were executed via a programme known as *Pembinaan Kesejahteraan Sosial Masyarakat Terasing/PKSMT Riau* ('Development of Social Welfare of the Isolated Tribal Communities of Riau'). However, the State also supplied cultural and moral guidelines for the unity of the

Indonesian people on fiscal organization and administration. As a result, autonomous development had to give way to State-led directed change. Individuals were obliged to support national development policies by totally subjecting themselves to the guidance, control and direction of the government. The argument for policies repudiating indigenous and horizontal political groupings is that indigenous peoples are inclined to 'primitive communalism' (or 'primitive communism') (Dove 1985: 29); their 'systems of mutual sharing and reciprocity' impair 'progressive initiatives towards surplus production, since hardworking individuals must share the fruits of their labours with the less energetic majority' (Colchester 1986: 92).

In 1994, a three-year *Inpres Desa Tertinggal* ('Presidential Instruction Programme for Less Developed Villages') was chartered. (*'Inpres'* means 'Presidential Directive', *'desa'* means 'village' and *'tertinggal'* means 'left out, remain behind or forgotten'.) The poverty of these villages was understood to be due to a lack of both economic and socio-cultural development. The implication was, as Rigg (1997: 49) points out, that poverty is something that exists when a people are not socio-culturally developed enough to participate in efforts for economic development, and that it is only through the incorporation of such 'forgotten' peoples into the development process that poverty can be tackled and eradicated. The state's reasoning was that communities such as the Orang Laut had somehow strayed from the main course of development and had to be guided back into it. Riau is predominantly a Muslim-Malay region. Since the Orang Laut are recognized to be indigenous Malays 'without a religion', the State worked vigorously in the name of development with the Muslim Malays to pressure them into what is considered a more progressive lifestyle. This progressive lifestyle involved observing Malay *adat* guided by local interpretations of Islam.

In 2000, the Department of Social Affairs was dissolved and the special programme for tribal people officially ended. However, government policies implemented by the Department of Social Affairs had already had significant impact. Furthermore, the supporters and associate partners of the defunct programme continue their efforts to address the peripheral status of the Orang Laut. In other words, the process of raising their living standards and incorporating them into mainstream Indonesian society has now been in force for many years.

Islamization and re-education programme on Pulau Nanga

There continues to be rigorous collaboration between the state and local offices in Riau to Islamize the Orang Laut. This stems from the time when the Department of Social Affairs officially came into operation in Tanjung Pinang. It assumed a particularly active role in orchestrating collective efforts by national and local Islamic missionary groups to visit Orang Laut communities. These visits came in various forms and with different strategies of persuasion.

In 1986, one such Islamization and re-education programme was implemented in Pulau Nanga. What follows is a case study of a particular group and setting that was subject to a programme of directed change. This programme had been in place prior to the commencement of my studies in the field, but members of the community recounted the launch of the event with much clarity.

Boat:
The officials contacted us and told us, 'People of Nanga Island, do not go fishing on this day because we are coming' . . . They came wearing their official uniforms. We just wore our ordinary clothes. There were about five or six officials including the *Penghulu* ('[Malay] village chief') from Karas Island. Upon their arrival, they asked for Bolong, our eldest brother. They asked him, 'What is your religion? Are you Muslims here?' Abang Bolong answered, 'No.' They then told us that it was *haram* ('forbidden') for us to live together without being married, that we should be Muslims, and that we should be married in the Muslim way. They said that they had come to officiate our marriages. They told Bolong to become a Muslim. Bolong agreed. Then, they asked us what our names were. We told them, 'Bolong, Siti Payung, Keladi and so on.' They completely disapproved of our names and said that they would give us new names. Bolong became Muhammad. Hence, on that day, we all became Muslims, were given new names, and got married. Our women wore makeup because we were all going to get married. The officials also took photographs of our marriage ceremony. The women wore their *baju kurung* ('long tunic') if they had one. Abang Bolong, Joya . . . many of us got married. We were told that we would be given a certificate, but that it was not ready yet. They also promised to give us clothes.

Cynthia:
I thought you were all married already.

Boat:
We were! It was not us who wanted to marry again. They wanted us to marry, so okay, we got married (laughs). We just went along.

Suri:
The officials wanted us to enter Islam. Do you understand why now? If we agreed, then we had to marry in the way they wanted us to marry. For those of us who did not want to enter Islam, they could not force us, but they said that we would not be the first to receive clothes, wood and so on from them.

Boat:
Formerly, I was a Christian and married in the Christian way. However,

my Christian marriage certificate fell into the sea and was lost. Therefore, I did not possess a marriage certificate when the government officials came to Nanga Island. As a result, they wanted me to marry again, so marry again I did (laughs)! Therefore now I have entered Islam. When I was visiting my sister Meri in Tiang Wang Kang, and the Christian missionaries visited us, I entered Christianity. When the officials came to Nanga Island and they told us that we had to pray to enter Islam, we simply followed exactly what they said. They were here for a long time.

Bolong:
After the visit, we heard a broadcast over our radios that government officials had been to Pulau Nanga, and that they had given us new homes, a religion, clothes, and rice. According to the broadcast, we were given a lot more things than we actually received.

Suri:
Just go along with the officials. There is nothing to lose.

The Orang Laut were given building material, but they were also coerced to erect their new houses on the shore, not over the sea. This was thought necessary and a symbolic shift from sea to land as it compelled them to abandon their seafaring life altogether in favour of a more mainstream, land-based and sedentary lifestyle. Even pile-dwellings in the sea were discouraged. Figure 6.1 shows how, at the start of the programme, most of them were supervised in building their houses on land. Other projects of directed change were implemented too. Measures to improve health conditions, including the insistence that everyone use soap and be involved in the national birth-control programme, were autocratically introduced. Parents were instructed to enrol their children for formal and religious education at a school on the neighbouring Malay island of Sembur. This was supposed to be a temporary arrangement while a school was being constructed on their island of Nanga. While children were obliged to attend school, adults were taught how to pray and women in particular were trained in housekeeping. To be respectable homemakers, the women were discouraged from going out to sea. They were given lessons in cooking, housekeeping and raising children. Furthermore, they were instructed that they should focus on selling snacks or cooked food if they desired to supplement their husbands' income. Alternatively, they could weave thatch, mats, hats and baskets, sew clothes or do other people's laundry. All of these options were to be explored instead of going fishing. In contrast to fishing, all these other activities were considered more respectable as they could be carried out either at home or within its proximity.

Material aid was also given to the community. To discourage their seafaring lifestyle, they were given small plots of land to grow vegetables. Each family was also given a house. Soap, veils, prayer mats, clothes, fishing nets, household utensils and food were also provided. An official, Bapak Mujiono, emphatically stated all of this in a conversation with me. Bapak Mujiono, also known simply as Pak Muji, was one of the officials present at the launch of the programme for directed change.

As soon as the officials and missionaries departed, several families started to dismantle their homes only to rebuild them over the sea or return to their houseboats. Thus began the exodus from land to sea. By 1991 (during the time of my fieldwork), as Figure 6.2 indicates, many families had either rebuilt or were in the process of rebuilding their homes over the sea. They were either living in them on a long-term basis or were oscillating between these dwellings and their boat-houses.

Several follow-up checks by government and religious bodies ensued after the launch of the programme. Initially, these visits were frequent, occurring once a month. Huge delegations comprising government officials and missionaries would arrive for a day's visit, or at the most for three days. During these visits, sermons on Islam were preached and material aid was handed out. Later, these visits became smaller and less frequent. Instead of a delegation, a lone government officer would appear, and the interval between visits widened to three or four months. During the course of my stay on Pulau Nanga in 1991, at least two follow-up events took place. Pak Muji, the government official who had been involved in the launch of the programme some five years earlier, was once again assigned to monitor the progress of the community.

On one of the two visits by Pak Muji in 1991, he was tasked with preparing the way for an official visit by a bigger delegation of officials and Muslim missionaries. On this occasion, he arrived with a woman whom he introduced as his 'wife'. However, Orang Laut gossip had it that this was but a pale cover-up for a wanton woman with whom he was simply having fun. When I probed them about their cynicism, they responded with a laugh, saying that Pak Muji would always be accompanied by a new lady who would inevitably be introduced as his wife.

The visit of Pak Muji and his wife lasted four days. The couple took great pains to explain the purpose of their mission.

> Pak Muji:
> There has been a government programme here in Pulau Nanga for the last 5 years. We come with Muslim missionaries to teach these Orang Laut how to be progressive and modern. We tell them that it is sheer backwardness to live in a boat. We give them planks of wood, nails and woven palm leaves, and set up new houses on land for them to live in. Just as important, we also teach them how to become Muslims so that they will have a religion and modernize. We are going to build a prayer house on this island for them. Over here . . . is going to be a school . . . We are also teaching the women how to set up a proper home . . . We have also given the men *songkok* (a rimless fezlike cap usually made of velvet) and the women veils for use especially at prayer times.

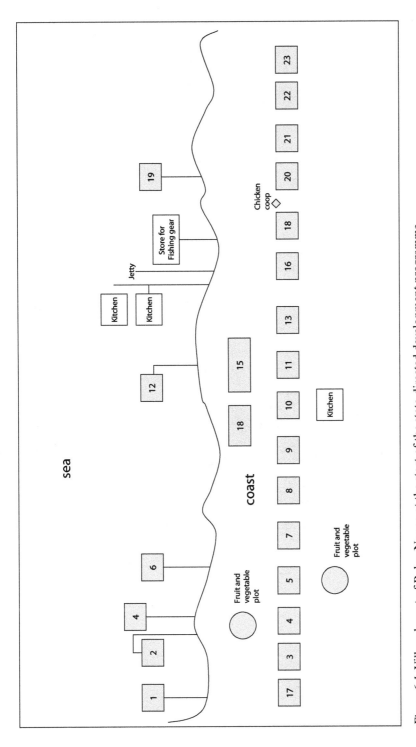

Figure 6.1 Village layout of Pulau Nanga at the start of the state-directed development programme.

Note: The numbers in Figure 6.1 correspond to the house numbers in Table 6.1

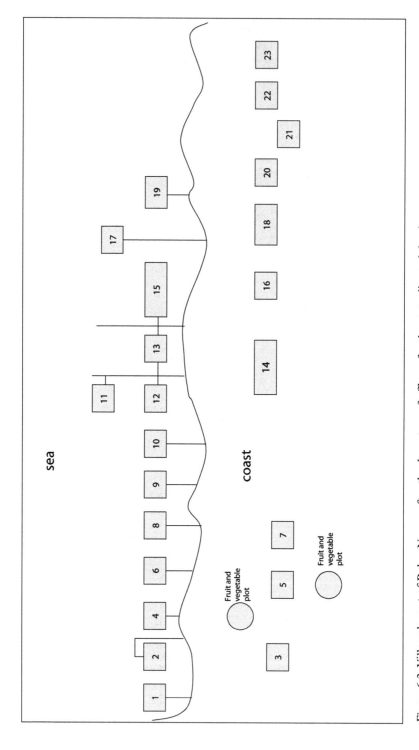

Figure 6.2 Village layout of Pulau Nanga after the departure of officers for the state-directed development programme.
Note: The numbers in Figure 6.2 correspond to the house numbers in Table 6.1

Table 6.1 Household composition of Pulau Nanga

House number	Name of head of house group	Family members	Generation structure/ depth	Type of residence
1	Sangau	Wife, 2 daughters and 2 sons	2	Semi-nomadic
2	Jiba	Wife, 1 daughter, 2 sons and 2 daughters from previous marriage	2	Semi-nomadic
3	Awang Bai	Wife (died before house was moved to position in Figure 6.2)	1	Sedentary
4	Budi	Wife, 3 daughters and 1 son	2	Nomadic
5	Larang	Wife	1	Nomadic
6	Silat	Wife and 1 son	2	Semi-nomadic
7	Vacant	–	–	–
8	Lampong	Wife, 3 daughters and 2 sons from deceased second wife	2	Sedentary
9	Lacet	Wife, 3 daughters (1 died before house was moved to position in Figure 6.2) and 1 son	2	Sedentary
10	Suri	Husband, 5 daughters and 2 sons	2	Sedentary
11	Ceco	Wife, 3 daughters (1 died before house was moved to position in Figure 6.2); husband of elder daughter; 1 grandson and 1 granddaughter	3	Sedentary
12	Zais	Wife, 1 daughter and 1 son (son died before house was moved to position in Figure 6.2)	2	Sedentary
13	Boat	Wife, 2 sons (1 died before house was moved to position in Figure 6.2), 1 daughter and 3 nephews from deceased sister	2	Sedentary
14	Kantor Sosial	No one wants to build it; incomplete construction	–	–
15	Bolong	2 wives, First wife childless; Second wife, 2 sons and 3 daughters; husband of eldest daughter and 1 grandson	3	Sedentary
16	Awang Go	Empty; not occupied at the time of my fieldwork	–	Nomadic
17	Atan	Single	1	Semi-nomadic

(*Continued Overleaf*)

Table 6.1 Continued.

House number	Name of head of house group	Family members	Generation structure/ depth	Type of residence
18	Pri	2 wives, 2 sons and 2 daughters from first wife; husband of eldest daughter and 1 grandson; 2 sons and 4 daughters from second wife	3	Sedentary
19	Den	Wife, 2 sons and 1 son from deceased first wife. 2 sons died during my fieldwork	3	Semi-nomadic
20	Kasim	Wife, 1 son and 1 daughter	2	Nomadic
21	Awang Asin	Empty; not occupied at the time of my fieldwork	–	Nomadic
22	Awang Chocho	Wife and 2 sons	2	Nomadic
23	Awang Keladi	Empty; not occupied at the time of my fieldwork	–	Nomadic

Pak Muji's wife:
[My husband and I] started out from Jakarta and then went to report at Pekanbaru before making our way here. Each Orang Laut household is given one hectare of land by the government to cultivate so that they will give up their seafaring lifestyle. My husband is building houses for the Orang Laut . . . Many did not want to live in houses on land even when they were given houses . . . Now, some are beginning to live on land. Even so, from time to time, they still return to sea. It is very difficult to get them to give up their old ways.

Every effort was made by the Orang Laut to keep Pak Muji and his female companion content during their visit. This was a strategy to dissuade him from forcing them to rebuild their houses on land or interfering too much with their daily lives. They considered it easier to handle Pak Muji because he was alone. Thus, they deliberately failed to notice that he was spending an inordinate amount of time gambling on neighbouring islands and, most importantly, they never broached questions about his many wives. The methods of their resistance to domination by state authorities, and even to Malay missionaries, are not outright challenges, as they know that such overt practices will not get them anywhere. Instead, they resist through 'dissimulation, calculated conformity, false compliance, feigned ignorance and slander' (Scott 1985: 29–34). These are 'weapons of the weak' (Scott 1985) that can be quickly and individually executed without any large-scale organizational or pre-meditated plan. Yet, these actions reap huge and immediate gains.

Pak Muji did nothing to complete the school building. Instead, he concentrated on supervising the construction of a jetty and collected 500 Rp. from each member of the community. He also ordered them to give the village and their houses a general clean-up. All of this was done in preparation for the proposed visit by the delegation of government officials and Muslim missionaries later. The jetty was to ensure that the visitors would have a landing and a sturdy walkway when they docked, and the money collected was meant for buying bottled soft drinks for the guests. Pak Muji told me that it was important that capped bottled drinks be served to ensure that the Orang Laut had not used their *ilmu* to tamper with them. Beverages served in a cup or glass were more exposed to the perceived threat of being poisoned by *ilmu*. Pak Muji also collected 'television tax' from everyone, even though there was only one television set on the entire island. Everyone was of the opinion that Pak Muji himself would pocket much of the money collected, but no one openly showed or uttered any resentment. They considered this a small price to pay to be left alone.

Following Pak Muji's visit, a delegation of government officials and student Muslim missionaries arrived for half a day. Days beforehand, a Malay headman from the neighbouring island of Sembur was notified by radio of the day of their arrival. He was instructed to relay the message to the Orang Laut and instruct them to remain on the island that day to receive the visitors instead of going fishing. Hours before the arrival of the delegation, they were, to everyone's amazement, instructed to again contribute money to buy bottled drinks for their guests, roll out ceremonial mats and wear their best clothes.

When the visitors arrived, bags of clothes were presented to Bolong and Ceco, the two eldest brothers. Although Bolong acted as the head of the community, Ceco, who was regarded by all as a more eloquent spokesman, assisted him. Unbeknown to the guests, Bolong instructed his siblings to remain silent and not raise any questions. Therefore, the more vocal siblings left to bathe or fish during the visit. This was not a to protest to Bolong's decision. Rather, this was because they agreed with their elder brother and reasoned that it would be better for them to observe from a distance. The visitors, like Pak Muji, were completely unaware of these carefully thought out and planned arrangements that were in fact passive acts of non-compliance. They were thus under the mistaken impression that they had procured total allegiance from the Orang Laut.

Upon arrival, the visitors were led to their places on the ceremonial mats, which had been rolled out under the shade of several coconut trees. The following is an excerpt of a lengthy opening speech delivered by the government official who led the delegation:

> Today, we are here with students from the Religious School in Tanjung
> Pinang to teach you how to pray in the correct Islamic way. The students
> will also deliver a message about Islam . . . In our records, there are over

50 children on Pulau Nanga. There is a school on the opposite island of Sembur. However, we understand that you are too busy to take your children there as it is far away. Therefore the Department of Social Affairs, the Office of Religious Affairs and the government will work together to build a school here to teach your children basic foundation subjects and Islam . . .

The school is built here for your convenience. You must send your children to school on a long-term basis and not stop them from attending school after a month or two or even after a year . . . It must not be the case that your children will attend school for a while and then stop until they produce the next generation of children and try to get them to go to school again . . . We have already been to Air Kelobi in Kelong to help the people there. We have set up a school for their children and they can now recite the *Qúran*. We have also built a prayer house there . . . When your children finish their education here, you can take them to school in Tanjung Pinang. When your children are smart in reading the *Qúran* or other subjects, they can return here to teach their younger siblings . . . We will send a teacher here to teach your children religion. This is now the season when it is difficult for you to go to sea, so we have brought cloth, clothes, and other things to help you.

After the speech, missionaries from the Religious School took centre stage to deliver sermons and lead everyone in prayers. Bolong and Ceco, as representatives of the community, donned their *songkok* and imitated their visitors in raising their hands during the prayer session. There were whispers and muffled laughter among the congregation. It was apparent to everyone that the two brothers were experiencing great discomfort as they were not used to praying. Their siblings whispered to me that Ceco was better than Bolong in putting up a public front.

As soon as the prayers were over, the leader of the delegation got up and instructed Bolong and Ceco to rise as well. They were also told to pose for the camera while boxes of material aid were given to them and their hands were shaken. As this was happening, several officials commented among themselves that the Orang Laut were 'pitiful but showing a bit of improvement'. After the ceremony, the visitors looked around the island briefly, but there was not much intermingling between hosts and guests. The visitors were curious about my presence and wondered why I would choose to be among the Orang Laut. No one seemed to understand why the Orang Laut would constitute an important topic of study or how I could possibly learn anything useful from them. I asked the visitors if they would also meet with the Orang Laut on the nearby islands of Teluk Nipah and Abang. An official responded to my query:

No. We were not given any instructions to go over there. Perhaps this is because the Orang Laut on Teluk Nipah are not Muslims. We did not even know there were Orang Laut on Pulau Abang. Are they Muslims?

The Orang Laut on the neighbouring island of Teluk Nipah were unhappy to see their rivals on the opposite island of Nanga receive gifts when no one from the visiting delegation was crossing over to give them anything. They had, however, previously received material aid from Catholic missionaries. From that point on, there has been a demarcation between Muslim and Catholic territories. This has inevitably led the Orang Laut to switch religious affiliations on the basis of whichever missionary group offers more material aid. The initial visits by government officials and missionary groups to launch a religious conversion and re-education programme were often accompanied with a great deal of material aid. However, as time passed, there was often a reduction in the flow of material aid. This often leads the Orang Laut, as in the case of the Pulau Nanga community, to complain bitterly among themselves and to me that since the first stages of help given by Pak Muji, he has become 'useless'. Yet, in spite of all the dissatisfaction and desire to be left alone to live their own lives, they are aware that as long as they remain officially recognized as having adopted a state-recognized religion, especially Islam, they can hope for visits and material aid. Although there are ideological differences and struggles between the Orang Laut and state officials and missionaries, the seeming willingness on the part of the former to 'go along' is a calculated conformity as 'there is nothing to lose' in terms of receiving material gifts.

It is noteworthy that although there are no longer officials from the now defunct Department of Social Affairs to carry out such visits, I discovered on a fieldtrip in 2001 that such activities aimed at changing the lifestyle of the Orang Laut are still actively promoted. One new group that has entered the scene is the multi-national corporations that have set up operations in Riau. Many of these corporations have been severely criticized by international and local pressure groups for their infringement upon the territory and resources of the local peoples. In an attempt to abate these criticisms and to show the local people that they would like to be good neighbours, some corporations have initiated programmes to *bantu* ('help') the local inhabitants to *maju* ('progress'), especially indigenous populations such as the Orang Laut. A special office or department is usually set up within these corporations to oversee the welfare of the indigenous people. Most surprisingly, I found these departments to be closely modelled upon the structure and agenda of the former Department of Social Affairs. Muslim men and women wearing the *songkok* and headscarf have been specially recruited, and programmes have been put into operation to re-educate communities such as the Orang Laut in religion, social habits and alternative sedentary livelihoods.

Nonetheless, the greatest pressures to Islamize the Orang Laut continue to occur on a more regular basis from the local Muslim-Malays of Riau. Particularly, Orang Laut children have felt these pressures. Malay children tease, bully and ostracize Orang Laut children with such intensity that even children who have expressed a desire to learn would rather give up any opportunity for formal education than attend school with Malay children.

Orang Laut adults face these pressures in ways that are not very dissimilar from their children. The Malays despise the Orang Laut. Through gossip about their backwardness, daily pressures mount on the Orang Laut to Islamize and change their lifestyle.

Lila:
There was a Malay couple in Pulau Abang. The wife was called Nilam, and the husband was called Padang. One day, they went to Pulau Petong to catch fish. At Pulau Petong, they met Subor, an Orang Laut. As they passed Subor's home, Nilam criticized his house and said that he was smelly. After their fishing trip, they returned to Pulau Abang. That night as they were sleeping, Nilam suddenly woke up at midnight. She left their home and left her husband, and went directly to Subor. Nilam's earlier criticism of Subor had caused her to be bewitched by him. Her husband just kept quiet because he was afraid. Her husband cried and wanted her to return but she did not return. The Orang Laut have no religion, and that is why they are unprogressive and practice black magic. Black magic is contrary to Islamic teaching, so we do not counter-attack. The devils are the friends of the Orang Laut.

Raja Ibrahim:
This is how unprogressive the Orang Laut are. When an Orang Laut dies, they dig a hole and place the body in it. Before they close the hole, they look for a long piece of bamboo to stick it into the dead person's mouth. Through the bamboo pole, they feed the dead person whatever was their favourite food in their lifetime. This is Orang Laut *adat*. It is so backward and coarse. At present . . . they are beginning to be more like us Malays and some have stopped observing this *adat* of feeding their dead. When they die . . . they do not mark their grave sites in the same manner as we do; instead, it is big stones or big trees that mark their grave sites.

Raja Rahim:
Do you know how unrefined and uneducated the Orang Laut are? When their children are born, they are given names in the following manner. The first thing that comes into the sight of the baby's parents upon its birth is the thing that the baby will be named after. Therefore, the name can be Pokok (tree), Ikan (fish) and so on. Now, a few Orang Laut have names with '*bin*' (son of) and so forth [in the manner of Malay *adat*]. Formerly, they just had single names like Bolong (black hole) and Payung (umbrella).

In an Orang Laut boat, there is usually a dog and a green bird. . . . When Malays go to the market, they know which fish have been caught by the Orang Laut. These will be fish bearing the marks of a harpoon. Malays will never buy fish caught by the Orang Laut because we Malays consider the Orang Laut and their fish unclean. What with a dog living in a boat, shitting and urinating – the fish will be unclean.

Intertwined in local interpretations of Islam is also the conception that being Malay, observing Malays *adat* and being Muslim are all the one and the same. Through their everyday gossip about the Orang Laut, the Malays clearly indicate that the Orang Laut are *tak/belum maju* ('not/not yet developed/progressive') Malays because they are *orang belum dalam Islam* ('people not yet in Islam'). The Malays have taken it upon themselves to enforce changes upon the Orang Laut, thus adding to the pressures already exerted by government officials.

Raja Rahim:
We are making efforts to help the Orang Laut progress. Together with the government, we teach them Islam, we construct houses on land for them to live in, and we teach the women how to look after their houses properly. Some Orang Laut women are now able to cook various dishes and make cakes like the kind we Malays eat. They have progressed considerably now. Some want to live in houses on land and some even possess their motor-boats. They are beginning to learn from us. Some women no longer go out to sea. Mark my words, it will not be long before the Orang Laut disappear. Their men and women will adopt lifestyles similar to us Malays. That will be good.

Undoubtedly, this is an echo of a statement made by Martono, the former Minister for Transmigration, during a meeting between the Department of Transmigration and the Inter-Government Group on Indonesia on 20 March 1985. Martono said groups such as the Orang Laut 'will in the long run disappear because of integration . . . and . . . there will be one kind of man' (Martono 1985: 11, 41 in Colchester 1986: 89).

7 The transformation of Orang Laut territories

The Growth Triangle

Introduction

Besides equating the modernization and development of Indonesia with the cultural refinement of its people, the Indonesian State also deems economic growth to be a defining factor in justifying the State's policy of modernization. In December 1989, the Indonesian government designated the region which had been home to the Orang Laut for centuries as a zone for development.[1] The Indonesian government signed an agreement with the governments of Singapore and Malaysia to pave the way for Riau to be part of a regional community or Growth Triangle programme (*Business Times*, 9 October 1990: 11). The aim was (and still is) to establish an economically integrated area with free movement of goods, services and people, making the entire area attractive as a united investment location for multi- and trans-national investors.

The joint partners of the Growth Triangle are now talking about the vast potential of the fishery industry. However, they are more interested in promoting high-technology ventures in aquaculture, seaweed and prawn farming. They are also promoting marine tourism, particularly the building of marina clubs and resort hotels in the more remote parts of the archipelago (Soh and Chuang 1990: 35). Golf courses have also been part of the master plan to turn the region into a tourist belt. Fresh water has also been earmarked for export.[2] Government and business entrepreneurs have declared that their aim is to 'diversify' Riau with 'three core thrusts': tourism, agri-business and industry (*Singapore Business* 1990: 34).

The Growth Triangle has been applauded by the governments of the three nation states, economists and trans-national corporations as an economic success. However, other stark realities are evident at the local level. Of the three points in the triangle, it is the landscape of Riau – and consequently the lives of the Orang Laut – that has been transformed most extensively. In view of this dramatic transformation, this chapter gives a concise description of how the planners, policy makers and investors visualize and map the Growth Triangle. I suggest that this chapter be read along with Chapter 4 in order to compare the community-based maps of the Orang Laut (especially Map 4.1) with that of the official mapping of the Growth Triangle. Reading these two chapters in juxtaposition clearly reveals that the Growth Triangle is forming a new landscape and supplanting older cultural and economic geographies. Also revealed is how the new maps of the state planners and investors marginalize and overlook the histories and geographies of the Orang Laut.

The changing landscape of Riau and the creation of the Growth Triangle also raises at least three important inter-related conceptual issues for consideration: (1) perspectives on globalization; (2) the concept of 'region states' (Ohmae 1995: 5) to service a 'spaceless' (Macleod and McGee 1996: 425) and 'borderless' (Chen and Kwan 1997) economy; and (3) the politics of mapping. It is therefore also relevant in this chapter to consider some leading theories of these three concepts in order to understand how and why projects to transform Riau are occurring and re-shaping the lives of the mariners.

In the case of Riau, the state has decided that new economic programmes within a global paradigm will be implemented to develop the region. To achieve this, the region must be transformed into a borderless zone known as the Growth Triangle.

Globalization, region states and maps

Globalization

There are at least two broad and contrasting perspectives on the processes of globalization. One perspective (e.g. Toffler 1990) argues that processes of globalization will result in an individualizing world of 'demassification' and 'customisation' marked by a heterogeneity driven by personal consumer choices (Cheater 1995: 117). A contrasting perspective (e.g. Appadurai 1996; Sassen 1996; Ohmae 1995) argues that processes of globalization are demolishing national boundaries and deterritorializing as well as denationalizing spaces, resulting in global interdependence.[3] According to this perspective, such processes will lead to homogenization of the world's socio-political values (e.g. Robertson 1990; Christie 1996; Vatikiotis 1996). It is this latter perspective that has inspired advocates of economic programmes such as the Growth Triangle. According to French critic Paul Virilio (1983), spatial relations will be re-arranged and restructured as a consequence of the technological, material and geopolitical transformations (e.g. Crang, Crang and May 1999). Territories will no longer be the stable and unquestioned actuality they once were. Rather than a given, the position and status of territories will now be questioned (Tuathail 2000). These assumptions have led to the current perception that there is indeed 'no sense of place' (Meyrowitz 1985) in our modern world.

Region states

These assumptions have inspired a rethinking of the concept of region and led to talks about the need to form inevitably borderless zones. Proponents of borderless zones (e.g. Castells 1996; Chen and Kwan 1997; Meyrowitz 1985; Sassen 1996) propose that all regional, national and international borders should be collapsed and all spaces should be realigned into region states and borderless Growth Triangles. It is argued that borderless zones will be more flexible in local political jurisdiction to accommodate trans-national corporate institutions which possess a combination of legal protections, controls and disciplinary regimes. The idea is that globalization will induce a situation of layered sovereignty in which the state retains overall control of its territory but, in specific instances, also allows corporate bodies to stipulate the terms for forming and regulating certain areas. Thus, within the state, its population, which is subjected to different regimes of value, will enjoy different kinds of rights, discipline, welfare and security, resulting in what is known as 'variegated citizenship' (Ong 1999: 215). The rationale behind this argument is that borderless Growth Triangles would better represent 'natural economic zones' (Ohmae 1995: 81) or 'extended metropolitan regions' (McLeod and MacGee 1996: 417) which conform more to flows of human economic activity than do political entities. This perspective speaks of 'space of flows' stemming from an inter-linked global economy. What is to be understood by this is that market linkages and investment flows are overriding artificial political constraints at national levels and transcending political borders at international levels. These proposed new zones can either lie within or across boundaries, and they possess synergies based upon complementary resources enabling successful linkages with the global economy. The challenge for those who uphold these views is thus to re-map areas and transform them into a 'global marketplace' (Ohmae 1995: 8).

The politics of mapping

Multi- and trans-national investors in collaboration with the Indonesian state are thus now busily engaged in re-mapping Riau to re-fashion it into a global estate of productive space for resource modification and ownership. This brings to the fore the politics of mapping.[4] Map reading is not only crucial for inferring the dynamics of political and economic power, but also for understanding other non-material interests in the political imagination (Cooke 2003: 266–267). It also draws our attention to debates concerning knowledge construction.

'Knowledge is not a simple echoing of facts, of confrontation with reality', as Gellner (1987: 162) observed. At every interstice of the global and local and of the centre and periphery, knowledge is constituted by pitching different systems of knowledge against each other, generating dichotomies. 'What constitutes knowledge?' and 'Who gets to define it?' are questions that concern issues of identity and representation. Nowhere else is this more clearly demonstrated than in the mapping of knowledge and the politics of mapping.

'Maps are not neutral instruments' but 'have both cadastral and political contexts' (Cooke 2003: 266). Official administrative maps represent a particular order of knowledge, and they can remake reality. Mapping is distinct from the act of map making; it is about power relations and prioritizing choices (Wood 1993: 36). The politics of mapping mirrors itself onto the creation of territories, the legitimization of territorial rule and the allocation and the realization of resource access rights. As those who have studied maps elucidate,

> All modern states divide their territories into complex and overlapping political and economic zones, rearrange people and resources within these units, and create regulations delineating how and by whom these areas can be used. These zones are administered by agencies whose jurisdictions are territorial as well as functional.[5]
>
> (Vandergeest and Peluso 1995: 387)

Administrative maps and mapping represent selective forms of knowledge useful for control, and this entails exercising exclusions. The intent of such 'tunnel vision' is that it distinctly brings into focus 'limited aspects of an otherwise far more complex and unwieldy reality' (Scott 1998: 11). Such simplification 'makes the phenomenon at the centre of the field of vision more legible and hence more susceptible to careful measurement and calculation ... selective reality is achieved, making possible a high degree of schematic knowledge, control and manipulation' (Scott 1998: 11). The attractiveness of administrative maps also lies in their generality and universality. The rules for drawing up such maps can be applied anywhere and everywhere, regardless of local contexts. They are maps to administer reforms in originally open spaces. Administrators perceive this administrative reform as promoting commercial and cultural progress, in addition to promoting rational citizenship (Scott 1998: 31–32). Basically, the more static, standardized and homogenous a social space is, the more comprehensible it is to state administrators and global business investors. This in turn expedites the ability of administrators and investors to exercise power and control to transform populations and landscapes into closed systems that present no challenges.

The Indonesian state is now exercising great power as chief architect in executing these agendas of standardizing resource and population maps of Riau for the global market. In this way, coastal and maritime tenure, settlement patterns and modes of livelihood are greatly re-fashioned. The state and multi-national investors are thus interested in portraying only one economically productive reality in their aim to develop Riau (see Map 7.1). The transformative power resides not in the map, of course, but in the power of the multi-national and trans-national corporations, which possess a monopoly on utilitarian simplification, and in the state, which possesses a monopoly on the legitimate use of force. In short, the genesis of the Growth Triangle is ostensibly a move towards borderless and multi-national zones of sovereignty. However, we could also view this state of affairs from another perspective. That is, today's globalized order of borderless worlds is a deliberate product of political choices and state decisions. Borderless worlds do not come into being merely by letting things take their own course. In fact, borderless worlds are political choices made within a context of assessment and enforced by the states concerned (Helleiner 1995: 150). Thus, contrary to the prevailing view that this is an era in which nation-states are evolving towards larger and borderless worlds, an examination of the Growth Triangle shows that the governments of the nation-states which embody the geography within these so-called borderless worlds are by no means presiding over the demise of their own borders. As reported in the Singapore-based *Business Times* (8 August, 1990),

> At present, the Growth Triangle is one only in concept, albeit endorsed at the highest levels by the three countries. In practice, there are three still separate sides, all a part of the Triangle only in that they have incorporated the concept into their respective development plans. Thus, Malaysian Prime Minister Datuk Sri Dr. Mahathir Mohamad recently said the Triangle fitted nicely into Johor State's development, while Indonesia saw the concept in the context of its own development master-plan for the Riau region, now being prepared [. . .] But the Triangle only becomes a reality, especially in the eyes of outside investors looking in, when it is jointly presented, marketed and operated as a distinct entity.

Zones such as the Growth Triangle are primarily created based on two factors. First, the individual states concerned acknowledge that they are in close geographical proximity to one another and that they are in fact in different stages of development. Second, each state considers itself to possess what it regards as a different sort of comparative advantage – be it in human, capital or material resources – over its neighbour state. Thus, the creation of a Growth Triangle is necessarily based upon divisions and differences. State borders are clearly the crucial markers of these. It is only when the individual states clearly see such divisions and differences that they are willing to form multi-lateral relationships and zones such as the Growth Triangle. A network of interdependency between these states is then formed to interweave their comparative advantages so as to increase the overall well-being of the individual states and the collective zone.[6] Thus, if anything at all, what can be said is that there is a 'paradoxical strengthening of state borders within . . . so-called borderless worlds' (Chou 2006b: 111) and that '[b]orderless worlds, in short, border worlds' (ibid.: 129)!

The notion of zones such as the Growth Triangle is thus relational, implying both difference and similarity. This is an eminent example of the relational nature of borders and their everyday practices of exclusion and inclusion. In this chapter, rather than concurring with the view of the Growth Triangle as a graduated zone of sovereignty, we see that it is actually a space marked by borders reflecting overlapping multiple realities.

The Growth Triangle and the master plan for Riau

The first attempt to re-think the region of Southeast Asia in this new conception of a borderless world occurred in 1989 with the mapping of the first Growth Triangle, variously known as the 'Singapore–Johor–Riau Triangle', 'SIJORI', the 'Southern Growth Triangle', 'JSR' (Johor–Singapore–Riau), the 'IMS' (Indonesia–Malaysia–Singapore) or the 'IMS-GT' (Indonesia–Malaysia–Singapore-Growth Triangle).[7]

The aim of business strategists, political cosmopolitans and other local state powers in linking these three geographically contiguous areas is to enable industries located within the Growth Triangle to enjoy advantages from two directions: (1) the efficient infrastructure and highly-skilled workers of Singapore; and (2) the cheaper unskilled labour and land, sea and other resources of Riau and Johor. From their perspective, this is a development strategy justified by the 'flying geese' theory of economists such as Akamatsu (1962), Chen (1989) and Yamasawa (1990). This theory argues for more openness within the Asia-Pacific region in the interest of sustained economic dynamism for everyone through the relocation of industries and investments from the more advanced countries to the less developed ones. In this way, the investing countries, such as Singapore, would be able to upgrade their industrial structure at cheaper costs and increase their profits, while the receiving countries, such as Indonesia and Malaysia, would be able to catch up with the industrialized countries.

The origin of the Growth Triangle lies in the fact that Singapore is dependant on outside sources for fresh water, low-cost land and labour – of these, water is the most crucial. For a long time, Singapore obtained these resources from Johor. In the mid-1980s, Singapore began to approach Indonesia as an alternative and additional resource. As relations between Jakarta and Singapore strengthened and joint initiatives in Riau increased, Singapore's relations with Johor were jeopardized. To solve this problem and to ensure co-operation from all sides, Singapore's First Deputy Prime Minister at the time, Goh Chok Tong, proposed the concept of a 'Triangle of Growth'.[8] The aim was to integrate Batam, Singapore and Johor in a collaborative and regional development plan.

The 'Growth Triangle' agreement signed between Riau, Singapore, and Johor is currently the most important economic programme that has been established for the transformation of the region of Riau. At the earliest stage of planning, Goh Chok Tong, then the Prime Minister of Singapore, announced that he was assembling 'a team of people who already know Indonesia, not just Singapore businessmen, multinationals, to survey the field and make an assessment of what can be done' (*Straits Times Weekly Edition*, 13 November 1991: 1). What is significant is that no mention of involving the local inhabitants of Riau in the planning process was ever made by any of the partners in the Growth Triangle.

The non-involvement of the local people of Riau in designing the Growth

Triangle should, however, come as no surprise at all. Through the lens of the Indonesian State itself, Riau is an archipelago of 'virgin' (Soh and Chuang 1990: 34) and 'small uninhabited islands' (*Jakarta Post*, 28 April 2000: 12) that span 235,000 square kilometres of unused territorial waters. These are 'idle resources' which must be 'put . . . to good use' (*Jakarta Post*, 28 April 2000: 12). Therefore, the plan is to 'rent' these islands to local and foreign investors (*Jakarta* Post, 28 April 2000: 12) so as to transform the island chain into the 'Caribbean of the East' (Wong and Ng 1991: 272). To lure investors to the area, pioneer status and tax breaks are awarded to anyone willing to invest, especially in the remote areas of the archipelago. Already, Singapore, Japan, Hong Kong, Australia and Taiwan are investing heavily in the region.

As a result, there has been a marked increase in the flow of goods, capital, information and people between Singapore and Johor and Singapore and Riau. Large-scale cross-border interactions started with media outlets, such as television from Johor and the 'private' FM 'Radio Zoo' from Batam, crossing their national borders. Population movements across the causeway between Singapore and Johor also escalated by about 17 per cent per annum (*Straits Times*, 24 August 1990). In 1989, visitors from Singapore to Johor spent a total of M$1 billion on their visits (*Straits Times*, 21 June 1990). With regard to border crossings between Batam and Singapore, 69,000 tourists entered Batam (mostly from Singapore) in 1980, and by 1985 they swelled to over 50,000 (*Batam Industrial Development Authority* 1991). These and many other initiatives between Singapore and Riau and Singapore and Johor have been operationalized in the past decades. All initiatives have been promoted by the government of Singapore as a reflection of the easing of national borders to enable the reaching out of the Singapore space-economy into Indonesia and Malaysia. The aim is for the flow of people and goods to be matched by an increasing level of cross-border capital flow.

In 1996, just prior to the Asian economic crisis of 1997, the economic disparities between the three countries were as shown in Table 7.1. Over and beyond monetary terms, these economic disparities have implications which may be tabulated as in Table 7.2.

Riau, the partner with the lowest GDP per capita and the lowest degrees of industrialization and urbanization, but the highest degree of rural subsistence, contributes to the economic complementarity within the Growth Triangle in the form of cheap land, labour and other resources.

The master plan for the Growth Triangle therefore re-maps Riau as a global investment and production zone, with the different islands to be re-fashioned in accordance with the projected needs and demands of global economy (see Map 7.1).

Targeting Riau as a resource zone for external interests has proceeded in a definite sequence: first on Batam Island, then eastwards to Bintan Island before moving westwards to include the Karimun islands and Sumatra and finally moving southwards to encompass the islands of Rempang and Galang.

Table 7.1 Gross Domestic Product (GDP) per capita in Indonesia, Malaysia and Singapore (1996)

Country	GDP per capita (1996 US$)
Indonesia	1,119
Malaysia	4,652
Singapore	31,787

Source: Hicklin, Robinson and Singh (1997)

All these islands have been home to the Orang Laut for centuries. What follows is a detailed account of the transformation of the landscape. This account shows the enormous impact that the transformation has had on the lives and livelihoods of the Orang Laut.

Significantly, Batam, the first site of spatial simplification, is the island nearest to Singapore and thus the most convenient for resource extraction. On this island, the Batamindo Industrial Park (BIP) has been built to house light industrial activities.[9] Three international ferry gateways have been introduced and another two are being planned. Also nearing completion are two larger domestic ferry terminals. The most heavily used roads will be upgraded to dual carriageways and a 'visa-free airport for foreign tourists' has been opened (Centre for Economic and Business Studies, Riau University and Promotion and Investment Board, Riau Province 2003: 24). In addition, six new bridges forming the Trans-Barelang highway were constructed to join Batam to the neighbouring islands of Rempang and Galang (ibid.: 20). Batam now has 20 star-rated hotels and five international golf courses. It has been explicitly stated by the Promotion and Investment Board Riau Province (ibid.: 26) that Batam is to become a recreation zone for international tourism, particularly for tourists from Singapore. An oil-service petroleum sector is also expanding at Sekupang and Batu Ampar, (<matrix-batam.hypermart.net/batamart.htm>, 28 November 1999). The comprehensive spatial simplifications that are occurring on Batam correlate with the huge amounts of foreign investment it has received.[10]

Eastwards, Bintan Industrial Estate (BIE) was constructed on the island of Bintan in 1993 to form a base for light industries (see Map 6.1). The first

Table 7.2 Relative degrees of industrialization, urbanization and rural subsistence

Country	Relative degree of industrialization	Relative degree of urbanization	Relative presence of rural subsistence
Indonesia	Low	Low	High
Malaysia	Medium	Medium	Medium
Singapore	High	High	Low

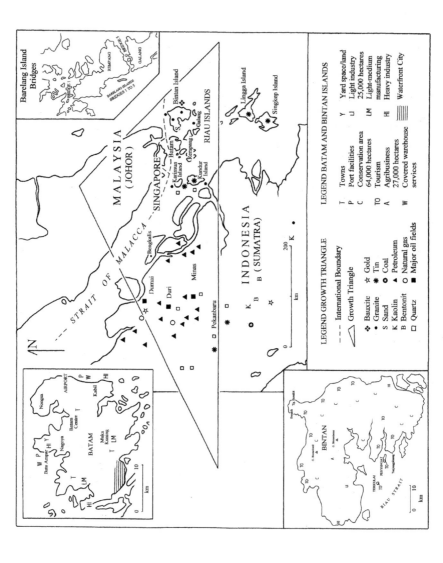

Map 7.1 The Growth Triangle and master plan for Riau as perceived by the Regional Investment Co-ordinating Board of Riau Province.

plant on the Lobam area, situated in the northwestern part of the island, opened in 1994 to house industries dealing with textiles, apparel, furniture, wood products, toys, plastic products, packaging, food processing and foot-wear. Ferry services from Singapore have been re-routed and a ferry terminal has been located at the entrance to the BIE to enhance its ability to trans-port both people and goods. Products are shipped to Singapore and then re-exported to the global market. In order to speed up customs formalities, Singapore and Indonesia have also jointly introduced a smart-card system which enables regular and selected travellers to cross the national boundaries of Singapore and Indonesia without having to present their passports for immigration clearance. At the same time, stricter border controls are enforced to keep certain other people, such as the Orang Laut, from making border-crossings. In addition, tourism and leisure facilities are already being built to take advantage of the 'unexplored natural beauty of the island' (Fukada 1997: 125). Luxury resorts, golf courses, a marine sports club and seafood restaurants have been constructed on a 23,000-hectare area in northern Bintan. A 20-year plan exists to transform a total of 26 areas on the island to attract tourists from Singapore, as well as from other regional and long-haul destinations. The master plan for Bintan Island also includes the construction of an 18-megawatt power station.

Since 1995, the four Karimun islands – Karimun Besar, Karimun Kechil, Kebil and Mudu – with their easy access to international shipping lanes in the Straits of Melaka, have been gradually transformed into a billion-dollar mar-ine complex, oil terminal and chemical supply base (Fukada 1997: 126–134). Fundamental to the project is a shipyard with the ability to accommodate oil tankers and container ships. Foreign investors have also been competing to quarry granite, sand and tin in the Karimun islands.

Southwards, in the islands of Rempang and Galang are the six new bridges that are now used to transport resources for export out of Riau to the global market. These bridges include the 644-metre Batam-Tonton bridge, the 420-metre Tonton-Nipah bridge, the 270-metre Setoko-Nipah bridge, the 365-metre Setoko-Rempang bridge, the 385-metre Rempang-Galang bridge and the 180-metre Galang-Galang Baru bridge (<www.batam.com/about/barelang.htm>).

These various ongoing projects in the Riau Archipelago are manifestations of modernization and development in the worldview of the Growth Triangle planners, who all share the vision of achieving quick economic returns. Investment in Riau to fulfil this vision is rapidly changing Riau's landscape. In a nutshell, the greater the amount of money invested in the region, the greater the spatial simplification imposed on the environment.

The impact on the Orang Laut

Displacement

To speak of the 'deterritorialization' of the Orang Laut's maritime world is actually to speak of generalized dismantling of an identity order complex that gives people, spaces and politics their meaning. A serious consequence of deterritorialization is the creation of a political map that would be, on the one hand, more integrated and connected, but on the other, more divided and dislocated as it tends to gloss over the reality of socio-political networks and economic systems which are already in place. 'The mapping of the Growth Triangle' is indeed 'like many other "official" mapping exercises' (Bunnell, Sidaway and Grundy-War 2006: 236) in that it necessarily selects only certain forms of knowledge for representation, involving acts of 'cartographic silencing', 'ideological filtering', 'discrimination' and '(mis)-representation' (Harley 2001). The Orang Laut are critically displaced in this new landscape and there is no space for them any longer unless they transform themselves or allow themselves to be transformed. Against the backdrop of this chapter, coupled with the Islamization process that is discussed in Chapter 6, the gravity of the situation is intensifying for the Orang Laut.

In the context of state-directed change to develop Riau for the global market economy, the Orang Laut feel that they have accrued few, if any, benefits. They feel that the projects of the Growth Triangle have had negative effects that far outweigh any intended benefits. They are distressed for many reasons. First, these projects have affected their local lives and living spaces to the extent that the continuity of entire communities has been threatened. Second, their strategic local knowledge, as well as modes of organization and participation which have for centuries sustained their livelihoods, have been repudiated. Third, landscapes which have formed the bases of their very existence have been wiped out.

For the Orang Laut, the mega infrastructural projects of the Growth Triangle in Bintan, Batam, Rempang, Galang and Karimun are in areas that impinge upon their seasonal routes. The projects are located in places where they obtain fresh water, find safe havens in storms, build short-term dwellings, repair their boats and harvest shellfish, sea cucumber and other littoral resources. Malay and European historical sources confirm that these islands have constituted a crucial part of their economy for centuries (see Sopher 1977: 345–388). Since the Growth Triangle came into effect, they have progressively lost this territory and all the resources therein. Many of the micro-environments within this territory have been irreversibly damaged and destroyed. Hence, even if the mega infrastructural projects were halted immediately, many of the damaged and destroyed micro-environments could not be reconstituted (Chou and Wee 2002: 341).

Vandergest and Peluso (1995: 389) point out that

People do not experience space as abstract, as they generally have no access to maps produced by militaries and government surveyors. Experienced territory or space is not abstract and homogenous, but located, relative, and varied. [The administrator's map] ignores and contradicts peoples' lived social relationships and the histories of their interactions with the land [and sea] . . . The lack of fit between lived space and abstract space has contributed to the instability of the territorial strategies of the modern state. States often have had to rely on open coercion against rural residents to implement territorial control. Even under such conditions, people often refuse to acknowledge the territories claimed by states for parks, protected production forests, and even state regulations on private property.

For centuries, the Orang Laut and the region of Riau have been sustained on the principle of diversity. However, the aim of the state and its investment partners to arrange these very diversities into simplified modern 'maps of standardisation' (Scott 1998: 12) has seriously impinged upon and modified the resources of the Orang Laut that are necessary for their very survival.

Modification of resources

In contrast to the modern maps of state administrators, the Orang Laut possess neither paper nor printed maps. Their knowledge of resource identification is transmitted orally. The historical significance of their indigenous maps lies in the fact that they have maintained an entire socio-economic system which has sustained their lives and living spaces. Just as significant is the fact that these orally transmitted and mental maps which reveal the discovered and potential wealth of the region's resources have yet to be fully documented. An important aspect of their concept of sea and coastal spaces is their ability to identify available and potential areas for fishing, trapping, charcoal making, worship, refuge, living, production, hunting, gathering and collecting construction material. They also possess a vast knowledge of marine species, edible and medicinal plants and detoxifying agents that currently have no market value. In addition, they possess extensive knowledge of multiple micro-environmental niches, including an in-depth understanding of tides, winds, currents, breeding grounds, danger zones, refuge areas and fresh water points (Chou 2006b: 127).

The master plan of the Indonesian government and its business entrepreneurs to re-map Riau (see Map 7.1) and re-define its resources so as to present it as a global investment zone does not in any way correspond to the resource identification maps of the Orang Laut (see Map 4.1). The result has been the wiping out of much of the region's diverse flora and fauna. From the state's perspective, such diversity – which is appreciated by the Orang Laut – constitutes a landscape of disorder and underdevelopment.

Resources such as water have also been transformed into higher-yielding revenue resources by bilateral investment projects of the states in the region. Water is now controlled by the state to serve what it defines as higher state interests. The Orang Laut are no longer allowed free use of fresh water that has always been available to them. This has come about because of an agreement signed on 28 August 1990 to serve the water needs of Singapore. In this agreement, the governments of Singapore and Indonesia also resolved to embark on joint development of tourism, water supply and industries in the Riau islands. This agreement explicitly states that Singapore and Indonesia will co-operate in the sourcing, supplying and distributing of water to Singapore and three fresh-water reservoirs are to be built on the island of Bintan alone (*Straits Times*, 29 August 1990: 1). One reservoir is to specifically serve the beach resorts on the north coast of Bintan, an area which has traditionally been inhabited by the Orang Laut. The second reservoir will be built in the middle of Bintan Island. The third reservoir will be a dam in Teluk Bintan ('the Bay of Bintan'), into which several rivers are to pour their fresh water. The amount of fresh water to be collected exceeds the needs of the Orang Laut and the other inhabitants of Bintan, who together form a total population of 90,000. Much of the water resources in Bintan will therefore now be channelled to meet the needs of other populations. As mentioned by the Indonesian Foreign Minister, the water will probably be pumped via Batam to Singapore (*Straits Times*, 10 April 1990).

Sand on Bintan Island is another resource that is now inaccessible for the Orang Laut. The island, which has been an important Orang Laut territory for centuries, is now facing serious environmental damage. Over 50 sand-quarrying companies, each with a government concession of five to 75 hectares for their operations, have shifted their sand quarrying from the sea to the island. The excavated sand is exported mainly to Singapore to support Singapore's construction sector and coastal reclamation projects. Sand is sold at S$1.50 (Rp. 7,500) per cubic metre to international brokers who then sell it to Singapore construction firms at S$15 per cubic metre. Singapore is estimated to need 1.8 million cubic metres of sand over the next seven years. The Orang Laut have suffered tremendous losses from these quarrying operations. The operations have led to land degradation, plant extinction, unmanaged waste from quarrying activities, serious damage to the marine environment and the subsidence of the smaller islets (*Jakarta Post*, 21 July 2003: 4).

In Karimun, six sand mining companies which were issued licenses in 2005 to export sand to Singapore destroyed the environments of several islands within just one year. Protests from local inhabitants forced a temporary stop in the mining operations (*Indo Pos*, 14 August 2006: 1 and 7; *Indo Pos*, 15 August 2006: 1 and 7). In spite of this emergency, the Deputy Karimun regent Aunur Rofiq decided to tone down the problem by stating, 'Actually, there isn't any island that is sinking, it is just that the reclamation activities have not been implemented properly. We will reevaluate the companies' commitment to maintaining the environment' (*Jakarta Post*, 18 August 2006: 8).

In such a collusion of interests among the Indonesian and Singapore governments and private conglomerates, local resources of the Orang Laut have been greatly modified, posing a serious challenge to the basic fabric of their survival. Of great concern is that the re-shaped landscape no longer leaves any room for them. They can find a place in the new landscape only if they proletarianize and allow themselves to be transformed.

Border prohibitions

It is paradoxical that in the contemporary revolutionary movement of reorganizing spatial relations to form more integrated and connected border-less worlds such as the Growth Triangle, zones have in fact become more divided and dislocated. That is, the development of borderless worlds does not contravene but actually hastens the simultaneous development of ever more modern worlds characterized by stark inequalities and divides.

The maritime world of the Orang Laut, which has been encroached upon by the Growth Triangle, provides a case in point. As the very concept of the Growth Triangle is premised on clear border divisions between nation-states, the power to define both the internal and external borders of such spaces and who shall be allowed to participate in such zones still primarily rests with each participating nation. Therefore, rather than effectuating a 'multi-national' and 'graduated zone of sovereignty' (Ong 1999: 215) in the maritime world of the Orang Laut, the Growth Triangle has in fact divided it into various different spatial polities. Each polity has come to embody a different sovereignty, knowledge domain and political economy. Each polity has also come to represent a different mode of resource allocation, production and consumption, in that order.

Borders, as Baud and van Schendel (1997: 222) say, 'rarely match the simplicity of their representation on maps'. They 'are simultaneously structures and processes, things and relationships, histories and events' (Donnan and Wilson 1999: 62). In these overlapping positions, the border functions as a marker and agent of both tradition and change. Borders inscribe the edge of a social system (Wallman 1978: 205). They are at the same time structures for policy making and tell of control and orders, and of the interface and identity lines between inside and outside, between 'us' and 'them' (Ross 1975). This important role of the border in the selectivity and reclassification of people, interest groups, territories and resources underscores why borders have also become a term of discourse in narratives regarding multiple reper-tories of identities and realities. They have become a screening mechanism to mark off members from non-members and acceptable participants from non-acceptable participants (Donnan and Wilson 1999: 15). In this perspective, the Orang Laut are seen by the Indonesian state and the other neighbouring states within the Growth Triangle as anomalous citizens who subvert state order and policies. In an apparent attempt to control the Orang Laut, the governments and their agents in the Growth Triangle have turned to the use of borders to forbid the Orang Laut to enter this supposedly global zone or move from one system to another within this zone (see Chapter 8). The Singaporean authorities pride themselves on having modernized and developed the island by either resettling Orang Laut communities into high-rise flats and re-categorizing them as 'Malays' or, alternatively, by excluding all of them from the area that is demarcated as the Republic of Singapore. Within the borders of the territory defined as the Republic of Indonesia, the Indonesian government considers the expert mariners an isolated, alien and backward people who form so-called 'pre-villages' or 'traditional villages'. The ultimate goal of the Indonesian government is, then, no different from that of Singapore: The Orang Laut must be transformed and resettled in urban settings.

Dead spaces

Another problem that has arisen in connection with forcing out the Orang Laut from the islands and coastal areas designated for the Growth Triangle is the collapse of an existing social order. Their removal from areas where they have lived and been custodians for centuries has led to the collapse of an informal but crucial network of local policing. As the Orang Laut and other permanent populations are replaced with transient migrants, opportunities have been created for the entry of roving bandits and pirates. This state of social chaos has even led a number of newly arrived entrepreneurs to abandon their planned operations in Riau.

Multi-national developers themselves have been affected by the environmental, social and economic impact of their projects. To pave the way for the construction of industrial parks and tourist resorts, local settlement sites had to be cleared, forcefully vacated and stripped of all life to provide vacant space for new developments. On Batam and Bintan, the two Riau islands that have seen the most extensive infrastructural development, the unexpected problem of 'dead spaces' (Scott 1998: 121) – spaces where there is no one around or which lack vital activity – has thus arisen. The clearance of the indigenous populations has meant that all other indigenous activities and resources for everyday living needs were also cleared. Many of the newly arrived migrants – managers and workers alike – who came to Batam and Bintan to set up their new homes and work in new enterprises found themselves without a supporting base for elementary survival. There were no permanent settlements to provide them with expert local knowledge and services. Many of the new arrivals have found they have had to incur massive and unexpected costs in setting up their bases in Riau. Many have had to build their own roads from their homes to other destinations and travel long distances just to obtain basic everyday needs. Tourists, who went to Riau expecting to see local cultures, found themselves isolated in beach resorts with nothing around the hotel except the beach, the sea and a golf course. Some of these beaches have even been artificially created to fill in uprooted mangrove coastlines. The only people whom these tourists met were the uniformed service personnel (Chou and Wee 2002: 353).

The notion of a borderless world is thus the ambition of the few who can afford to dream of becoming homogeneous; the majority find it a constant struggle just to be alive. The conflicts that the Growth Triangle has imposed on the living and production spaces of the Orang Laut is a sobering example of deliberately enforced wider differentiation which has favoured the privileged and penalized the weak. The reformers of Riau aim to portray only one reality in their maps. However, in order to do this, the old – in this case, the Orang Laut – must be obliterated, first from consciousness and then in reality; hence the predicament of the Orang Laut.

8 Conclusion

Introduction

External agendas have been imposed by foreign agencies on the Orang Laut maritime world since the arrival of European colonizers in the sixteenth century. The current pressures for change imposed by the state, the Muslim Malay community and the forces of globalization are part of a continuous historical process that began centuries ago. The difference lies in the rate and scale of these new changes in the living and production spaces of the Orang Laut. This chapter considers the current predicaments of the Orang Laut, in addition to those already raised in the previous chapters. First, it examines the consequences of resettlement and regrouping programmes. Second, it examines the double impact of these changes on the lives of women.

Disputes over redefining the tenure of territory

Following the downfall of Suharto in 1998, there have been numerous demonstrations and protests by thousands of people all over Riau. For three decades of his regime, the local inhabitants felt it was unsafe to voice any grievance over land loss and unfair compensation. Since 1998, however, the people have been releasing their pent-up frustrations in vocal and increasingly violent ways. They are now demanding adequate compensation for land that has been taken from them. In 1993, displaced villagers in Batam were given 50 rupiah (then US$0.23) for each square metre of the floor space of their houses. As the villagers were designated as 'squatters', they did not receive compensation for their land and resources outside their houses. Between 1994 and 1995, villagers on the north coast of Bintan were displaced and failed to receive the promised compensation. Their problems have intensified because of the desire of Singaporeans to own houses. Plots of land are now supplied by Riau, especially from Batam island, which is just across the Strait from Singapore. Although Indonesian law makes it problematic for foreigners to own land, there are many proxy purchases (*Business Times*, 5 September 1990: 1). Within three days of the launch of the southern part of Batam's S$1 billion Waterfront City Project in June 1990, 80 per cent of the

644 residential units were sold to Singaporeans. A few other examples of Batam property projects that have attracted property speculators include the S$30 million First City Complex, the S$50 million Batam Centre and the S$20 million Shangri-la Garden Residential Project. The first two projects are located in Batam Centre and sold most of their 500 bungalows and semi-detached houses on 11.6 hectares of land and 670 housing units on 31 hectares of land respectively to Singaporeans. The Shangri-la Garden Project sold most of its 107 bungalows to Singaporeans (*Business Times*, 8 August 1990: 2). The consequences of this property speculation are manifold: displacement of the Orang Laut from places that have been their territory for centuries; dramatic escalation in property prices in Batam; and an increasing number of squatters as migrants from other Indonesian islands arrive in search of jobs but are unable to find affordable housing. It was reported that in '1997 alone, an estimated 3,000 spontaneous migrants arrived in Batam in search of jobs in tourism and manufacturing' (Chang 2004: 4). These are but a few examples of what has ignited the rage of the local inhabitants who are now claiming, in almost every case, that their land was acquired forcibly and without adequate compensation.

In the context of these articulated grievances, the absence of a written historical tradition among the Orang Laut has made it almost impossible for them to stake any claim over territorial rights. This has been the source of many clashes between foreign business entrepreneurs and local inhabitants, including the Orang Laut. The lack of written proof is something which distresses many people who have yet to understand the importance and relevance of oral histories in the modern world. The newcomers can neither understand how a people can lay any claim to a territory based simply on an oral tradition nor how such claims can have the power to override the legal rights they have obtained from the Indonesian state.

In January 2000, the inhabitants of Bintan disputed the joint land ownership of Indonesian Salim Group and the Singaporean Bintan Resort. The inhabitants demonstrated and demanded restitution for a Suharto-era deal in which they had received little compensation for their land that was sold to business entrepreneurs. The clash resulted in widespread cancellations of hotel reservations, disruption in the tourist industry and an undermining of the confidence of foreign investors. It was reported that the problem was eventually resolved in closed-door negotiations, but the protest had spread to the neighbouring Bintan Park, yet another joint venture by the Indonesian Salim Group and several Singaporean companies. 13 people were injured and the Park was forced to close for a day. This resulted in losses amounting to US$200,000 (*Straits Times* 21 January 2000; Cohen 2000: 24; Chang 2004: 4). In the midst of all the raging disputes, the historical significance of Bintan, home to the Orang Laut for centuries, was simply forgotten.

In June 2003, Pelalawan Regency in Sumatra sued Asia's largest pulp producer, PT Riau Andalan and Paper (RAPP), for water pollution, destruction of protected rainforests in Kepungan Sialang and neglect of local communities. According to my coastal informants, the waters in the area were so polluted by the waste generated from the pulp mills that maritime life was greatly affected. Food supply and sources of income for local people were drastically diminished. Fresh water points were so contaminated that skin diseases rose at an alarming rate. Results of a field investigation by the regency revealed that 95 per cent of the raw material supplied to the pulp and paper mills was sourced outside of the company's industrial estates, infringing sometimes on protected forests. The spokesman for the locals said that the company's industrial estates could not meet its demand for logs as it was producing 1.7 million tons of pulp and paper per year, up from the previous 1.3 million tons. The increase had forced the company to source more logs from outside its own timber estates and natural forests. RAPP denied knowing anything but decided to donate packages of financial aid to several groups after the demonstrations. The company had long been protected by security forces and had strong links with the former President Suharto. This was not the first protest staged by the locals against RAPP. The Betawi people in North Sumatra had earlier staged violent protests against Toba Pulp Lestari, a sister company, over widespread environmental damage caused by deforestation and air and water pollution, including the world-famous Lake Toba (*Jakarta Post*, 17 July 2003: 5). PT Riau Andalan Pulp and Paper is now trying to improve its reputation by sponsoring a Riau reforestation programme that was launched by President Susilo Bambang Yadhoyono in Jakarta in April 2006 (*Jakarta Post*, 14 August 2006: 8).

As clashes between foreign investors and local inhabitants escalate, there are some signs of awareness among a few of the newcomers that they must make some attempt to honour indigenous claims of territory. However, the investors are not without problems. As a chief executive officer of a multi-national company despaired,

> First of all, we cannot understand why there can be two systems of law here. The State of Indonesia has given us the right to come here and use these territories, but those who claim to be the local Malays are claiming that this is their land. Initially, we persisted in carrying on with our enterprise, but the local villagers organized protests and blocked the entrance points to our headquarters here. What could we do? They were women and children. We got our local representatives to go out and talk to them, and to advise them to move away quietly. However, they persisted. They finally moved off, and we thought the problem was solved. But this just kept happening again and again. Each time we have to stop a day of operations because of such protests, we lose US$1 million.
>
> So, okay, we thought, this cannot go on. We will talk to the local people and find out what we have to do to allow us to coexist peacefully with them. One group comes and tells us that according to their customary law, they cannot sell us the land. However, they are allowed to lend it to us for a certain fixed period and we have to pay them an amount as

stipulated by their law. This arrangement is not even regarded as renting us the land, but lending it to us. Even then, there are spaces we are not even allowed to touch. Can you imagine how impossible this is for us? Right smack in the middle of our operations there may suddenly be a patch of sacred land we cannot set foot on. All right, so we agree to the terms of this group. Before long, another group comes up with another story about another area where we have ongoing operations, and another group and another group. It is just endless. How can we know if these people are honest or not? They may just be opportunists. They see us giving compensation to one group, and they decide to take advantage of the situation too. Whose story is to be believed? Which story is the correct one? Who can authenticate and verify these stories? Can anyone even tell us how to identify an indigenous Malay? Everyone seems to be giving me different definitions. This has to stop somewhere or it will be impossible for us to carry on any operation here.

What must also be said at this point is that paying compensation to settle claims does not solve all possible problems. In rare cases where resettlement and compensation fees have been awarded to the affected Orang Laut, these payments have only been made to men. This is based on the assumption that men are the heads of households and sole title-holders of the tenure of territory. Women are not recognized in this matter. Correspondingly, in cases where programmes for the retraining of skills are offered to those who have been affected by resettlement schemes, women are again not included. It is no surprise that, from the perspective of the Orang Laut, the Growth Triangle has also become a structure and symbol of differentials in gender and status.

With the escalation of problems and conflicts between the investors of the multi-corporations in Riau and the local inhabitants, I was approached by a pulp and paper company in 2003 for help. The company – whose name I will not disclose here – had discovered from the coastal inhabitants that I had been doing research in Riau for many years and had established a very warm relationship with the Orang Laut. The company executives proposed that they would allow me access to all their confidential files containing information on problems that they were facing with the local inhabitants. My task would then be to use my 'anthropological skills' to 'talk to the local people to persuade them to stop giving so many problems' and 'to be on [the] side [of the company], rather than against [them]'. The executives who spoke to me believed that the grievances and demands of the locals were 'irrational'. They emphasized that, in view of the 'millions of dollars' that they could lose each day, whenever disputes disrupted their operations, they were not keen on a longitudinal study of the problems but instead required a solution that could be accomplished in as little time as possible with 'immediate results'. The executives who met with me were divided in their opinions concerning the 'copyrights' of the proposed 'findings' from the field. Some were of the opinion that I could hold the rights to publish the 'findings' as 'some of the problems [were] already known internationally' and it would only be a matter of time before all other problems would be publicly known anyway. Others thought that permitting me the rights would only 'open the Pandora's box' to even more problems. Needless to say, it was beyond my conscience even to consider such an undertaking for the company. Nonetheless, in the course of my meetings with representatives of the company, I was able to obtain some insights into their operations in Riau.

Loss of identity and sense of belonging

The Indonesian state authorities kept records on the programme to resettle Orang Laut. Up to the early 1990s, it has been reported that 19 per cent of an estimated 5,000 Orang Lauts on the islands of Singkep, Lingga and Galang have voluntarily moved to resettlement sites. Resettlement programmes were carried out in Singkep, Lingga and Galang during the periods of 1982 to 1983, 1985 to 1986 and 1988 to 1989 respectively (Bupati Kepala Daerah Tingkat II Kepulauan Riau 1988: 6; Ketua Bappeda Tingkat II Kepulaun Riau 1990/91: 2). Each resettled Orang Laut family consisted of an average of five members (Lenhart 1997: 588). Planning for the resettlement of the Orang Laut of the sub-district of East-Bintan began in 1989; 30 families were finally resettled in 1993 (Camat Pembantu Baputi Wilayah IV Bintain 1989). In 1986, plans to move all the Orang Laut of the Batam area en masse to a large resettlement site on Bertam island began (Walikotamadya Kepala Wilayah Kotmadya Administratif Batam 1986). In the following year, the first 14 of the 100 planned houses were built on the island to begin the exodus (Forum Komunikasi dan Konsultasi Sosial (FKKS) Batam 1987: 4–5); Institut Teknologi Bandung, Fakultas Teknik Sipil dan Perencanaan, Jurusan Teknik Arsitektur 1988). In 1991, another 24-house resettlement site neighbouring Bertam and facing Lingke Island was opened (*Kompas* 18 May 1991: 14; 22 September 1991: 8). Although no other recent concrete figures for the resettlement of Orang Laut communities are available, Indonesian state authorities maintain that figures continue to rise.

Critical problems have resulted in the resettlement programmes for displaced Orang Laut. Some of these problems have already been discussed in Chapter 7, but there are others. Due to a lack of awareness among resettlement officers, members of rival Orang Laut groups have been resettled on the same island. Clashes with serious consequences between these rivals are everyday occurrences. During the course of my fieldtrip in 1994 to the communities on Teluk Nipah and Nanga Island, my Orang Laut friends confided that they had heard of plans to move all the different groups to a single island, making way for an intended highway that was to stretch from Batam Island to Senyentong. Such plans cause great apprehension among the Orang Laut. In their view, social harmony is a factor to which the planners of modernization and development programmes have not given much consideration. They are also concerned that concentrating too many Orang

Laut on any one island would surely overburden the carrying capacity of the area in terms of its marine and fresh-water resources.

Another basic problem is that many Orang Laut find it difficult to adapt to a land-based lifestyle. The fact that they have been accustomed to living together in small groups of kinsmen makes it difficult for them to adjust to living with several other groups on one resettlement site. Very little inter-action takes place between the new neighbours that crowd on one resettle-ment site, and even less interaction occurs between the different Orang Laut groups who now make up one resettlement community. Resettlement officers are faced with endless problems in trying to co-ordinate and involve all of them in activities concerning the community as a whole. A sense of communal spirit has not developed because the rival groups cannot agree on any one person to serve as official representative.

Gendering of borders and spaces

From the perspective of the Orang Laut, the maritime world which they inhabit is considered to be both living and working spaces for all men, women and children.[1] All have shared these spaces and all have had equal access to all resources for their sustenance. In their work and other daily activities, there are no demarcations of male and female spaces.

The houseboat functions as a living space as well as a site of production. It can be owned by men and women or both. The ideals and realities of the relationships between men and women are clearly reflected in the spatial structure of their houseboats. Neither spatial barriers nor gendered spaces exist at home or in the work place. What is usually demarcated as the private sphere or home and the public sphere or workplace merges into a similar spatial context: the houseboat which men and women share and of which each has equal control.

Fishing is a family-oriented activity, and men and women work together as partners in the entire production process. Forming a man–woman partner-ship to fish for daily subsistence and to earn an income from selling off surpluses is so significant that, as described in Chapter 5, their productivity as a fishing couple is central to their identity as Orang Laut. The women also own their own spears and harpoons, and it was not uncommon during the course of my fieldwork for the women to show and demonstrate to me the use of these.

The non-gendered and shared spaces of the Orang Laut bear many broad implications, especially for the women. First, the equal rights to mobility enable them to have similar access to information and knowledge of all spaces. Second, equal access to spaces allows women to apply their know-ledge in effective ways. Third, shared spaces are intricately connected with equal advantages to the rights of ownership, inheritance and claims. Fourth, there is equal access to and control of economic and income-generating activities so that the women have equal access to earned income.

The tradition of non-gendered spaces is now being challenged by the strengthening of state borders within the so-called borderless world of the Growth Triangle programme. The Orang Laut are presently caught in a web of institutionalized spatial and gender segregation, and the impact has fallen particularly hard on the women. I will shortly discuss two examples which highlight this impact. The first shows the gendering of spaces within the zone that has since been demarcated to be the Province of Riau, Indonesia; the second presents the problems that the Orang Laut, and especially the women, are confronted with when they attempt to cross the border between Indonesia and Singapore.

Within the demarcated area that constitutes the Province of Riau, the Orang Laut are being challenged by the Indonesian state's perception of how their life and living spaces should be ordered. From the state's perspective, it is clear that the Orang Laut possess different and undesirable interpretations of spatial arrangements, property ownership and labour division. Therefore, the state finds it necessary to re-define what it perceives to be proper places for Orang Laut men and women. In the name of development, as discussed in Chapter 6, the state and the Muslim Malays are pressuring the Orang Laut into what they regard as a more progressive lifestyle.

The pressures to observe Malay and Islamic rules are bringing extensive changes to the spatial alignments for Orang Laut men and women. The Orang Laut explain that the building of a place of worship and observing the stipulated regular daily prayers would only cause them to be bound to land. Fishing, as they explain, involves irregular hours and it is therefore difficult for them to observe the rituals of prayer and carry enough fresh water for pre-prayer cleansing purposes in their boat-houses.

The impact of modernization and development programmes has also been felt in other areas. The Orang Laut's shared and non-gendered spaces in their maritime world and their houseboats are regarded by the Muslim Malays as unprogressive. The Orang Laut are pressured to demarcate all things pertaining to being out at sea (e.g. boats and fishing gear) as falling within the domain of the men. They are likewise pressured to demarcate all things pertaining to the land-based house and the responsibilities of maintaining it as falling within the domain of the women. This inevitably implies that all fishing activities should only be carried out by men, it being improper for women to be out at sea.

Marriage is the expected norm for all adult members in the Malay community. Once married, the Malays observe their custom and, as Malay conduct dictates, carry out a clear division of labour according to sex in their households. For the Malays, the association of women with the house accounts for much of the boundary differences between themselves and the Orang Laut. Of course, the most visible difference is that Malay women do not go fishing. Any woman who expresses an interest in doing so is frowned upon as a naughty and wanton woman. It is no wonder that the officers implementing the programme of development on Pulau Nanga (see Chapter 6) were adamant in propagating the idea that a woman's responsibilities and rightful place remain within the perimeter of the house itself. The men should never involve themselves in household chores and fishing is solely for the men. They go out to sea to fish to provide for the household's main – if not only – source of income. At most, women might occasionally engage in line fishing, the catch of which is for home consumption and not for trade. When women actually go to the sea, they seek safety in numbers by forming huge and lively parties. Failure to do so would subject them to accusations of being wanton. Such trips are considered leisure outings which can and must only last for a few hours in broad daylight.

In the light of all this, Orang Laut women in communities selected for programmes of directed change are pressured to observe what is perceived as their rightful place in the home on land. This has resulted in some of them taking a less active role in fishing or becoming producers of income for the family. Facing mounting pressures to Islamize, the Orang Laut are confronted with the problem of re-casting their gender relations.

The impact of denying women access to certain spaces has also taken on an added dimension with border controls between states. The maritime world of the Orang Laut is no longer an open space in which to travel, nor is it a production site that is accessible to both men and women. Den and Boat, both of whom are Orang Laut men from Nanga island, explain:

Den:
I bring in sea cucumbers to [the Pasir Panjang Barter Trading Station in] Singapore. The customs officers of Singapore will stop us and ask what we bring in – sea cucumbers or marijuana. When we tell them that we are bringing in sea cucumbers, they will contact our bosses in Pasir Panjang before they allow us onto land. In Pasir Panjang, there are many Chinese and Indian middlemen who are our bosses.

Boat:
We bring fish for the people of Singapore.

Den:
I have cards for entry into Singapore here with me.

Cynthia:
Do you have to pay for these documents?

Den:
S$30 ... these permits [to enter Singapore] were made in Singapore. Formerly, our children were allowed to accompany us into Pasir Panjang, Singapore. Now, they no longer allow our children in. Women are not allowed in ... I do not know why. I got these documents a long time ago – a few years back. These papers are no longer valid. The permit

allows us to be in Singapore for two days. If we arrive today, we will have to leave the next day.

The documents issued by the Port Authority of Singapore to the Orang Laut state that the latter are only allowed to be within the premises of the Pasir Panjang Barter Trading Centre 'for a period of 48 hours' and that the Orang Laut 'are not permitted to leave this area for any other part of Singapore, except for Indonesia on the day of their departure'. The Orang Laut are also 'required to report to an immigration officer for examination at the time they leave Singapore'. It is significant that the Singaporean port authorities will not grant Orang Laut women and children the necessary documents to cross the borders from Indonesia to Singapore. Iyang, Den's wife, and Den explain the situation:

Iyang:
From Belakang Padang . . . [we used to] enter Singapore. [Now] Den goes in alone. I cannot enter Singapore with him so I wait in Belakang Padang for him . . . I have never been [to the Pasir Panjang Barter Trading Station], but I have heard Den talk about it.

Den:
I row a small boat into Singapore. If we were to start rowing from here [Nanga island], we could reach Singapore within two days. If we make our way to Singapore from Belakang Padang, it is even nearer. If we start off at ten o'clock in the morning, we can reach Singapore by 3 p.m. It takes only a day . . . If there are no strong winds, we can get there even faster.

As discussed in Chapter 5, many Orang Laut bring their maritime products to the Barter Trading Station in Singapore. This is because, as Den explained, a kilogramme of the high grade sea cucumber *nabi* would fetch S$45 in Singapore, while the same would only fetch from 5,000 Rp. to 8,000 Rp. (approximately S$2 to S$4) in Indonesia. Remarkably, the entire production process of harvesting sea cucumbers and other maritime products is carried out by men and women together until they reach the border between Indonesia and Singapore. At that point, their working partnership is forcibly halted. Orang Laut women are not allowed to cross the border to carry out the sale of their products. This has substantially altered the working partnerships of Orang Laut couples and their incomes. Hitherto, women always participated in the marketing of their products. Further, many Orang Laut feel that women are better negotiators and can thus bring in higher yields.

In short, the Orang Laut are re-classified when they attempt to cross a border, and Orang Laut women in particular are classified as unacceptable. As noted by Donnan and Wilson, 'all people who cross international borders must negotiate not only with the structures of state power that they encounter', but also with 'new frameworks of social status and organisation with their concommitant cultural ideals and values' (1999: 108).

Thoughts for present and future research

The Orang Laut live in a region which has changed dramatically since the 1990s. On the one hand, the re-mapping of Riau through development programmes has created new possibilities for the region and its inhabitants to attain higher levels of co-operation with the global community. On the other hand, new political formations and economic processes for the region have resulted in dramatic shifts in geopolitical and socio-political relationships.

As trans-national shifts of capital, equipment and technical know-how usually flow from higher-growth to lower-growth countries, natural and, sometimes, human resources tend to move in the opposite direction, feeding into the higher-growth countries. This has had massive consequences for the Orang Laut and other lower-growth communities whose livelihoods depend on the resources that are flowing out. By re-mapping Riau within such a context for generating wealth, we are simultaneously confronted with questions, such as 'What is being exchanged for what?', 'For whose gain is it?' and 'At whose expense is this wealth generated?'

Yet, ambitions to re-map Riau to increase its competitive edge in the global market economy cannot be averted. Increasingly pervasive are the forces of globalization with which rural subsistence economies must co-exist if they are to survive at all. This book presents the ethnography of the Orang Laut and highlights their concepts of appropriation and production. Inadequate concepts of what constitutes appropriation and production have resulted in the Orang Laut being seen as roving hunters and foragers engaged in an extractive economy. If they are such hunters and foragers, it follows that they do not have the slightest concern for the production and reproducion of resources. These are unwarranted perceptions which challenge us to reconsider indigenous concepts of appropriation and production. The implication is that these concepts are inextricably tied to other issues.

First, reconsideration should lead us to a better understanding of the Orang Laut claim to tenure of territory. Second, reconsideration would enable us to become aware of the fact that the Orang Laut are actually keen partners in the production and reproduction of resources. Their conception of coalescing the spirit of people with places and things has convinced many that their claims to tenure lack relevance in pursuing economic concerns and modern-day progress. Critics argue that such claims of tenure are merely spiritual concerns. As Ingold (1986: 139) aptly states, 'implicit in this formulation is the notion that tenure must be either religious (as in the aboriginal conception) or economic (as in the Western conception), but cannot exhibit both aspects simultaneously.' This dichotomy rests on 'a very narrow view of the economy' (ibid.: 14).

To argue for a refinement of concepts is to re-present an entire social system of principles that shape the rights to a network of tenure of territories and distribution of resources. The distribution of property, of course, has fundamental political implications. Anthropology, as the scientific authority for the interpretation of cultural 'others', can play an important role in initiating programmes to recognize cross-determining discourses and opportunities for consent from all concerned to set up new institutional arrangements. The poignant reminder from the Orang Laut is that if they 'do not fish', they will have 'nothing to eat'.

The quest for conceptual clarity is as relevant to anthropologists as it is to state planners and business entrepreneurs. In view of ambitions to promote marine tourism in Riau and the concern to include local inhabitants in this programme, as stressed by the former Governor of Riau, it is important to consider ways whereby Orang Laut contribution and participation can be included in the realization of transforming Riau into what the state planners and business entrepreneurs alike hope to be the Caribbean of the East. As two Singaporean economists have already noted, 'in order to create an authentic rural environment and enhance the attractiveness' of the area as a tourist resort, 'a larger permanent settlement is necessary' (Kumar and Lee 1991: 16). How can the gap be bridged?

The multiple realities involved in re-mapping Riau for rapid modernization and growth can therefore either form scenarios of marginalization and disempowerment for its own people or, if the development programme is adequately and sensitively implemented, empower its people and place them in the global community.

Appendix

Orang Laut fishing gear

The fishing gear illustrated below can be viewed at the National Museum, Singapore and the University Museum of Archaeology and Anthropology, Cambridge, United Kingdom.

Scratchers

These are instruments for collecting shellfish in muddy and rocky areas during low tide.

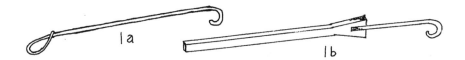

1a: *Kais ketam* ('crab scratcher') is used to get *ketam bakau* ('mangrove crab') from mangrove swamps and rocky crevices. *Ketam bakau* fetches a higher price than other species of crab. The end of the hook is curved gently to avoid damaging the crab.

1b: *Kais gamat/nabi* ('sea cucumber scratcher') is for collecting various types of sea cucumbers, the commonest of which are *gamat* and *nabi*. The handle is much longer than the *kais ketam*, enabling the user to hook a sea cucumber while remaining in an upright position.

Spears and harpoons

The *serampang* ('spear' or 'harpoon') refers to any sharp-pointed instrument used for spearing fish, including spears with several sharp prongs. It is the most distinctive fishing equipment used by the Orang Laut.

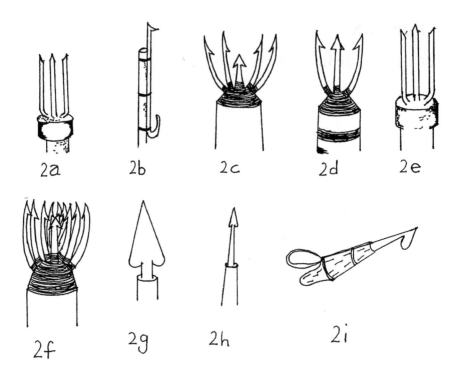

2a: *Seligi* ('javelin') is the name of a simple wooden lance in the pre-settlement era. It is the prototype of all *serampang* used today.

2b: *Panah ikan* ('fishing arrow') is for catching big fish weighing up to 30 kg, such as rays and groupers. The arrow is released by tension of an elastic band along the length of the wooden shaft.

2c: *Serampang tua* ('big spear'), also known as *serampang lima mata* ('five-pronged spear') or *serampang sotong* ('cuttlefish spear'), is the most common of all spears used by the Orang Laut.

2d: *Serampang tiga mata tua/ketam* ('three-pronged spear'), also known as *serampang ketam* ('crab spear'), is used in inner tidal zones for catching crab or sea cucumber.

2e: *Serampang udang* ('shrimp spear') is for catching shrimp at night.

2f: *Serampang mata sepuluh* ('ten-pronged spear') is for spearing large fish. It is also said to be an effective weapon in a fight as it is capable of fatally wounding an enemy.

2g: *Lembing tombak babi hutan* ('spear to kill wild boar') is also an effective weapon in combat.

2h: *Tiok* ('spear') is for collecting sea cucumbers such as *gamat* and *nabi*.

2i: *Mata tali angku* ('harpoon with line for turtle') is a single-barb harpoon for catching large fish and, specially, turtles. The iron head is set in a wooden holder which fits into a long wooden shaft. A line is attached to the wooden holder, allowing the catch to be played out if necessary.

Hooking devices

Weighted hooks are used for *moden mata pancing* ('modern hand line') fishing. A hook is identified by size and usually also by the name of the fish it is used to catch. Hooked fish fetch a higher price than speared ones as their bodies are not damaged.

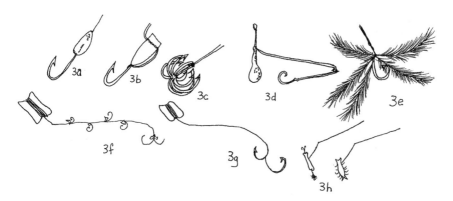

3a: *Moden mata pancing ikan pinang* ('modern hand line for catching starry pigface bream').

3b: *Moden mata pancing ikan mentimun* ('modern hand line for catching blue and yellow snapper').

3c: *Candat lebam* ('jig for small fish') or *candat udang* ('jig for shrimp').

3d: *Ranggung pancing/ton-ton* ('swivel line)' for catching *ikan kerapu* ('grouper').

3e: *Pancing tunda* ('troll line') for catching *ikan parang* ('dorab'), usually camouflaged by chicken feathers or sisal hemp and towed by a moving boat.

3f: *Rawai ikan selar* ('multi-hooked line for catching mackerel').

3g: *Keranggong/tali tunda* ('tow line') for catching mackerel.

3h: *Udang sungei* ('river shrimp') and *Udang pasir* ('sand shrimp') are hooks camouflaged in hand-carved wooden shrimps for catching a wide variety of fish. The hooks are now widely commercialized in plastic form painted in various luminous colours.

Trapping structures

Many types of trapping devices are used. Most can be handled by a single person. The bait, usually suspended or fastened inside, can be pieces of any kind of tough-skinned fish like shark, ray or cuttlefish.

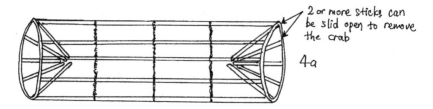

4a: *Injap* (trap with spikes turning inward to allow easy entry but make regress impossible) is for catching crabs. It is made of bamboo strips held together by three or five weavings of rattan. A pole is stuck through the trap into the sea bed to hold it fast and the visible top serves as a marker of its location.

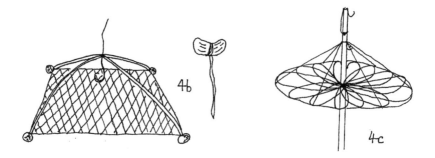

4b: *Bintur/Bento tali* ('crab trap'/'crab trap with curved strings or ropes') is weighted down by sinkers attached to the corners of the netting.

4c: *Bintur/Bento kayu* ('crab trap'/'crab trap with curved wood') is set in shallow water by pushing the pointed staff into the sea bed until the trap touches bottom.

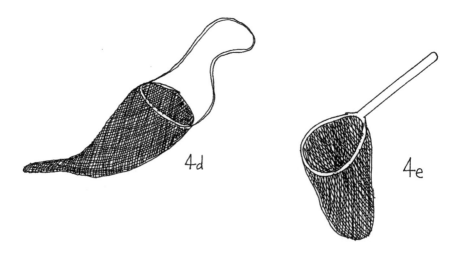

4d: *Pongjeng* ('conical bag') is usually hand-woven and tied to a boat with the bag immersed in water. It keeps the catch alive until it is time for them to be sold or eaten.

4e: *Penyedo/penyendok* ('hand-woven scoop') is for scooping the catch from the sea.

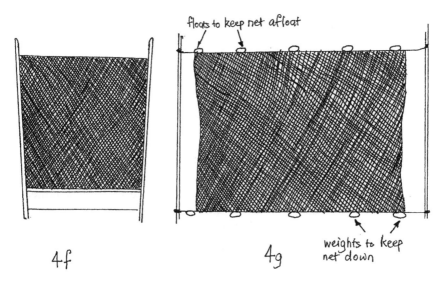

4f: *Sondong* ('shrimp net') is a net extended by opening a two-pronged fork fastened at the ends of a cross bar. The user walks slowly in a shallow river at low tide and scoops up shrimp, small fish and cuttlefish.

4g: *Jaring tansi* ('nylon dragnet') is bought and used by some Orang Laut today and is held by two persons in two boats.

Glossary

abhi şeka	rites to consecrate a ruler
adat	Malay customary law or a body of Malay customs and traditional law
Akuan Sa'ti	'Spirit of the Supernatural or Divine Power'; *sa'ti*, a Sanskritic word usually spelt *sakti* in Malay, refers to supernatural or divine power
Akuan Sidi	'Sprit of Effective Charms'
alam dalam di'l	world within ourselves
Alam Melayu	'Malay World'
alam secara makro, alam yang luas	macrocosm
asal	origin
asli Melayu	indigenous Malay
atap	leaves of a stemless palm, Eugeissona; thatched roofing made of palm leaves
bagi	to share or to divide
bagian	sharing
bahasa	language
bahu	shoulder
bangun	to get up or rise
baju kurung	long tunic
bantu	to help
batin	headman or leader who regulates the customary law of a group and is usually considered to have esoteric knowledge and spiritual power; literally 'spiritual' in Malaysia and Indonesia
Bendahara	Grand Vizier; traditionally the principal official in a Malay polity after the sultan; also means 'treasurer'
berat	heavy
Bhinneka Tungal Ika	'Unity in Diversity'

bin	son of
bolong	black hole; also used by the Orang Laut as person's name
buang hantu laut	throw out the sea spirit
Bughung Putih Yono Belayeh	'White Bird of the Changing Colours'
bujang	bachelor
burung hantu	bird spirit
cap burung	[currency with the] seal of a bird
Celates; Orang Selat	Portuguese terms for sea nomads in the Strait of Malacca and Riau
cocok	to suit, fit or match
comek	variety of cuttlefish
daging	meat
daulat	spiritual prowess; strictly defined as 'sovereignty', but used to refer to the charismatic quality of the ruler that symbolizes his power
derhaka	treason towards a Malay ruler
dosa	sin or sinful
dugong	sea cow
duit sen	money [in the form of] cents or coins
dukun	indigenous medical practitioner; shaman
daulat	strictly defined as 'sovereignty', but used to refer to the charismatic quality of the ruler that symbolizes his spiritual prowess
gadis	maiden; virgin
gagang	long wooden or bamboo shaft
gamat	variety of sea cucumber
hak ulayat	traditional communal property rights
hantu	spirit; ghost
hantu darat	land spirit
hantu laut	sea spirit
haram	forbidden
harga	cost and price
Hikayat Hang Tuah	'*The Story of Hang Tuah*'
hulubalang	military officer; warrior; war leader
ikan	fish
ilmu	all-encompassing knowledge, including scientific and common
ilmu itam	black magic
inpres	presidential directive
Inpres Desa Tertinggal	'Presidential Instruction Programme for Less Developed Villages'
ipar	in-law

itam	black
jaga	to guard; to preserve; to look after
jaga tanah	look after (their) territory
jampi	magic formula, spell or incantation
jaring	dragnet; seine
jati diri	identity
jual	sell
kajang	mat woven from palm frond used for boat-coverings, roofs, drying fish, etc.
kampong kita	our village
kasi	to give
kemantan	shaman
kepala	head
kerahan	traditional or corvée services
keris	wavy double-bladed dagger
ketua kampong	village headman
Ketuhanan Yang Maha Esa	'One True God'
keturunan	descent
kong	cast a spell over
koto bumi selebe dulang, koto langit sekombang payung	'the earth is as broad as a tray, the sky is as an unfurled umbrella'
kurang halus	less refined
labang nyao, nyawa, lobang semangat	soul hole
Laksamana	Admiral
lari	to elope, run off
maju	progress
mandala	arena; area in a circle
masyarakat terasing	isolated communities
mata dugong	'eyes of the sea cow', but used to refer to the tear drops of the sea cow
merompak	commit piracy
nabi	variety of sea cucumber that is higher priced than *gamat*
nafiri	reed pipes
nenek hantu	grandmother spirit
nobat	drums
orang asli	indigenous people
orang belum dalam Islam	'people not yet in Islam'
orang darat	land-dwelling people
Orang Kaya	'rich people'; Malay nobles
orang kerahan	nobility's vassals
orang lain	outsider; other people

Orang Suku Laut	'people of the sea'; different groups with a distinctive maritime culture
pakai layah	with [the seal of] a sail
Pancasila	the five basic moral principles of the Republic of Indonesia
Pancing	hand-line fishing
pangkalan asal	original starting point
pangkalan tambatan	initial mooring-point
Panglima	title of a leader of a military force
Parameswara	title of a ruler
payung	umbrella
pembangunan	development; (re)construction
pembangunan masyarakat suku terasing	development of tribal communities
pembangunan nasional	national development
pembinaan kesejahteraan masyarkat terasing	development of welfare of the isolated tribal communities
pembinaan kesejahteraan sosial masyarakat terasing Riau	development of social welfare of the isolated tribal communities of Riau
Penghulu	Malay village chief
persumpahan	place of oath-taking
piara	adopt
pisang raja	large sweet banana, literally king's bananas
Pohon Alam	'World Tree'
pokok	tree
pondok	a hut
pra-desa	pre-village
pulau	island
punya	to possess, own or have rights to
raja	king; lord
rakyat	subject
Rakyat Johor	subjects of Johor
Rakyat Laut	sea subjects
Rencana Pembangunan Lima Tahun	'Five-Year Plan of Development'
roh	soul; spirit
rumah	house
sama sama kuat	similarly powerful or strong
sampan	boat
Sangaji	title of a ruler
saudara	brother, sister or cousin of the same generation
saya punya	my own; I own
Sejarah Melayu	'Malay history'; specifically refers here to the Malay Annals

semangat	soul; spirit
sembah di teluk tanjung	worship or to pay homage and obeisance in bays and capes
serampang lima mata	five-pronged spear
sialang	bee-hived tree *Koompassia excelsa* or *tualang*
songkok	rimless fez-like cap usually made of velvet
Suku Sampan	'Boat People'
suku, suku-suku	group(s)
suku-suku terasing	isolated groups; official Indonesian term for isolated ethnic groups
swadaya	'self-supporting'; used here as an Indonesian official term to refer to traditional villages
swakarya	'self-developing' (of projects without government funding); used here as an Indonesian official term to mean transitional villages
swasembada	'self-supporting' or 'self-sufficient'; used here as an Indonesian official term to mean developed villages
tak/belum maju	not or not yet developed or progressive
tanah	territory; land
tempat kita juga	our place too
tempuling	single-pronged harpoon
tukar	exchange
uang dollar	dollar money
uang kopi	coffee money
upacara menebar beras	ritual of scattering of uncooked rice
yang laut pulang ke-laut, yang darat balik ke-darat	'what is of the sea returns to the sea, and what is of the land goes home to the land'

Notes

1 Orang Laut

1 Lenhart's estimates are based on the following sources: Bupati Kepala Daerah Tingkat II Kepulauan Riau (1988); Departemen Sosial Republik Indonesia (1994); Ketua Bappeda Tingkat II Kepulauan Riau (1990/91) and Walikotamadya Kepala Wilayah Kotamadya Administratif Batam (1986). Lenhart (2002: 297) also estimates that approximately half of the total Orang Laut population are still nomads.

2 As the Chinese in Indonesia have for long been regarded as an 'alien population' (Siddique and Suryadinata 1982: 677) by the Indonesian government, some Chinese have married the Orang Laut in an attempt to gain Indonesian citizenship.

3 Collins (1995: 186–191) has compiled an excellent bibliography on the works relating to the study of Orang Laut language.

4 Parts of this section draw on Chou 1997.

5 In 1945, Sukarno participated in drafting the first Constitution of the nation of Indonesia. In 1949, he became the president of the Republic of the United State of Indonesia and on 17 August 1950 president of the Republic of Indonesia. After the coup of 1965, he handed over power to Suharto via the letter of 11 March 1966 (Vickers 2005: 225).

6 See Article 51 of the Constitutional Law of the Netherlands Indies. It constitutes the foundation of the colony's agrarian legislation. The first three paragraphs date from 1854 and the rest were included in 1870.

3 History and culture change

1 According to Andaya, 'An important relationship between the court and the Orang Laut was . . . maintained successfully until the death of the last scion of this ancient dynasty in 1699' (1975: 47).

4 The inalienable gift of territory

1 Parts of the discussion for this section are drawn from Chou (2003: 76–77).

7 The transformation of Orang Laut territories

1 Parts of the discussion for this chapter are drawn from Chou and Wee (2002: 318–363).

2 The Indonesian Co-ordinating Minister for Trade and Industry and Ministry of Industry stressed that half of Bintan's water would be reserved for domestic

consumption, but this will in fact be mainly reserved for the tourist resorts (Wee and Chou 1997: 534–535).

3 This is expressed, for instance, in regular inter-continental travel, the use of plastic cards that negate national currency boundaries and the use of fax, e-mail and the internet for the free flow of information.

4 The discussion on mapping is drawn from Chou (2006c: 241–256).

5 Through the politics of mapping, a state can enforce measures to classify, alter, regulate and eliminate areas, individuals and resources within its territorial boundaries (Sack 1986: 21–22). The mapping of territories by the state is based on 'abstract space' (Vandergeest and Peluso 1995: 388). That is to say, spaces are viewed as 'homogenous', 'uniform' and 'linear' dimensions that can be divided into discrete units. Any piece is equivalent to another and they can be compared easily. The fabrication of these abstract and comparable units allows any one area to be located within the larger national space and within the even larger global space (Anderson 1991: 173; Vandergeest and Peluso 1995: 388). Anderson (1991: 173) directs our attention to the way in which mapping situates space in a global geometrical grid defined by the precise calculation of longitudes. On a basis of 'totalizing classification', it squares off seas, land and all things in it into measured boxes for surveillance.

6 The sovereignty of the state in defining borders – even in borderless Growth Triangle zones – can be seen in the proceedings during the recent meeting of the foreign ministers of the Association of Southeast Asian Nations (ASEAN), where the ASEAN Troika was established. The latter was 'a rapid reaction ad hoc committee that was composed to deal with trans-border problems occurring between member states' as an answer to the 'challenges of increased interdependency in the region' (*Jakarta Post*, 25 July 2000: 1). It was agreed that in order to resolve any sort of problem, it was necessary for all participants to rest on the premise of knowing 'precisely' what each member state's 'respective position' is. This assumes that transnational problems could be solved 'without compromising ASEAN's cardinal principle of non-intervention' in 'domestic' issues (ibid.: 1).

7 Other Growth Triangles have been proposed since. They include: (i) the Northern Growth Triangle comprising the northern states of Malaysia, northern Sumatra and southern Thailand; (ii) the Brunei-Indonesia-Malaysia-Philippines East ASEAN Growth Area or BIMP–EAGA; and (iii) The Indochina Economic Zone encompassing Burma (Myanmar), Laos, Thailand, Cambodia, Vietnam, Hong Kong and Yunnan Province of China.

8 Goh Chok Tong became Singapore's Prime Minister in November 1990.

9 Initially, the development was known as Batam Industrial Park. However, to establish a brand name, it was changed to 'Batamindo'. This name is a composite of Batam and the first two syllables of Indonesia (see Fukada 1997: 126).

10 From 1988 to 1990, Batam enjoyed foreign direct investments of more than US$680 million. By the mid-1990s, these investments exceeded US$1,000 million (or US$1 billion) (Grundy-Warr, Peachey and Perry 1999: 311). According to Singapore Trade Development Board figures, in 1996, Batam's share of foreign investment constituted 20 per cent of that received by Riau Province and two per cent of that received by Indonesia as a whole (Grundy-Warr, Peachey and Perry 1999: 311).

8 Conclusion

1 Parts of this section draw on Chou 2006b.

Bibliography

Achmedi (1972) 'Isolated ethnic groups in Indonesia', *Indonesia* 16, p. 20.

Akamatsu, K. (1962) 'A historical pattern of economic growth in developing countries', *The Developing Economies, Preliminary Issue*, March–August, pp. 3–25.

Alexander, G. (1973) *The Invisible China: The Overseas Chinese and the Politics of Southeast Asia*, New York, Macmillan.

Andaya, L. (1974) 'The structure of power in seventeenth century Johor', in Anthony Reid and Lance Castles (eds.), *Pre-colonial State Systems in Southeast Asia: The Malay Peninsula, Sumatra, Bali-Lombok, South Celebes*, Kuala Lumpur, Malaysian Branch of the Royal Asiatic Society, Monograph 6, pp. 1–11.

Andaya, L. (1975) *The Kingdom of Johor 1641–1728: Economic and Political Developments*, Kuala Lumpur, Oxford University Press.

Andaya, B. (1997) 'Recreating a vision: daratan and kepulauan in historical context', in Cynthia Chou and Will Derks (eds.), 'Riau in transition', *Bijdragen Tot de Taal-, Land-en Volkenkunde*, 153,4e: 483–508.

Anderson, B.R. (1972) 'The idea of power in Javanese Culture', in Claire Holt (ed.), *Culture and Politics in Indonesia*, Ithaca, NY, Cornell University Press, pp. 1–69.

Anderson, B. (1991) *Imagined Communities*, London, Verso.

Anderson, J. (1965) [1824] *Political and Commercial Considerations Relative to the Malayan Peninsula and the British Settlements in the Straits of Malacca; with an introduction by J.S. Bastin*, Singapore, Royal Asiatic Society, Malayan Branch.

Appadurai, A. (1996) *Modernity at Large: Cultural Dimensions of Globalization. Public Worlds*, Vol. 1, Minneapolis, London, University of Minnesota Press.

Barnard, T.P. (2001) 'Texts, Raja Ismail and violence: Siak and the transformation of Malay identity in the eighteenth century', *Journal of Southeast Asian Studies*, 32,3: 331–342.

Barnard, T.P. (2003) *Multiple Centres of Authority: Society and Environment in Siak and Eastern Sumatra, 1674–1827*, Leiden, Koninklijk Instituut voor Taal-, Land- en Volkenkunde Press.

Barnard, T.P. (2007) 'Celates, Rayat-Laut, pirates: the Orang Laut and their decline in History', *Journal of the Malaysian Branch of the Royal Asiatic Society* 80,2: 33–49.

Barnes, R.H. (1995) 'Introduction', in R.H. Barnes, Andrew Gray, and Benedict Kingsbury (eds.), *Indigenous Peoples of Asia*, Ann Arbor, Michigan, Association for Asian Studies, Inc. Monograph and Occasional Paper Series 48, pp. 1–12.

de Barros, J. (1777) *Da Asia. Dos Feitos que os Portuguezes fizeram no descubrimento e conquista dos mares e terras do Oriente. Decade segunda, parte II*. Lisbon.

Basic Agrarian Law of 1960. Government of Indonesia, *State Gazette* No. 104.

Batam Industrial Development Authority (1991) *Guide for Investors.*

Baud M. and van Schendel, W. (1997) 'Toward a comparative history of borderlands', *Journal of World History*, 8,2: 211–242.

Benjamin, G. (1989) 'Achievements and gaps in orang asli research', *Akademika* 35,7: 7–46.

Benjamin, G. (2002) 'On being tribal in the Malay World', in G. Benjamin and C. Chou (eds.), *Tribal Communities in the Malay World: Historical, Cultural and Social Perspectives*, Singapore and the Netherlands, Institute of Southeast Asian Studies and International Institute for Asian Studies, pp. 7–76.

Bodenhorn, B. (1989) ' "The Animals Come to Me, They Know I Share": Inupiaq Kinship Changing Economic Relations and Enduring World Views on Alaska's North Slope', unpublished Ph.D. thesis, Cambridge, University of Cambridge.

Bowen, J.R. (2003) *Islam, Law and Equality in Indonesia: An Anthropology of Public Reasoning*, Cambridge, Cambridge University Press.

Bradley, D. and Harlow, S. (1994) 'East and South-East Asia', in C. Moseley and R.E. Asher (eds.), *Atlas of the World's Languages*, London and New York, Routledge, p. 166.

Brown, C.C. (1970) *Sejarah Melayu or Malay Annals: An Annotated Translation*, Kuala Lumpur, Oxford University Press. First in 1952 *Journal of the Malayan Branch of the Royal Asiatic Society* 25,2 and 3: 1–276.

Bryant, R.L. (1998) 'Resource politics in colonial South-east Asia', in Victor T. King (ed.), *Environmental Challenges in South-east Asia*, Richmond, Curzon Press, pp. 29–51.

Bunnell, T., Sidaway, J.D. and Grundy-Warr, C. (2006) 'Introduction: re-mapping the "Growth Triangle": Singapore's cross-border hinterland', *Asia Pacific Viewpoint* 47,2: 235–240.

Bupati Kepala Daerah Tingkat II Kepulauan Riau (1988) 'Laporan bupati kepala daerah tingkat II kepulauan Riau', mimeographed, Tanjung Pinang, Kantor Bupati.

Burling, R. (1965) *Hill Farms and Padi Fields*, New Jersey, Prentice Hall.

Camat Pembantu Baputi Wilayah IV Bintan (1989) 'Pandangan umum bagi pembinaan dan pengembangan Suku Laut', mimeographed, Kijang, Kantor Camat.

Cameron, J. (1882) 'Landing of Raffles in Singapore by an eye-witness', (translation), *Journal of the Straits Branch of the Royal Asiatic Society* 10: 285–286.

Campo, J. N.F.M. (2003) 'Discourse without discussion: representations of piracy in colonial Indonesia 1816–25', *Journal of Southeast Asian Studies* 34,2: 199–214.

Cashdan, E. (1983) 'Territoriality among human forager: ecological modes and an application to four Bushman groups', *Current Anthropology* 24: 47–66.

Castells, M. (1996) *The Informational City: Economic Restructuring and Urban Development*, Oxford, Blackwell.

CCCIL (Collected Commentary and Cases on Indonesian Law) (1988), Sydney, University of Sydney.

Center for Economic and Business Studies (PUSEB), Riau University in cooperation with Promotion and Investment Board, Riau Province (eds.), (2003) *Policy and Investment Opportunities in Riau Province of Indonesia*, Pekanbaru Riau, Promotion and Investment Board Riau Province.

Chang, T.C. (2004) 'Tourism in a 'borderless world: The Singapore experience', *Asia Pacific Issues* 73 (May): 1–8.

Chau, J. (1911) *Chau Ju-kua: His Work on the Chinese and Arab Trade in the Twelfth and Thirteenth Centuries, entitled Chu-fan-chï.* Translated from Chinese and annotated by Friedrich Hirth and W.W. Rockhill, St. Petersburg, Printing Office of the Imperial Academy of Sciences.

Cheater, A.P. (1995) 'Globalisation and the new technologies of knowing: anthropological calculus or chaos?' in Marilyn Strathern (ed.), *Shifting Contexts: Transformations in Anthropological Knowledge*, London and New York, Routledge, pp. 117–130.

Chen, E.K.Y. (1989) 'The changing role of the Asian NICs in the Asian-Pacific region towards the year 2000', in Miyohei Shinohara and Fu-chen Lo (eds.), *Global Adjustment and the Future of Asian-Pacific Economy*, Tokyo, Institute of Developing Economies, pp. 207–231.

Chen, E. and Kwan, C.H. (1997) *Asia's Borderless Economy: The Emergence of Sub-regional Zones*, St. Leonards New South Wales, Allen and Unwin.

Chou, C. (1995) 'Orang Laut women of Riau: an exploration of difference and the emblems of status and prestige,' in L. Summers and W.D. Wilder (eds.), 'Gender and the Sexes in the Indonesian Archipelago', *Indonesia Circle*, 67: 175–198.

Chou, C. (1997) 'Contesting the tenure of territoriality: the Orang Suku Laut', in C. Chou and W. Derks (eds.), 'Riau in transition', *Bijdragen Tot de Taal-, Land- en Volkenkunde*, 153, 4e: 605–629.

Chou, C. (2003) *Indonesian Sea Nomads: Money, Magic, and Fear of the Orang Suku Laut*, London and New York, RoutledgeCurzon, Taylor & Francis Group and International Institute for Asian Studies.

Chou, C. (2005) 'Southeast Asia through an inverted telescope: maritime perspectives on a borderless region', in P.H. Kratoska, R. Raben and H.S. Nordholt (eds.), *Locating Southeast Asia: Geographies of Knowledge and Politics of Space*, Singapore and Athens, Singapore University Press and Ohio University Press, pp. 234–249.

Chou, C. (2006a) 'Research trends on Southeast Asian sea nomads', *Kyoto Review of Southeast Asia*, 7: 1–11.

Chou, C. (2006b) 'Borders and multiple realities: the Orang Suku Laut of Riau, Indonesia', in A. Horstmann and R.L. Wadley (eds.), *Centering the Margin: Agency and Narrative in Southeast Asian Borderlands*, New York and Oxford, Berghahn Books, pp. 111–134.

Chou, C. (2006c) 'Multiple realities of the Growth Triangle: mapping knowledge and the politics of mapping', *Asia Pacific Viewpoint*, 47,2: 241–256.

Chou, C. and Wee, V. (2002) 'Tribality and globalization: the Orang Suku Laut and the "Growth Triangle" in a contested environment', in Geoffrey Benjamin and C. Chou (eds.), *Tribal Communities in the Malay World: Historical, Cultural and Social Perspectives*, Singapore and the Netherlands, Institute of Southeast Asian Studies and International Institute for Asian Studies, pp. 318–363.

Christie, C.J. (1996) *A Modern History of Southeast Asia: Decolonization, Nationalism and Separatism*, London, Tauris.

Chuleeporn, V. (2002) 'Power relations between the Orang Laut and the Malay kingdoms of Melaka and Johor during the fifteenth to seventeenth centuries' in S. Chutintaranond and C. Baker (eds.), *Recalling Local Oasts: Autonomous History in Southeast Asia*, Chiang Mai, Silkworm Books, pp. 143–166.

Cohen, M. (2000) 'Chorus of discontent', *Far Eastern Economic Review*, 163–7: 24–25.

Colchester, M. (1986) 'Unity and diversity: Indonesian policy towards tribal peoples', *The Ecologist*, 16–2/3: 89–98.

Colchester, M. (1989) *Pirates, Squatters and Poachers: The Political Ecology of Dispossession of the Native Peoples of Sarawak*, Kuala Lumpur, Survival International and Institute of Social Analysis.

Colchester, M. (1991) 'A future on the land? Logging and the status of native customary land in Sarawak', *Ilmu Masyarakat* 19, April–June: 36–45.

Colchester, M. (1995) 'Indigenous peoples' rights and sustainable resource use in South and Southeast Asia', in R.H. Barnes, A. Gray and B. Kingsbury (eds.), *Indigenous Peoples of Asia*, Ann Arbor, Michigan, Association for Asian Studies, Inc. Monograph and Occasional Paper Series 48, pp. 59–76.

Collins, J.T. (1995) *Bibliografi Dialek Melayu di Pulau Sumatera*, Kuala Lumpur, Dewan Bahasa dan Pustaka Kemeterian Pendidikan Malaysia.

Colombijn, F. (2003) 'When there is nothing to imagine nationalism in Riau', in P.J.M. Nas, G.A. Persoon and R. Jaffe (eds.), *Framing Indonesian Realities: Essays in Symbolic Anthropology in Honour of Reimar Schefold*, Leiden, Koninklijk Instituut voor Taal-, Land-en Volkenkunde Press, pp. 333–370.

Combes, F. 1987 [1667]. *Historia de las islas de Mindanao, Joló y sus adyacentes*, Madrid, Retana W.E. and P. Pastells.

Cooke, F.M. (2003) 'Maps and counter maps: globalised imaginings and local realities of Sarawak's plantation agriculture', *Journal of Southeast Asian Studies* 34,2: 265–284.

Crang, M., Crang, P. and May, J. (eds.) (1999) *Virtual Geographies*, London, Routledge.

Cribb, R. (2000) *Historical Atlas of Indonesia*, Surrey, Curzon.

Day, A. (2002) *Fluid Iron: State Formation in Southeast Asia*, Honolulu, University of Hawaii Press.

de Bruyn Kops, G.F. (1855) 'Sketch of the Rhio-Lingga archipelago', *Journal of the Indian Archipelago and Eastern Asia* 9: 96–108.

Dentan, R.K. (1975) 'If there were no Malays who would the Semai be?', in J.A. Nagata (ed.), *Pluralism in Malaysia: Myth and Reality*, Leiden, E.J. Brill, pp. 50–64.

Departemen Sosial (1981) *Program Memorandum Pembinaan Masyarakat Terasing*, Jakarta, Departemen Sosial.

Departement Sosial (1986) *Data dan Informasi Pembinaan Masyarakat Terasing*, Jakarta, Departemen Sosial.

Departemen Sosial Republik Indonesia (1994) *Masyarakat Terasing dalam Angka*, Jakarta, Direktorat Jenderal Bina Kesejahteraan Sosial, Direktorat Bina Masyarakat Terasing.

Dietrich, S. (1998) ' "We don't sell our daughters": a report on money and marriage exchange in the township of Larantuka (Flores, E. Indonesia)', in T. Schweizer and D.R. White (eds.), *Kinship, Networks, and Exchange*, Cambridge, Cambridge University Press, pp. 234–250.

Direktorat Bina Masyarakat Terasing (1990) Peta Suku dan Permasalahan Masyarakat Terasing di Indonesia, Jakarta, Departemen Sosial RI.

Direktorat Bina Masyarakat Terasing (1994/95) *Data dan Informasi Pembinaan Masyarakat Terasing*, Jakarta, Departemen Sosial R.I.

Donnan, H. and Wilson, T.M. (1999) *Borders: Frontiers of Identity, Nation, and State*, New York, Berg.

Dove, M. (1985) 'The Agroecological Mythology of the Javanese and the Political Economy of Indonesia', *Indonesia* 39.

Dunn, F.L. and Dunn, D.F. (1984) 'Maritime adaptations and exploitations of marine resources in Sundaic Southeast Asian prehistory', in P. van de Velde (ed.), *Prehistoric Indonesia: A Reader*, Dordrecht, Floris Publications Holland, pp. 244–271.

Edhie D. (1993) 'Masyarakat tradisional di pedalaman ("masyarakat terasing")' in Mubyarto (ed.), *Riau Nenatap Masa Depan*, Yogyakarta, Aditya Media, pp. 31–50.

Erni, C. and Stidsen, S. (2006) 'Indonesia', in Stidsen, S. (eds.), *The Indigenous World 2006*, Copenhagen, The International Work Group for Indigenous Affairs, pp. 300–308.

Ekachai, S. (1990) *Behind the Smile: Voices of Thailand*, Bangkok, Thai Development Support Committee.

Esser, S.J. (1951) *Peta Bahasa-bahasa di Indonesia*, Djakarta, Djawatan Kebudajaan, Kementerian Pendidikan, Pengadjaran dan Kebudajaan.

Evans, G. (1995) 'Central highlanders of Vietnam', in R.H. Barnes, A. Gray and B. Kingsbury (eds.), *Indigenous Peoples of Asia*, Ann Arbor, Michigan, Association for Asian Studies, Inc. Monograph and Occasional Paper Series 48, pp. 247–272.

Ferrand, G. (1913) *Relations de Voyages et texts géographiques arabes, persans et turks relatifs à l'Extrême-Orient du VIII au XVIIIe siècles*, Paris, Ernst Leroux, 2 vols.

Foley, W.A. (1983) 'Sumatra language atlas of the Pacific area, Part II, Japan area, Philippines and Formosa, mainland and insular South-East Asia', in S.A. Wurm and S. Hattori (eds.), *Language Atlas Pacific Area*, Canberra, The Australian Academy of the Humanities in collaboration with The Japan Academy, p. 38.

Forrest, T. (1780) *A Voyage to New Guinea, and the Moluccas from Balambangan*, London, G. Scott.

Forrest, T. (1792) *A Voyage from Calcutta to the Mergui Archipelago, Lying on the East Side of the Bay of Bengal*, London, G. Scott.

Forum Komunikasi dan Konsultasi Sosial (FKKS) Batam (1987), 'Proyek pemukiman "Suku Laut" di kepulauan Riau', mimeographed, Sekupang, FKKS.

Fox, R. (1966) *Kinship and Marriage: An Anthropological Perspective*, London, Penguin Books.

Fukada, J. (1997) 'Appendix to chapter 4: the Riau islands: development in progress' in E.K.Y. Chen and C.H. Kwan (eds.), *Asia's Borderless Economy: The Emergence of sub-regional zones*, Sydney, Allen and Unwin, pp. 124–135.

Garnet, M. (1975) *The Religion of the Chinese People*, New York, Harper and Row.

Gatot Soeherman, H. (1993) 'Sambutan dewan nasional Indonesia untuk kesejahteraan sosial', in Koentjaraningrat (ed.), *Masyarakat Terasing di Indonesia*, Jakarta, Gramedia, pp. ix–x.

Gellner, E. (1987) *Culture, Identity and Politics*, Cambridge, Cambridge University Press.

Gibson-Hill, C.A. (1973) 'The Orang Laut of Singapore River and the sampan panjang', *150th Anniversary of the Founding of Singapore, Journal of the Malayan/ Malaysian Branch of the Royal Asiatic Society*, Commemorative Reprint, 42, I: 121–134. First in *Journal of the Malayan/ Malaysian Branch of the Royal Asiatic Society*, 25, I: 161–174.

Gray, A. (1995) 'The indigenous movement in Asia', R.H. Barnes, A. Gray and B. Kingsbury (eds.), *Indigenous Peoples of Asia*, Ann Arbor, Michigan, Association for Asian Studies, Inc. Monograph and Occasional Paper Series 48, pp. 35–58.

Grundy-Warr, C., Peachey, K. and Perry, M. (1999) 'Fragmented integration in the

Singapore-Indonesian border zone: Southeast Asia's "Growth Triangle" against the Global Economy', *International Journal of Urban and Regional Research*, 1 June, pp. 304–328.

Hamilton, A. (2002) 'Tribal people on the southern Thai border: internal colonialism, minorities, and the state', in G. Benjamin and C. Chou (eds.), *Tribal Communities in the Malay World: Historical, Cultural and Social Perspectives*, Singapore and the Netherlands, Institute of Southeast Asian Studies and International Institute for Asian Studies, pp. 77–96.

Harley, J.B. (2001) 'The new nature of maps: essays in the history of cartography', in Paul Laxton (ed.), *Edited Collection of J.B. Harley Writings*, Baltimore, Maryland and London, The John Hopkins Press.

Hart, K. (1986) 'Heads or tails? Two sides of the coin', *MAN*, 21,4: 637–656.

Heine-Geldern, R. (1923) 'Südost Asien', in G. Buschan (ed.), *Illustrierte Völkerkunde*, Bd. II, Stuttgart: Schrecker and Schröder, pp. 689–968.

Helleiner, E. (1995) 'Great transformations: a Polayanian perspective on the contemporary global order', *Studies in Political Economic*, 48.

Hicklin, J., Robinson, D. and Singh, A. (1997) *Macroeconomic Issues Facing ASEAN Countries*, Washington, D.C., International Monetary Fund. Also available at <www.imf.org/external/pubs/nft/macro/OVERVIEW.HTM>

Hill, A.H. (trans.) (1973) 'The founding of Singapore described by 'Munshi Abdullah', *150th Anniversary of the Founding of Singapore, Journal of the Malayan/ Malaysian Branch of the Royal Asiatic Society*, Commerative Reprint, pp. 94–111. First in *Journal of the Malayan/ Malaysian Branch of the Royal Asiatic Society*, 28, III: 125–132, 137–149.

Hitchcock, M. (1996) *Islam and Identity in Eastern Indonesia*, Hull, The University of Hull Press.

Hobart, M. (1993) 'Introduction: the growth of ignorance', in M. Hobart (ed.), *An Anthropological Critique of Development: The Growth of Ignorance*, London, Routledge, pp. 1–30.

Hooi, C. (1957) 'Piracy and its Suppression in Malayan Waters 1800–1867', unpublished academic exercise, University of Malaya.

Indo Pos, (Jakarta) 'Penambangan pasir harus dihentikan', 14 August 2006, pp. 1 and 7.

Indo Pos, (Jakarta) 'Pulau hilang, batas pun melayang: pasir dikeruk, pajak nunggak', 15 August 2006, pp. 1 and 7.

Ingold, T. (1986) *The Appropriation of Nature: Essays on Human Ecology and Social Relations*, Manchester, Manchester University Press.

Institut Teknologi Bandung, Fakultas Teknik Sipil dan Perencanaan, Jurusan Teknik Arsitektur (1988) 'Pengembangan permukiman Suku Laut di Pulau Bertam Kepulauan Riau: kerangka acuan', mimeographed, Bandung, Institut Teknologi.

Ismani (1985) 'Rice culture, viewed from myths, legends, rituals, customs, and artistic symbolism relating to rice cultivation in Indonesia', *East Asian Cultural Studies* 24,1–4: 117–130.

Jacobsson, B. (2005) *Conquest of the Forest: Rice Rituals Among the To Pamona in Central Sulawesi (Indonesia)*, Göteborg, Department of Social Anthropology, Göteborg University.

Jakarta Post, (Jakarta) 'ASEAN Troika to cross borders', 25 July 2000, p. 1.

Jakarta Post, (Jakarta) 'Indonesian giant sued for illegal logging', 17 July 2003, p. 5.

Jakarta Post, (Jakarta) 'Reforestation program launched in Riau', 14 August 2006, p. 8.

Jakarta Post, (Jakarta) 'Islands in focus: sand exports to S'pore stopped', 18 August 2006, p. 8.

Jacoby, N.H., Nehemkis, P. and Eells, R. (1977) *Bribery and Extortion in World Business: A Study of Corporate Political Payments Abroad*, New York and London, Macmillan Publishing Co., Inc. and Collier Macmillan Publishers.

James, K. (8 August 1990) *Business Times*, (Singapore), 'Triangle of golden opportunity', p. 2.

James, K. (9 October 1990) *Business Times*, (Singapore), 'Catching the boat of opportunity', p. 11.

Jessup, H.I. (1990) *Court Arts of Indonesia*, New York, The Asia Society Galleries, New York in association with Harry N. Abrams, Inc., New York.

Kähler, H. (1960) *Ethnographische und linguistische Studien über die Orang Darat, Orang Akit, Orang Laut und Orang Utan im Riau-Archipel und auf den Inseln an der Ostküste von Sumatra. Veröffentlichungen des Seminars für Indonesische- und Südseesprachen der Universität Hamburg. No. 2*. Berlin: D. Reimer.

Ketua Bappeda Tingkat II Kepulauan Riau (1990/91), 'Bahan rapat ketua bappeda tingkat II kepulauan Riau tentang pembinaan Suku Terasing', mimeographed, Tanjung Pinang, Bappeda.

King, V.T. (1993) *The Peoples of Borneo*, Oxford, Blackwell.

King, V.T. (1999) *Anthropology and Development in South-east Asia: Theory and Practice*, Kuala Lumpur, Oxford University Press.

Kompas, (Jakarta), '24 KK Suku Laut dimukimkan di permukiman terapung' 18 May 1991, p. 14.

Kompas, (Jakarta), 'Percobaan untuk Suku Laut', 22 September 1991, p. 8.

Kumar, S. and Lee Tsao Yuan (1991) 'A Singapore perspective' in Lee Tsao Yuan (ed.), *Growth Triangle: The Johore-Singapore-Riau Experience*, Singapore, Institute of Southeast Asian Studies and Institute of Policy Studies, pp. 1–36.

Kwa, C.G. (2005) 'From Temasek to Singapore: locating a global city-state in the cycles of Melaka Straits history', in J.N. Miksic and C. Low (eds.), *Early Singapore 1300s–1819: Evidence in Maps, Text and Artefacts*, Singapore, Singapore History Museum, pp. 124–146.

Leach, E. (1989) *Culture and Communication: The Logic by which Symbols are Connected: An Introduction to the Use of Structuralist Analysis in Social Anthropology*, Cambridge, Cambridge University Press.

Lenhart, L. (1997) 'Orang Suku Laut ethnicity and acculturation', in C. Chou and W. Derks (eds.), 'Riau in transition', *Bijdragen Tot de Taal-, Land- en Volkenkunde*, 153, 4e: 577–604.

Lenhart, L. (2002) 'Orang Suku Laut identity: the construction of ethnic realities' in G. Benjamin and C. Chou (eds.), *Tribal Communities in the Malay World: Historical, Cultural and Social Perspectives*, Singapore and the Netherlands, Institute of Southeast Asian Studies and International Institute for Asian Studies, pp. 293–317.

Li, T. (1999) 'Marginality, power and production: analyzing upland transformations' in T. Li (ed.), *Transforming the Indonesian Uplands: Marginality, Power and Production*, Amsterdam, Harwood Academic Publishers, pp. 1–44.

Lian, K.F. (2001) 'The construction of Malay identity across nations: Malaysia, Singapore and Indonesia', *Bijdragen tot de Taal-, Land- en Volkenkunde* 157,4: 861–879.

Logan, J.R. (1847) 'The present condition of the Indian archipelago', *Journal of the Indian Archipelago and Eastern Asia* 1: 1–21.

Logan, J.R. (1849a) 'The piracy and slave trade of the Indian archipelago', *Journal of the Indian Archipelago and Eastern Asia* 3: 581–588; 629–636.

Logan, J.R. (1849b) 'Malay amoks and piracies. What can we do to abolish them?' *Journal of the Indian Archipelago and Eastern Asia* 3: 463–467.

Logan, J.R. (1850) 'The piracy and slave trade of the Indian archipelago', *Journal of the Indian Archipelago and Eastern Asia* 4: 45–53; 144–162; 400–410; 617–628; 734–746.

Lui, F. (1985) 'An equilibrium of queuing model of bribery', *Journal of Political Economy* 93,4: 760–781.

McKinnon, J. and Wanat Bhrukasari (eds.) (1986) *Highlanders of Thailand*, Singapore, Oxford University Press.

Macleod, S. and McGee, T.G. (1996) 'The Singapore-Johore-Riau growth triangle: an emerging extended metropolitan region', in F. Lo and Y. Yueng (eds.), *Emerging World Cities in Pacific Asia*, Tokyo, New York, Paris, United Nations University Press, pp. 417–464.

Maier, H. (1988) *In the Center of Authority*, Ithaca, Southeast Asia Program Publications, Southeast Asia Program, Cornell University.

Mariam, M.A. (1984) 'Orang Baru and Orang Lama: Ways of Being Malay in Singapore's North Coast', unpublished B.Soc.Sc. (Hons.) dissertation, National University of Singapore.

Mariam, M.A. (2002) 'Singapore's Orang Selat, Orang Kallang, and Orang Selat: The Last Settlements' in G. Benjamin and C. Chou (eds.), *Tribal Communities in the Malay World: Historical, Cultural and Social Perspectives*, Singapore and the Netherlands, Institute of Southeast Asian Studies and International Institute for Asian Studies, pp. 273–292.

Mauss, M. (1954) *The Gift*, translated by Ian Cunnison, Glencoe, IL.: Free Press.

Meyrowitz, J. (1985) *No Sense of Place: The Impact of Electronic Media on Social Behaviour*, New York, Oxford University Press.

Moch N. Kurutawan (21 July 2003) *Jakarta Post* (Jakarta) 'Bintan island facing serious environment damage', p. 4.

Mohamed Shahrom bin Mohamed Taha (2003) 'The Orang Laut: 19th Century British Perceptions of a Changing Community', unpublished Bachelor of Arts and Social Sciences (Hons.) dissertation, National University of Singapore.

Mohamed Suffian bin Hashim, Tan Sri (1972) *An Introduction to the Constitution of Malaysia*, Kuala Lumpur, Government Printers.

Netscher, E., (1854) 'Beschrijving van een gedeelte der residentie Riouw', *Tijdschrift voor Indische Taal-, Land- en Volkenkunde* 2: 108–270.

Newbold, T.J. (1839) *Political and Statistical Account of the British Settlements in the Straits of Malacca, viz., Pinang, Malacca and Singapore: With a History of the Malayan States on the Peninsula of Malacca*, London, John Murray, 2 vols.

Nicholas, C. (1990) 'In the name of the Semai? The state and Semai society in Peninsular Malaysia', in L.T. Ghee and A.G. Gomes (eds.), *Tribal Peoples and Development in Southeast Asia*, Selangor, University of Malaya.

Ohmae, K. (1995) *The End of the Nation State: The Rise of Regional Economies*, New York, The Free Press.

Ong, A. (1999) *Flexible Citizenship: The Cultural Logics of Transnationality*, Durham, N.C., Duke University Press.

Persoon, G. (1988) 'Isolated groups or indigenous peoples: Indonesia and the international discourse', *Bijdragen Tot de Taal-, Land- en Volkenkunde*, 154,2: 281–304.

Pires, T. (1944) *Suma Oriental: An Account of the East from the Red Sea to Japan Written in Melaka 1512–1515*, Vol. II, translated by Cortesão, London, Hakluyt Society.

Porath, N. (2002) 'Developing indigenous communities into *Sakais*: South Thailand and Riau' Settlements' in G. Benjamin and C. Chou (eds.), *Tribal Communities in the Malay World: Historical, Cultural and Social Perspectives*, Singapore and the Netherlands, Institute of Southeast Asian Studies and International Institute for Asian Studies, pp. 97–118.

Republic of Indonesia, *Repilita* VI, (1994/95–1998/99), Indonesia's Sixth Five-Year Development Plan: A Summary, Jakarta, Department of Information.

Rigg, J. (1997) *Southeast Asia: The Human Landscape of Modernization and Development*, London and New York, Routledge.

Robertson, R. (1990) 'Mapping the global condition: globalisation as the central concept', in M. Featherstone (ed.), *Global Culture: Nationalism, Globalisation and Modernity*, London, Sage, pp. 15–30.

Ross, J.K. (1975) 'Social borders: definitions of diversity', *Current Anthropology*, 16,1: 53–72.

Sack, R.D. (1986) *Human Territoriality: Its Theory and History*, Cambridge, Cambridge University Press.

Safian, H. (1999) *Ensiklopedia Sejarah dan Kebudayaan Melayu*. Jilid 1, Kuala Lumpur, Dewan Bahasa dan Pustaka, Kementerian Pendidikan Malaysia.

Sahlins, M.D. (1968) *Tribesmen*, Englewood Cliffs, New Jersey, Prentice Hall Inc.

Sahlins, M.D. (1974) *Stone Age Economics*, London, Tavistock Publications.

Salmond, A. (1984) 'Nga huarahi o te ao Maori: pathways in the Maori world', in S.M. Mead (ed.), *Te Maori, Maori Art from New Zealand Collections*, New York: Harry N. Abrams, Inc, pp. 109–137.

Sama Bajau Studies Newsletter. No. 1. (1995), Kyoto, Centre for Southeast Asian Studies.

Sama Bajau Studies Newsletter. No. 2. (1996), Kyoto, Centre for Southeast Asian Studies.

Sandbukt, Ø. (1982) 'Duano Littoral Fishing: Adaptive Strategies within a Market Economy', unpublished Ph.D. thesis, Cambridge, University of Cambridge.

Sandbukt, Ø. (1984) 'The sea nomads of Southeast Asia: new perspectives on ancient traditions', *Annual Newsletter of the Scandinavian Institute of Asian Studies*, 17: 3–13.

Sassen, S. (1996) *Losing Control? Sovereignty in an Age of Globalization*, New York, Colombia University Press.

Sather, C. (1995) 'Sea nomads and rainforest hunter-gatherers: foraging adaptations in the Indo-Malaysian archipelago', in P. Bellwood, J.J. Fox and D. Tryon (eds.), *The Austronesians: Historical and Comparative Perspectives*, Canberra, Department of Anthropology, Research School of Pacific and Asian Studies Publication, Australian National University, pp. 229–268.

Sather, C. (1998) 'Sea nomads, ethnicity, and "otherness": the Orang Suku Laut and Malay ethnicity in the Straits of Melaka', *Soumen Antropologi* 23,2: 20–36.

Sather, C. (1999) 'The Orang Laut', *Occasional Paper* No. 5, *Academy of Social Sciences (AKASS) Heritage Paper Series*, Penang, Academy of Social Sciences.

Schefold, R. (1970) 'Divination in Mentawi', *Tropical Man: Yearbook Department of Social Research (Formerly Anthropology Department), Royal Tropical Institute Amsterdam* 3: 10–87.

Schot, J.G. (1882) 'De Battam archipel', *De Indische Gids*, 4,2: 25–54; 161–188; 470–479; 616–625.

Schot, J.G. (1883) 'De Battam archipel', *De Indische Gids*, 5,1: 205–11; 463–84.

Schot, J.G. (1884) 'Het stroomgebied der Kateman, bijdrage tot de kennis van Oost-Sumatra', *Tijdschrift voor Indische Taal-, Land- en Volkenkunde* 29: 555–581.

Scott, J.C. (1985) *Weapons of the Weak: Everyday Forms of Peasant Resistance*, New Haven, CT and London, Yale University Press.

Scott, J.C. (1998) *Seeing Like A State: How Certain Schemes to Improve the Human Condition Have Failed*, New Haven and London, Yale University Press.

Shields, R. (1991) *Places on the Margin: Alternative Geographies of Modernity*, London, Routledge.

Siddique, S. and Suryadinata, L. (1982) 'Bumiputra and pribumi: economic nationalism (indigenism) in Malaysia and Indonesia', *Pacific Affairs*, 54,4: 662–687.

Singapore Business (December 1990) 14:2, p. 34.

Skeat, W.W. and Blagden, C.O. (1906) *Pagan Races of the Malay Peninsula*, Vol. I, London, MacMillan and Co.

Skeat, W.W. and Ridley, H.N. (1900) 'The Orang Laut of Singapore', *Journal of the Straits Branch of the Royal Asiatic Society* 33: 247–250.

Smedal, O.H. (1989) *Order and Difference: An Ethnographic Study of Orang Lom of Bangka, West Indonesia*, Oslo, Department of Social Anthropology, University of Oslo.

Soh, S. (5 September 1990) *Business Times*, (Singapore), 'Who can buy residences in Batam, and who can't', p. 1.

Soh, S. and Chuang, Peck Ming (1990) 'Riau: an investor's guide to the 3,000-island province', *Singapore Business*. [Special theme issue].

Sopher, D.E. [1965] (1977) *The Sea Nomads: A Study of the Maritime Boat People of Southeast Asia*, Singapore, National Museum Publication.

St. John, H. (1853) *The Indian Archipelago: Its History and Present State*, Vol. I, London, Longman, Brown, Green, and Longmans.

St. John, S. (1849) 'Piracy in the Indian Archipelago', *Journal of the Indian Archipelago and Eastern Asia* 3: 251–266.

Stoler, A.L. (2001) 'Comment', in André Burguière and Raymond Grew (eds.), *The Construction of Minorities: Cases for Comparison Across Time and Around the World*, Ann Arbor, MI, The University of Michigan Press, pp. 165–170.

Stoll, D. (1982) *Fishers of Men or Founders of Empire?* London, Zed Books.

Straits Times, (Singapore), 'Singapore team finds water potential in Riau', 10 April 1990.

Straits Times, (Singapore), 21 June 1990.

Straits Times, (Singapore), 'Call to forge growth triangle with Thailand and Indonesia', 24 August 1990, p. 20.

Straits Times, (Singapore), 'S'pore, Jakarta seal Riau deal', 29 August 1990, p. 1.

Straits Times, (Singapore), 'Bintan protest', 21 January 2000.

Straits Times Weekly Edition, (Singapore), 13 November 1991, p.1.

Survival International (1987) *Thailand: Akha Expelled*, London.

Tagliacozza, E. (2000) 'Kettle on a slow boil: Batavia's threat perception in the Indies outer islands 1870–1910', *Journal of Southeast Asian Studies* 31,1: 70–100.

Tambiah, S.J. (1976) *World Conqueror and World Renouncer*, Cambridge and New York, Cambridge University Press.

Tapp, N. (1986) 'The minorities of southern China: a general overview', *Journal of the Hong Kong Branch of the Royal Asiatic Society*, 26: 102–114.

Tarling, N. (1962) *Anglo-Dutch Rivalry in the Malay World, 1780–1824*, London and St. Lucia, Cambridge University Press and University of Queensland Press.

Tarling, N. (1963) *Piracy and Politics in the Malay World; A Study of British Imperialism in Nineteenth-century South-east Asia*, Melbourne, F.W. Cheshire.

Tenas, E. (2002) 'The Orang Petalangan of Riau and their forest environment', in G. Benjamin and C. Chou (eds.), *Tribal Communities in the Malay World: Historical, Cultural and Social Perspectives*, Singapore and the Netherlands, Institute of Southeast Asian Studies and International Institute for Asian Studies, pp. 364–383.

Thomson, J.T. (1847) 'Remarks on the Sletar and Sabimba Tribes', *Journal of the Indian Archipelago and Eastern Asia* 1: 341–351.

Thomson, J.T. (1851) 'Description of the eastern coast of Johor and Pahang, and adjacent islands', *Journal of the Indian Archipelago and Eastern Asia* 1: 341–351.

Toffler, A. (1990) *Powershift*, New York, Bantam Books.

Trocki, C.A. (1979) *Prince of Pirates: The Temenggongs and the Development of Johore and Singapore, 1784–1885*, Singapore, Singapore University Press.

Tsing, A. L. (1993) *In the Realm of the Diamond Queen: Marginality in an Out-of-the-Way Place*, Princeton, New Jersey, Princeton University Press.

Tsing, A.L.(1999) 'Becoming a tribal elder, and other green development fantasies' in T. Murray Li (ed.), *Transforming the Indonesian Uplands: Marginality, Power and Production*, Amsterdam, Harwood Academic Publishers, pp. 159–202.

Tuathail, G.Ó. (2000) 'Geopolitics and globalization: the changing world political map', *GEOPOLITICS*, 4,2. Also available as 'Borderless worlds? Problematizing discourses of deterritorialization' at <http://www.majbill.vt.edu/geog/faculty/toal/papers/Borderless.htm>

Turner, A. (1997) 'Cultural survival, identity and the performing arts of Kampar's Suku Petalangan', in Cynthia Chou and Will Derks (eds.), 'Riau in transition', *Bijdragen Tot de Taal-, Land- en Volkenkunde*, 153, 4e: 648–672.

Turner, V. (1969) *The Ritual Process*, London, Routledge Kegan Paul.

Urry, J. (1981) 'A View from the West: Inland, Lowland and Islands in Indonesian Prehistory', unpublished paper presented at the 51st ANZAAS Congress, Brisbane.

Valentijn, F. (1724). *Oud en Nieuw Oost-Indië, Vervattende een Naaukeurige en Uitvoerige Verhandelinge van Nederlands Mogenthevd in die Gewesten*, vols. 3 and 5, Dordrecht-Amsterdam.

Van Gennep, A. (1960) *The Rites of Passage*, translated by M.B. Vizedom and G.L. Caffee, London and Henley, Routledge and Kegan Paul.

Vandergeest, P. and Peluso, N.L. (1995) 'Territorilization and state power in Thailand', *Theory and Society* 24: 385–426.

Vatikiotis, M.R.J. (1996) *Political Change in Southeast Asia: Trimming the Banyan Tree*, London, Routledge.

Vickers, A. (2005) *A History of Modern Indonesia*, Cambridge, Cambridge University Press.

Virilio, P. and Lotringer, S. (1983) *Pure War*, New York, Semiotext(e).

Voorhoeve, P. (1955) *Critical Survey of Studies on the Languages of Sumatra*, Koninklijk Instituut Voor Taal-, Land- en Volkenkunde Bibliographical Series 1, 'S-Gravenhage, Martinus Nijhoff.

Walikotamdya Kepala Wilayah Kotamadya Administratif Batam (1986) 'Tinjauan umum Suku Laut dalam kotamadya administratif Batam', mimeographed, Sekupang.

Warren, J. (1981) *The Sulu Zone 1768–1898: The Dynamics of External Trade, Slavery, and Ethnicity in the Transformation of a Southeast Asian Maritime State*, Singapore, Singapore University Press.

Warren, J. (1998). *The Sulu Zone: The World Capitalist Economy and the Historical Imagination*, Amsterdam: VU University Press.

Wallman, S. (1978) 'The boundaries of "race": processes of ethnicity in England', *MAN*, 13,2: 200–217.

Wee, V. and Chou, C. (1997) 'Continuity and discontinuity in the multiple realities of Riau', in C. Chou and W. Derks (eds.), 'Riau in transition', *Bijdragen Tot de Taal-, Land- en Volkenkunde*, 153, 4e: 525–541.

Weiner, A.B. (1985) 'Inalienable wealth', *American Ethnologist: The Journal of the American Ethnological Society*, Vol. 12,2: 210–227.

Weinstock, J.A. (1979) Land Tenure Practices of The Swidden Cultivators of Borneo, Master of Science dissertation, Ithaca, NY, Cornell University.

Wessing, R. (1997) 'Nyai Roro Kidul in Puger: Local applications of a myth', *Archipel*, Vol.53: 97–120.

Wilkinson, R.J. (1959) *A Malay-English Dictionary (romanised)*, London, Macmillan.

Wolters, O.W. (1967) *Early Indonesian Commerce: A Study of the Origins of Srivijaya*, Ithaca, NY, Cornell University Press.

Wolters, O.W. (1975) *The Fall of Srivijaya in Malay History*, Kuala Lumpur, Oxford University Press.

Wolters, O.W. (1979) 'Studying Srivijaya', *Journal of the Malaysian Branch of the Royal Asiatic Society*, 52,2: 1–32.

Wolters, O.W. (1999) *History, Culture, and Region in Southeast Asian Perspectives*, Ithaca, NY and Singapore: Southeast Asia Program Publications, Southeast Asia Program, Cornell University and The Institute of Southeast Asian Studies.

Wong, Poh Kam and Ng, Chee Yuen (1991) 'Singapore's internationalization strategy for the 1990s', in S. Siddique and Ng Chee Yuen (eds.), *Southeast Asian Affairs 1991*, Singapore, Institute for Southeast Asian Studies, pp. 267–276.

Wood, D. (1993) *Power of Maps*, London, Routledge.

Yamazawa, I. (1990) *Economic Development and International Trade: The Japanese Model*, Honolulu, HI, East-West Center, Resource Systems Institute.

Yampolsky, P. (1996) 'Introductory notes to Melayu music of Sumatra and the Riau Island', Smithsonian Folkways CD recording, *Music of Indonesia* series 2.

Websites

<matrix-batam.hypermart.net/batamart.htm>, accessed on 28 November 1999.
<http://www.riau.go.id/penduduk/penduduk.php>, accessed on 25 November 2002.

Index

9 780415 626231